Yoshiki Oshida, Takashi Miyazaki
Bone-Grafting Biomaterials

Also of interest

Magnesium Materials.
From Mountain Bikes to Degradable Bone Grafts
Yoshiki Oshida, 2021
ISBN 978-3-11-067692-1, e-ISBN (PDF) 978-3-11-067694-5

Biomaterials and Engineering for Implantology.
In Medicine and Dentistry
Yoshiki Oshida and Takashi Miyazaki, 2022
ISBN 978-3-11-074011-0, e-ISBN (PDF) 978-3-11-074013-4

Materials for Medical Application
Edited by: Robert B. Heimann, 2020
ISBN 978-3-11-061919-5, e-ISBN (PDF) 978-3-11-061924-9

Artificial Intelligence for Medicine.
People, Society, Pharmaceuticals, and Medical Materials
Yoshiki Oshida, 2021
ISBN 978-3-11-071779-2, e-ISBN (PDF) 978-3-11-071785-3

Nickel-Titanium Materials.
Biomedical Applications
Yoshiki Oshida and Toshihiko Tominaga, 2020
ISBN 978-3-11-066603-8, e-ISBN (PDF) 978-3-11-066611-3

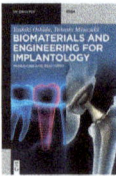

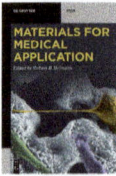

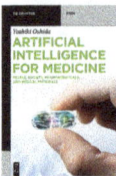

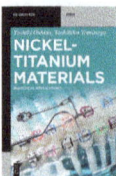

Yoshiki Oshida, Takashi Miyazaki

Bone-Grafting Biomaterials

—

Autografts, Hydroxyapatite, Calcium-Phosphates,
and Biocomposites

DE GRUYTER

Authors
Prof. Yoshiki Oshida
School of Dentistry
University of California San Francisco
513 Parnassus Ave
San Francisco CA 94143-0430
USA
yoshida@iu.edu

Dr. Takashi Miyazaki
Miyazaki Dental Clinic
3F Nishimami Plaza 1-5-1
639-0222 Kashiba-City, Nara Prefecture
Japan
miyarin3366@gmail.com

ISBN 978-3-11-113666-0
e-ISBN (PDF) 978-3-11-113669-1
e-ISBN (EPUB) 978-3-11-113889-3

Library of Congress Control Number: 2024930627

Bibliographic information published by the Deutsche Nationalbibliothek
The Deutsche Nationalbibliothek lists this publication in the Deutsche Nationalbibliografie;
detailed bibliographic data are available on the Internet at http://dnb.dnb.de.

© 2024 Walter de Gruyter GmbH, Berlin/Boston
Cover image: above: alex-mit/iStock/Getty Images Plus; below: PhonlamaiPhoto/iStock/Getty Images Plus
Typesetting: Integra Software Services Pvt. Ltd.
Printing and binding: CPI books GmbH, Leck

www.degruyter.com

Preface

Once a patient is evaluated as a candidate for implant recipient (free from all risky contraindications for implant treatments), the patient will face various stages of treatment, including pre-, intra-, and post-treatments, as well as operational procedures. Referring to attached figure, once the placed implant(s) is biologically fused to receiving hard/soft tissues (i.e., establishment of osseointegration), the patient is allowed to use implant(s) under normal occlusal function or ordinary daily activities. Thus the successfully functioning implant(s) is expected to exhibit a quite long-term service (aka, longevity) to which success rate and survival rate are contributed. The long-term servicing implant(s) would, in general, enhance patient's health-related quality of living (HRQoL) or dental health related quality of living (DHRQoL). Of course, to this end, a great corporation and responsibility should be demanded from patients, which should include daily hygiene management and well-organized maintenance checkup schedule.

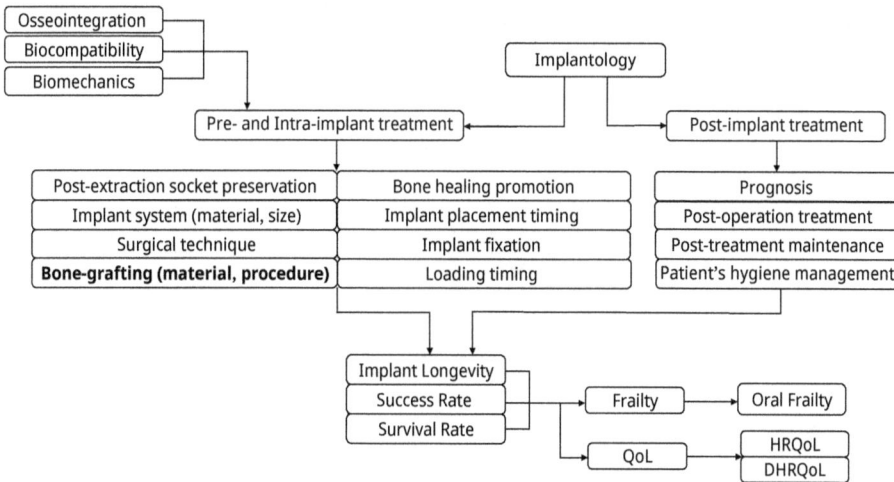

While QoL concept has been considered as a sort of a subjective evaluation from each implant recipient in either orthopedic treatment or dental treatment, the longevity, success rate, and survival rate should be evaluated in terms of direct or indirect objective, based on scientific examinations and/or observations, as well as clinical case data. Demographically, the term "longevity" is a synonym for the life expectancy; in the same way, longevity is normally used in medical and dental fields. Especially, the longevity of placed orthopedic joint replacement implants and dental implants are principally subjected to be discussed. The longevity is controlled by several crucial factors, which should include pre-operation procedures (e.g., appropriate implant system selection, post-extraction treatment for dental implant, etc.), intra-operation procedures (e.g., surgical skill and infection management, etc.), and post-operation issues

https://doi.org/10.1515/9783111136691-202

(e.g., poor hygiene and uncontrollable diabetes or an increased body-mass index or missing of preventive cares from both physicians and patient, etc.).

As seen in the figure, bone-grafting (with materials and procedures) is positioned in the pre-implantation and/or intra-implantation stages. The subsequent occurring success rate and survival rate are strongly relied on appropriate selection of bone-grafting materials as well as proper choice of grafting technique.

In this book, we will be discussing types and procedures of bone-grafting, anatomy, and physiochemistry of natural bone tissue, type of bone-grafting materials, supportive devices for bone-grafting procedures, and technical sensitivity of bone-grafting methods.

Contents

List of nomenclatures

AAAA	autolyzed antigen-extracted allogeneic
ABBM	anorganic bovine bone mineral
ABG	autogenous block graft
ACP	amorphous calcium phosphate
ADSCs	adipose-derived stem cells
ALP	alkaline phosphatase
AM	additive manufacturing
ARST	alveolar ridge split technique
BAG	bioactive glasses
BBA	bone bank allografts
BCC	body-centered cubic
BCP	biphasic calcium phosphates
BCT	biomechanical computed tomography
BHA	bovine-derived hydroxyapatite
BJ	binder jetting
BMD	bone mineral density
BMI	body-mass index
BMP	bone morphogenetic protein
BRS	bioresorbable scaffolds
BRU	bone remodeling units
CAP	cold atmosphere plasma
CBCT	cone-beam-computed tomography
CCP	cubic close packed
CDHA	calcium-deficient hydroxyapatite
CGF	concentrated growth factor
CMC	carboxymethylcellulose
CPBS	chitosan-poly(butylene succinate)
CPC	calcium phosphate cements
cpTi	commercially pure titanium
CS	chitosan
CSTi	cancellous structured titanium
μCT	micro-computer tomography
DBBM	deproteinized bovine bone mineral
DBM	demineralized bone matrix
DCPA	dicalcium phosphate anhydrous
DCPD	dicalcium phosphate dihydrate
DEXA	dual energy X-ray absorptiometry
DF	fractal dimension
DFDBA	Demineralized FDBA
DMLS	direct metal laser sintering
DHRQoL	dental health-related quality of living
EBM	electron beam melting
EBSS	Earle's balanced salt solution
ECM	extracellular matrix
EDL	electrical double layer
EGF	epidermal growth factor
EPC	endothelial progenitor cells

https://doi.org/10.1515/9783111136691-204

FBR	foreign body reaction
FCC	face-centered cubic
FDA	US Food and Drug Administration
FDBA	freeze-dried bone allograft
FDM	fused deposition modeling
FFI	full-length femur imaging
FGF	fibroblast growth factor
FTIR	Fourier transform infrared
GAG	glycosaminoglycan
GBR	guided bone regeneration
GFs	growth factors
GTR	guided tissue regeneration
HA	hydroxyapatite
HAS	hip structural analysis
HCP	hexagonal closed packed or hexahedron
HEPES	(4-(2-hydroxyethyl)-1-piperazineethanesulfonic acid)
hFOB	human fetal osteoblastic
HIV	human immunodeficiency virus
hMSC	human mesenchymal stem cells
HR-pQCT	high-resolution peripheral quantitative computed tomography
HRQoL	health-related quality of living
HV	hydroxy-valerate
IAN	inferior alveolar nerve
IGFs	insulin-like growth factors
ISQ	implant stability quotient
L-PBF	laser powder bed fusion
MCPA	monocalcium phosphate anhydrous
MCPM	monocalcium phosphate monohydrate
mCT	micro-computed tomography
MDCT	multidetector computed tomography
MIS	minimally invasive surgery
MNGCs	multinucleated giant cells
MSCs	mesenchymal stem cells
NBCM	native bilayer collagen membrane
NIH	National Institutes of Health
OC	osteocalcin
OCP	octacalcium phosphate
OCT	octahedron
OHRQoL	oral health-related quality of living
OPF	osteoperiosteal flap
OPG	osteoprotegerin
P3HB	poly(3-hydroxybutyrate)
PASS	primary closure, angiogenesis, space maintenance, stability of wound (principle)
PCL	polycaprolactone
PDA	polydopamine
PDGF	platelet-derived growth factor
PE	polyethylene
PEEK	poly ether ketone
PEG	polyethylene glycol

PET	poly(ethylene terephthalate)
PGA	poly(glutamic acid)
PHB	poly hydroxyl butyrate
PHEMA	polyhydroxyethyl methacrylate
PLA	polylactide or polylactic acid
PLGA	polylactideglycolide copolymer or copoly(lactic-glycolic acid
PLLA	poly(L-lactic acid)
PMMA	polymethylmethacrylate
PMN	polymorphonuclear
PP	polypropylene
PPE	polyphosphoester
PPF	poly(propylene fumarate)
PPF	periosteal pocket flap (technique)
PRF	plasma-rich in fibrin
i-PRF	injectable form of PRF
L-PRF	leukocyte PRF
PRP	platelet-rich plasma
PSU	poly sulfone
PTFE	polytetrafluoroethylene
d-PTFE	high-density PTFE
e-PTFE	expanded PTFE
PTH	parathyroid hormone
PTMC	polytrimethylene carbonate
PU	polyurethanes
PVA	poly(vinyl alcohol)
PVC	polyvinyl chloride
QCT	quantitative computed tomography
REMS	radiofrequency echographia multi-spectrometry
RGD	Arg-Gly-Asp amino acid
SBF	simulated body fluid
SEM	scanning electron microscopy
SF	silk fibroin
SLM	selective laser melting
SLS	selective laser sintering
TBS	trabecular bone score
TCP	tricalcium phosphate
TE	tissue engineering
TEA	triethanolamine
TGF	transforming growth factor
THA	total hip arthroplasty
TKA	total knee arthroplasty
TSA	total shoulder arthroplasty
TTCP	tetra-calcium phosphate
UHMWPE	ultra-high molecular weight polyethylene
VEGF	vascular endothelial growth factor
VFA	vertebral fracture assessment

1 Introduction

In this chapter, we will discuss the interrelationship of bone-grafting materials and their applications in both medicine and dentistry implantology. As seen in Figure 1.1, bone-grafting materials are categorized in a global term of the biomaterials and further classified into, generally, natural bone-grafting materials, synthetic bone-grafting materials, and supporting membrane structures. In an area of application, traumatized bone healing is the most important process in both orthopedic implants and dental implants via osteointegration.

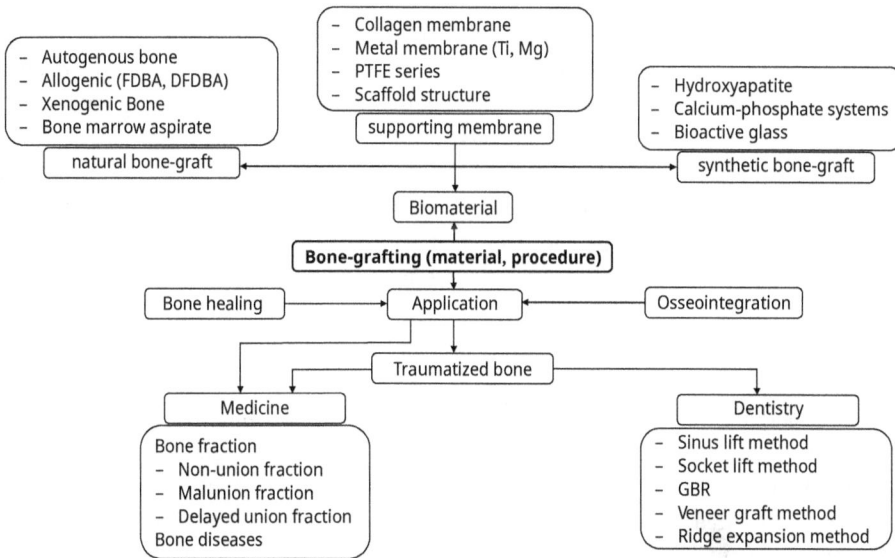

Figure 1.1: Interrelationship of bone-grafting materials with applications in both medical and dental fields. (FDBA: freeze-dried bone allograft; DFDBA: demineralized FDBA; PTFE: poly-tetra-fluoroethylene; GBR: guided bone regeneration).

1.1 Success rate and survival rate

The most common of these replacement joints are artificial knees and hips, which constitute almost 90% of the worldwide demand for joint implants. As a result of various technological breakthroughs, other extremity joint implants for ankles, shoulders, elbows, hands, feet, and jaws are increasingly more common as well [1, 2]. It is reported that more than 7 million Americans are living with a knee or hip implant and the number is increasing rapidly every year [3]. It was also reported that (i) in 2011, orthopedic surgeons performed 306,000 total hip replacements in the US alone and (ii) also in 2011, doctors also performed an additional 50,600 revision procedures to

https://doi.org/10.1515/9783111136691-001

replace previously implanted artificial hips [3]. Even in these circumstances, the expected longevity of placed orthopedic implants is more than 20 years [2, 4] or between 15 and 20 years [5].

The placed implants are constantly subjected to hostile environments including biomechanics, biotribology, and biotriocorrosion [6]. In the case of joint implants, tribological action in biological environment would produce the wear debris, which might be harmful to surrounding soft/hard vital tissues. These are challenges to material scientists as well as surface engineers. Actually, several remarkable R&D outcomes have been recognized to prolong lifespan expectation of placed implants. Shot peening or laser peening onto orthopedic implant surfaces has been carried out to develop beneficial surface-negative residual stresses [2, 7–12]. Recently structural integrity of materials has been manipulated to facilitate osseointegration. Such new methods can include: (i) controlled porosity of implant surface skin by the ion-assisted polymerization process to create bio-functional 3D-printed Ti implants or by selective laser melting to create porous titanium implants with enhanced bone-mimicking mechanical properties [13] or additive manufacturing technology [14] and (ii) functional gradation from core to skin of the implants [6, 10, 15], in which there is a descending gradation of mechanical strength from core to skin; reversely, a descending slope of biological characteristics from skin to core.

It was reported that around 120 million people in the United States alone are missing at least one of their natural teeth and an incredible 36 million or more people have no teeth at all (namely, edentulous) [16]. Normally, dental implant treatment includes three components that technically make up a single dental implant; these different components must all be considered when determining how long an implant-supported restoration will last. They are an implant main body (most of which is immersed into bone or augmented bone), abutment, and prosthesis. Most sources put the average lifespan of a dental implant at around 25 years or more; however there are also some sources that say implant posts can be permanent [17]. There are also reports confirming the 25-year longevity [18, 19]. The implant-supported restoration, in general, may need to be replaced approximately every 10–15 years since the constant forces of chewing and biting will eventually wear down the exterior surface thereof [17, 20].

Implants are often evaluated in terms of either success rate or survival rate, as described before. There is a definitive difference between these two terms [1]. The term "success" is used if a particular implant meets the success criteria it is being evaluated with, while the term survival simply implies that the implant exists in the body or mouth and appears not to include evaluated biofunction. However, the survival rate has been treated as a longevity indicator. Singh et al. [21], examining patients with total shoulder arthroplasty (TSA), reported that (i) 2,207 patients underwent 2,588 TSAs, with 63% of patients with underlying diagnosis, (ii) 212 TSAs were revised during the follow-up, and (iii) at 5-, 10- and 20-year implant survival rates were 94.2%, 90.2%, and 81.4%, respectively. It was reported that THA (total hip arthroplasty) achieves excellent technical outcomes with 10-year survival exceeding 95%, 25-year implant survival greater than 80%, and significant benefits for pain, mobility, and physical function [22, 23]. Bae et al. [24] performed 224 revision TKAs (total knee arthroplasty) in 194 patients from September 1990 to June 2009 and reported that (i) the 5-, 8-, and 10-year survival rates were 97.2%, 91.6%, and 86.1%, respectively, (ii) re-revision TKAs were performed on 20 knees because of infection (seven knees), loosening (six knees), polyethylene wear (six knees), and periprosthetic fractures (one knee), and (iii) the long-term survival rate of revision TKA was satisfactory, but careful attention is necessary to detect the late failure.

Historically, we had three important international conferences on acceptable criteria for dental implants [25, 26]. During the NIH Harvard Conference (1978), the following five criteria were set forth as acceptable success rate: (1) less than 1 mm of movement in each direction is allowed, (2) X-ray observed transmission images cannot serve as a reference standard, (3) bone resorption of less than 1/3 of the vertical height of the implant is acceptable, (4) no untreatable gingivitis, no inflammation, nor infection, no damage to adjacent teeth, no paresthesia or hypoesthesia, and (5) should function for 5 years in more than 75% cases. In 1986, the conference organized by Albrektsson had reached a consensus for acceptable criteria: (1) upon examination, individual unconnected implant should not move, (2) no X-ray penetration image around the placed implant, (3) vertical interim bone-sorption over time after 1 year after implant placement should be less than 0.2 mm, and (4) no persistent or irreversible signs or symptoms due to the implant (pain, infection, nerve paralysis, paresthesia, mandible injury, etc.), and (5) under the above conditions, a 5-year success rate of 85% shall be the lowest success criterion. At the latest conference (at Toronto, 1988), the followings were determined as acceptable success criteria: (1) the implant supports a functional and aesthetic superstructure that satisfies both the patient and the implantologist, (2) no pain, discomfort, sensory changes, or signs of infection caused by the implant, (3) when clinically examined, (4) the average annual vertical absorption after the start of the function should be less than 0.2 mm. During the Toronto Osseointegration Conference, the "more than 85% survival rate for 5 years, more than 80% after 10 years" was also determined, leading to further material development including commercially pure titanium (cpTi) as well as surface modification technologies.

The term "survival" is defined as the condition in which the implant remains in the mouth. If you have peri-implantitis but the implant is not removed, the placed implant can be considered alive (survive). On the other hand, the term "success" is recognized as a condition in which there are no subjective symptoms and no findings of peri-implantitis and there are four major factors influencing the success rates of placed implants. They include: (1) correct indication and favorable anatomic conditions (bone and mucosa), (2) good operative technique, (3) patient cooperation (oral hygiene), and (4) adequate superstructure design and fabrication [6].

Referring to Figure 1.2, Nishinaka [27] tries to differentiate survival rate and success rate, comparing two distinctive outcomes two years after implant surgery, in which (A) represents a development of peri-implantitis while (B) exhibits an excellent prognosis.

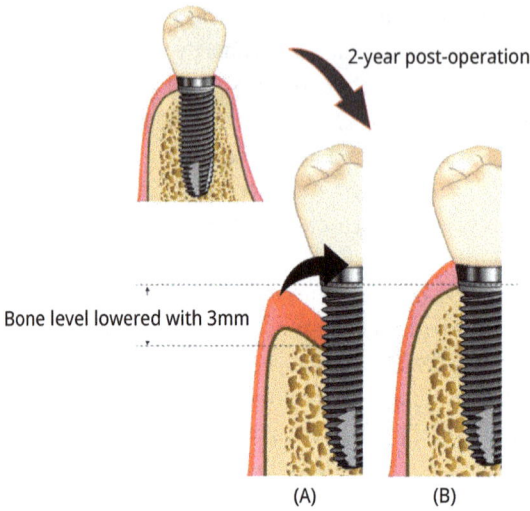

Figure 1.2: Two outcomes after two years after implant surgery: (A) developed peri-implantitis and (B) excellent prognosis [27, with kind permission of Dr. Nishinaka, Japan].

In either case, there was no movement in the placed implant, and the patient was able to eat without any problems with only a slight discomfort in the gingiva of the implant as a subjective symptom. It was observed that as for the objective findings of (A), the gums were inflamed due to infection, bleeding and pus were observed, and the bone level was about 3 mm lower than when it first functioned due to the spread of inflammation. In light of the criteria for success established at the Toronto Conference, although the implant was not movable and there was no functional problem, infection originating from the implant was observed, and vertical bone resorption was also 3 mm (if no abnormality is found, about 0.4 mm is the Toronto Conference standard if 2 years have passed), so it was evaluated as a failure on the basis of the Toronto standard. In the case of (B), it functioned in the mouth without falling out during the 2-year functional period, so the survival rate (2 years) can be evaluated to be 100%. Of course, regarding case (B), since it meets the success criteria established at the conference, it can be said that the success rate in the second year is 100%. Thus, there is a qualitative difference between the success rate and survival rate of implant treatment. In other words, survival is an indicator that does not reflect the health status of the implant [27].

As the living standard of the population improves, dental restoration has become the definitive therapy for most dental defects. Implants have been recognized as the "third set of teeth," since they are beautiful, comfortable, and have good chewing efficiency, making them feel like natural teeth [28]. Large-scale studies have reported that the long-term survival rates of implants are between 93.3% and 98% [29, 30], indicating that dental implants are an effective treatment for edentulousness. Busenlechner et al. [29] placed, from 2004 to 2012, a total of 13,147 implants in 4,316 patients at the Academy for Oral Implantology in Vienna and computed the survival rates after eight years of follow-up and assessed the impact of patient- and implant-related risk factors. It was found that (i) overall implant survival was 97%, independent of various factors of implant length, implant diameter, jaw location, implant position, local bone quality, previous bone augmentation surgery, or patient-related factors including osteoporosis, age, or diabetes mellitus. Krebs et al. [30] placed, between April 1991 and May 2011, 12,737 implants in 4,206 patients for a variety of clinical indications and reported that the Kaplan-Meier cumulative survival rate [31] was 93.3% after 204 months. Artzi et al. [32] differentiated between the survival and success definitions of functional hydroxyapatite-coated implant prosthesis by evaluating a total of 248 implants (62 patients), 5–10 years in function. It was reported that (i) the accumulative survival rate after 5 and 10 years was 94.4% and 92.8%, respectively, (ii) accumulative success rates were 89.9% and 54%, respectively, (iii) implants 13 mm and 15 mm in length (97.9% and 96.4%, respectively) had the highest survival rate, which was higher over implants 8 mm and 10 mm in length (75% and 88.2%, respectively), and (iv) the survival rate of 4-mm diameter implants compared with 3.25-mm was 96.5% and 90.3%, respectively, concluding that (v) a distinguishable observation between survival and success rate was noted particularly in long-term observations and (vi) implant length and diameter have an influence on the survival rate.

Moraschini et al. [33] evaluated the survival and success rates of osteointegrated implants determined in longitudinal studies that conducted a follow-up of at least 10 years and reported that survival rates were 94.6%, indicating that osteointegrated implants are safe and present high survival rates and minimal marginal bone resorption in the long term. Thirty-five placed implants in 19 patients were subjected to assess the success and survival rates and it was reported that (i) there was a success rate of 74% after definitive prosthetic rehabilitation, while six implants showed bone loss of between 2 mm and 4 mm, being classified as satisfactory survival, which was 100%, (ii) there was no relationship between the success and/or survival rate and any of the parameters evaluated, (iii) four implants presented with peri-implant mucositis, while peri-implantitis was observed in two implants, and (iv) regarding the definitive restorations, 17 prostheses were classified as successful, while there were complications in eight prostheses [34]. Evaluating 590 patients with 990 implants, Cochran et al. [35] reported that: (i) the majority of implants were 10 mm and 12 mm long (78.7%) and were placed in type II and III bone (87%) and 73% of the implants were placed in the mandible, and 27% were placed in the maxilla, (ii) the cumulative survival rate was 99.56% at 3 years and 99.26% at 5 years, and (iv) the overall success rate was 99.12% at 3 years and 97.38% after 5 years. Another study on 185 patients with 271 implants reported that: (i) three implant failures were recorded, resulting in a cumulative survival rate of 98.6% after 5 years post-loading, (ii) at 5-year follow-up, the mean crestal bone loss was −0.28 ± 0.60 mm, and (iii) over 99% of patients reported satisfaction with the restoration as excellent or good [36]. Studying 1,078 cases

(601 males and 477 females) with 2,053 implants, it was found that after implantation, 1,974 implants were retained, and the early survival rate was 96.15% [28]. Del Fabbro et al. [37] conducted a systematic review to evaluate the survival rates of immediately loaded implants after at least five years. Besides implant failure, the amount of marginal bone loss around implants and the complication type were assessed. Thirty-four prospective studies with at least 5-year follow-up, published between 2007 and 2017 were included. A total of 5,349 immediately loaded implants in 1,738 patients were analyzed. It was reported that: (i) the mean weighted implant survival was 97.4%, (ii) cumulative survival rate of implants placed in the mandible was significantly higher than for the maxilla, and (iii) no significant difference in failure rate was found among the types of prosthesis employed, suggesting that (iv) immediate loading of implants appears to have long-term predictability and success rate under well-defined circumstances.

Unlike orthopedic implants, survival and success rates are sensitively influenced by several unique parameters, including size of implant main body, immediate loading or delayed loading, maxilla or mandible, or year span. Papaspyridakos et al. [38] compared survival rates between short implants (≤6 mm) and longer implants (>6 mm) after periods of 1–5 years in function. It was found that: (i) a short implant has higher variability and lower predictability in survival rates compared to longer implants and (ii) the mean survival rate was 96% (range: 86.7% ~ 100%) for short implants, and 98% (range 95% ~ 100%) for longer implants. The possibility of using short implants was indicated as a valid alternative in selected cases where bone quantity precludes the use of longer implants, which would require potentially extensive bone-grafting that increases invasiveness as well as morbidity of the treatment and treatment time. Especially for the posterior mandible where vertical ridge augmentation tends to be a challenging procedure with guarded predictability, the use of short implants seems to offer an excellent alternative [38]. Di Martino et al. [39] performed a registry study to query survival rates, hazard ratios, and reasons for revision of different stem designs in THAs after developmental dysplasia of the hip. A regional arthroplasty registry was inquired about cementless THAs performed for hip dysplasia from 2000 to 2017. Patients were stratified according to stem design in tapered (TAP; wedge and rectangular), anatomic (ANAT), and conical (CON), and divided on the basis of modularity (modular, M; non-modular, NM). In total, 2,039 TAP stems (548 M and 1,491 NM), 1,435 ANAT (1,072 M and 363 NM), and 2,287 CON (1,020 M and 1,267 NM) implants were included. Survival rates and reasons for revisions were compared. It was obtained that: (i) the survival rates (using any revision surgery as an endpoint) showed that the NM-ANAT group achieved the best results, whereas the NM-CON cohort achieved the worst outcomes at long-term follow-up, and (ii) specifically, the NM-ANAT cohort achieved a survival rate of 98.4% (95%), whereas 94.4% of the NM-CON stems survived at 10 years (95%). Based on these findings, it was concluded that (iii) NM-CON stems showed the highest risk of failure, especially high rates of cup aseptic loosening, (iv) NM-CON implants were not more prone to dislocations and stem aseptic loosening, and (v) clinical comparative studies are required to investigate the causes of NM-CON failures, which may be due to abnormal acetabular morphology or imperfect restoration of the proximal biomechanics.

As to location effects, it was mentioned that the survival rate of implants in 10 ~ 15 years is about 90% in the upper jaw and about 94% in the lower jaw [40]. Furthermore, Arakawa et al. [41] reported that the probability of successful osseointegration for maxilla anterior was 84.1%, mandible anterior 96.9%, maxilla posterior 78.9% and mandible posterior 95.9%, suggesting that: (i) mandible exhibits higher probability of osseointegration than maxilla, (ii) posterior shows lower probability than anterior, and (iii) survival rate at 10-year post-implantation for maxilla was 96.4% while it was 94.0% for the mandible [41].

Reasons why either success rate or survival rate has been decreasing by years, as we have seen in the above, can be expatiated by some irreversible factors. Implant material per se and its structural integrity would deteriorate constantly during the usage, although its rate might be relatively slow. At the same time, implant receiving hard/soft tissue of the implant recipient is weakened and embrittled with much faster speed than the material's weakening process, mainly due to natural aging and some unexpected age-related disease (local or systemic) as well as frailty. Antiaging or lookism and the like could be included in these concerns. As indicated in Figure 1.1, all these concerns are strongly affected by how bone-grafting (materials selection as well procedures) is effectively and efficiently managed and achieved.

1.2 Bone-grafting, bone augmentation, and bone regeneration

Bone-grafting can be considered as a medical treatment to enable living tissues to promote bone healing. A certain type of biomaterial is transplanted into a bony defect or bony structure with less quantity and/or poor quality. Such transplanted grafting material can be either natural or synthetic material alone or in a composite form or hybrid form with other materials [42, 43]. In articles relevant to orthopedic implants and dental implant treatment, there are different yet similar terms used: i.e., bone-grafting, bone augmentation, bone reconstruction, and bone regeneration, with unclear differentiations of one from the others.

Bone augmentation, also referred to bone-grafting, is a procedure typically needed when the current bone mass is too thin or soft unsuitable for successfully incorporating dental implants (in a sense of osteointegration). In dental implantology, there is an important treatment known as a ridge augmentation. The main difference between bone-graft and ridge augmentation is the entirety of the process. A dental bone-graft is used to help stimulate and encourage bone healing after a tooth has been pulled. A ridge augmentation is to fully form and shape the newly grown bone after the graft has been completed. The alveolar ridge itself is a special jawbone surrounding teeth roots. When any tooth or teeth get removed, this leaves an empty socket in the alveolar ridge bone. The particular surgical procedure has more to do with improving the alveolar ridge's shape and size. It prepares the ridge area to better receive and likewise retain any form of dental prosthesis to be added to the site [44]. A common use of bone-grafting is for ridge augmentation, which can recapture the natural contour of gums and jaw after the loss of a tooth as a result of trauma, congenital abnormalities, infection, or periodontal disease. Achieving an ideal amount of gum and bone as a support to surrounding restorations or implants may require hard and soft tissue reconstruction [44]. Graft should have several stages of healing in the following sequences: inflammation (necrotic debris stimulates chemotaxis) → osteoblast differentiation (differentiates from mesenchymal precursor cells) → osteoinduction (stimulation of osteoblast and osteoclast function) → osteoconduction (bone forms around the new scaffold) → remodeling (continual process for years) [45].

Bone reconstruction is considered as regeneration of lost bone. The reconstruction of bone defects relies upon the reconstitution of bone in the area. The basic concepts of fracture bone healing are presented to better understand the mechanism of bone-grafting. Conventional bone-graft techniques are quite effective in reconstructing defects, including those up to 25 cm in length, provided an adequate vascular bed is available to allow vascularization of the grafts. If an adequate milieu is not available, then the use of a pedicle or free vascular bone-graft can be employed. Just as skin and muscle flaps have provided the solution to soft-tissue loss, vascularized bone-grafts provide the technique to deal with large bone defects not amenable to conventional treatment. Allografts additionally have a place in skeletal reconstruction but require appropriate management and consideration of host immune response [46].

Bone regeneration is a complex, integrated physiological process of bone formation, which can occur during the healing stage of fracture bone, and is employed in continuous bone remodeling throughout adult life [47]. Currently, there are quite different techniques to augment the impaired or "insufficient" bone-regeneration process, including the "gold standard" autologous bone-graft, free fibula vascularized graft, allograft implantation, and use of growth factors, osteoconductive scaffolds, osteoprogenitor cells, and distraction osteogenesis [47].

References

[1] Negm SAM. Implant success versus implant survival. Dentistry. 2016, 6, 1–5; doi: 10.4172/2161-1122.1000359.
[2] Orthopedic Implants. Part 1 – Surface finishing enhances component life, function. 2021; https://roslerblog.com/2021/05/25/orthopedic-implants-part-1-surface-finishing-enhances-component-life-function/#:~:text=The%20longevity%20of%20these%20implants,attached%20to%20the%20respective%20bones.
[3] Can implant longevity be extended? Int Congr Joint Reconstruc. 2016; https://icjr.net/articles/can-implant-longevity-be-extended.
[4] How long do metal implants last in the body? The question you should ask before implant placement. 2020; https://monib-health.com/en/post/55-how-long-do-metal-implants-last.

[5] How Long do Hip Replacements Last? https://www.coreorthosports.com/how-long-do-hip-replacements-last/.

[6] Oshida Y, Miyazaki T. Biomaterials and engineering for implantology – in medicine and dentistry. De Gruyter STEM Series. 2022.

[7] Wagner L, Gregory JK. Improve the fatigue life of titanium alloys. Part I. Adv Mater Processes. 1994, 146, 35–36.

[8] Wagner L, Gregory JK. Improve the fatigue life of titanium alloys. Part II Adv Mater Processes. 1994, 146, 50–53.

[9] Oshida Y, Daly J. Fatigue damage evaluation of shot peened high strength aluminum alloy. In: Meguid SA (Ed.)., Surface Engineering. Elsevier Applied, NY. 1990, 404–16.

[10] Oshida Y. Bioscience and Bioengineering of Titanium Materials. Elsevier, NY. 2007.

[11] Dane CB, Hackel LA, Daly J, Harrison J. High power laser for peening of metal enabling production technology. Mater Manuf Process. 2000, 15, 81–96.

[12] Orthopedic Implants – Shot Blasting Improves Longevity. 2022; https://roslerblog.com/2022/03/29/orthopedic-implants-part-6-shot-blasting-improves-longevity/.

[13] Croes M, Akhavan B, Sharifahmadian O, Fan H, Mertens R, Tan RP, Chunara A, Fadzil AA, Wise SG, Kruyt MC, Wijdicks S, Hennink WE, Bilek MMM, Yavari SA. A multifaceted biomimetic interface to improve the longevity of orthopedic implants. Acta Biomater. 2020, 110, 266–70.

[14] van Hengel IAJ, Gelderman FSA, Minneboo M, Weinans H, Fluit AC, van der Eerden BCJ, Fratila-Apachitei LE, Apachitei A, Zadpoor AA. Functionality-packed additively manufactured porous titanium implants. Mater Today Bio. 2020, 7, 10060; doi: https://doi.org/10.1016/j.mtbio.2020.100060.

[15] Kurtin K Using Nanotechnology to Boost the Lifespan of Medical Implants. 2010; https://today.uconn.edu/2010/01/using-nanotechnology-to-boost-the-longevity-of-medical-implants/.

[16] Feurstein S. Do dental implants last forever? Long Dent Implants. 2020; https://www.mkdentist.com/do-dental-implants-last-forever.

[17] How Long Do Dental Implants Last? 2021; https://newteethchicagodentalimplants.com/how-long-do-dental-implants-last/.

[18] Everything you need to know about the longevity of dental implants. Mt Laurel Dent Implant Center. 2020; https://www.mtlaureldental.com/longevity-dental-implants.

[19] Rigby JC. The Lifespan of Dental Implants. 2020; https://www.jerseysmiles.com/lifespan-of-dental-implants.

[20] Bidez MW, Misch CE. Clinical biomechanics in implant dentistry. In: Misch CE (Ed.)., Dental Implant Prosthetics. 2nd Ed, Elsevier-Mosby. 2014; doi: 10.1016/B978-0-323-07845-0.00005-1.

[21] Singh JA, Sperling JW, Cofield RH. Revision surgery following total shoulder arthroplasty: Analysis of 2,588 shoulders over 3 decades (1976–2008). J Bone Joint Surg Br. 2011, 93, 1513–17.

[22] Prime MS, Palmer J, Khan WS. The national joint registry of England and wales orthopedics. 2011, 34, 107–10.

[23] Smith GH, Johnson S, Ballantyne JA, Dunstan E, Brenkel IJ. Predictors of excellent early outcome after total hip arthroplasty. J Orthop Surg Res. 2012, 7, 13–15.

[24] Bae DK, Song SJ, Heo DB, Lee SH, Song WJ. Long-term survival rate of implants and modes of failure after revision total knee arthroplasty by a single surgeon. J Arthroplasty. 2013, 28, 1130–34.

[25] Akagawa Y. Implant history in the last 100 years and the evolution of osseointegrated implant. JICD. 2020, 51, 28–33; https://www.icd-japan.gr.jp/pub/vol51/08-vol51.pdf.

[26] Zarb GA, Albrektsson T. Consensus report: Towards optimized treatment outcomes for dental implants. J Prosthet Dent. 1998, 80, 641; doi: 10.1016/s0022-3913(98)70048-4.

[27] Nishinaka H. How long does an implant last? -Survival rate and success rate-; https://nishinaka-dental.com/2020/08/04/survival-rate-of-implant/.

[28] Yang Y, Hu H, Zeng M, Chu H, Gan Z, Duan J, Rong M. The survival rates and risk factors of implants in the early stage: A retrospective study. BMC Oral Health. 2021, 21, 293; doi: https://doi.org/10.1186/s12903-021-01651-8.

[29] Busenlechner D, Fürhauser R, Haas R, Watzek G, Mailath G, Pommer B. Long-term implant success at the Academy for Oral Implantology: 8-year follow-up and risk factor analysis. J Periodontal Implant Sci. 2014, 44, 102–08.

[30] Krebs M, Schmenger K, Neumann K, Weigl P, Moser W, Nentwig G-H. Long-term evaluation of ANKYLOS® dental implants, Part I: 20-year life table analysis of a longitudinal study of more than 12,500 implants. Clin Implant Dent Relat Res. 2015, 17, e275–86.

[31] Goel MK, Khanna P, Kishore J. Understanding survival analysis: Kaplan-Meier estimate. Int J Ayurveda Res. 2010, 1, 274–78.

[32] Artzi Z, Carmeli G, Kozlovsky A. A distinguishable observation between survival and success rate outcome of hydroxyapatite-coated implants in 5–10 years in function. Clin Oral Implants Res. 2006, 17, 85–93.

[33] Moraschini V, Poubel LDC, Ferreira VF, Dos Sp Barboza E. Evaluation of survival and success rates of dental implants reported in longitudinal studies with a follow-up period of at least 10 years: A systematic review. Int J Oral Maxillofacial Surg. 2015, 44, 377–88.

[34] Bandeira de Almeida A, Prado Maia L, Demoner Ramos U, Luís Scombatti de Souza S, Bazan Palioto D. Success, survival and failure rates of dental implants: A cross-sectional study. J Oral Health Rehab. 2017; https://www.dtscience.com/success-survival-and-failure-rates-of-dental-implants-a-cross-sectional-study.

[35] Cochran D, Oates T, Morton D, Jones A, Buser D, Peters F. Clinical field trial examining an implant with a sand-blasted, acid-etched surface. J Periodontol. 2007, 78, 974–82.

[36] Beschnidt SM, Cacasi C, Dedeoglu K, Hildebrand D, Hulla H, Iglhaut G, Trennmair G, Schlee M, Sipos P, Stricker A, Ackermann K-L. Implant success and survival rates in daily dental practice: 5-year results of a non-interventional study using CAMLOG SCREW-LINE implants with or without platform-switching abutments. Int J Implant Dent. 2018, 4, 33; doi: https://doi.org/10.1186/s40729-018-0145-3.

[37] Del Fabbro M, Testtori T, Kekovic V, Goker F, Tumedei M, Wang H-L. A systematic review of survival rates of osseointegrated implants in fully and partially edentulous patients following immediate loading. J Clin Med. 2019, 8, 2142; doi: 10.3390/jcm8122142.

[38] Papaspyridakos P, De Souza A, Vazouras K, Gholami H, Pagni S, Weber H-P. Survival rates of short dental implants (≤6 mm) compared with implants longer than 6 mm in posterior jaw areas: A meta-analysis. Clin Oral Implants Res. 2018; https://onlinelibrary.wiley.com/doi/epdf/10.1111/clr.13289.

[39] Di Martino A, Castagnini F, Stefanini N, Bordini B, Geraci G, Pilla F, Traina F, Faldini C. Survival rates and reasons for revision of different stem designs in total hip arthroplasty for developmental dysplasia: A regional registry study. J Orthop Traumatol. 2021, 22, 29; doi: https://doi.org/10.1186/s10195-021-00590-y.

[40] Cumulative implant survival rate and causes of shortened lifespan; https://www.cidjp.org/implant-life-span.html.

[41] Arakawa H, Kuboki T, Kanayama M, Sonoyama W, Kojima S, Yatani H, Ueno T, Takagi S, Sugawara T, Mano T, Matsumura T. A clinical follow-up study on acquisition and maintenance of osseointegration on relation to risk factors for osseodisintegration. J Jpn Soc Oral Implant. 2002, 15, 66–74.

[42] Elsalanty ME, Genecov DG. Bone grafts in craniofacial surgery. Craniomaxillofacial Trauma Reconstr. 2009, 2, 125–34.

[43] Zhao R, Yang R, Cooper PR, Khurshid Z, Shavandi A, Ratnayake J. Bone grafts and substitutes in dentistry: A review of current trends and developments. Molecules. 2021, 26, 3007; doi: 10.3390/molecules26103007.

[44] Madigan MP. Bone Grafts and Ridge Augmentation; https://www.madiganperio.com/bone-grafts-and-ridge-augmentation.

[45] Bone Grafting; https://www.orthobullets.com/basic-science/9011/bone-grafting.

[46] Bieber EJ, Wood MB. Bone reconstruction. Clin Plast Surg. 1986, 13, 645–55.

[47] Dimitriou R, Jones E, McGonagle D, Giannoudis PV. Bone regeneration: Current concepts and future directions. BMC Med. 2011, 9, 66; doi: https://doi.org/10.1186/1741-7015-9-66.

2 Bone-grafting treatment

Bone-grafting is one of the most frequently used procedures in traumatology, orthopedics, and oral and maxillofacial surgery, with the intent to form new bone tissue at the target area (e.g., skeletal defect, atrophy region, a space between bones to be fused) [1]. Bone-grafting is employed in multiple surgical areas, which could include orthopedic surgery, neurosurgery, plastic surgery, and dental surgery. Bone-grafts fill bony voids or gaps of the skeletal system that may be the result of surgically created osseous defects (e.g., tumor resection), related to osteonecrosis, or the result of trauma [2]. Almost two million bone-grafting procedures are performed worldwide every year [3]. In the United States, over 500,000 bone-grafting procedures are performed annually, making bone-grafts second only to blood transfusion as the most common tissue transplantation [4]. It is also reported that annually, half a million patients require bone repair intervention in the US and Europe [5]. The global annual expenditure in bone fractures and orthobiologics is estimated at US$5.5 billion and US$4.7 billion, respectively, whilst the total cost of bone repair-related expenditure is estimated at US$17 billion per year [6].

Bone is known for its self-healing abilities [7]. The healing of bone fractures is a remarkable repairing process, resulting in the complete reconstruction of the tissue achieving its original form and functionality [8]. Bone healing is a well-orchestrated process and for most minor fractures, a mechanical fixation of the damaged bone region is sufficient for successful convalescence [9]. Critical-sized bone defects are regularly treated by autograft and allograft transplantation. However, such treatments require harvesting bone from patient donor sites, with often limited tissue availability or risk of donor site morbidity [9]. It is said that if a defect reaches a critical size (≥2.5 cm) [10, 11], the endogenous regenerative capacity of bone tissue is insufficient for self-repair [12]. Critical-sized bone defects caused by diseases such as osteogenesis imperfecta, osteoarthritis, osteomyelitis, osteoporosis, or conditions related to infection or induced by wear, still remain crucial key challenges to be addressed in clinical practice [11, 13]. Porter et al. [13] mentioned that besides illnesses, trauma and tumors can lead to a critical-sized bone defect. The gold standard treatment involves autografts (bone taken from the patient's own body) and allografts (bone tissue taken from a donor) [7]. Even if successful, challenges like the limited supply of autografts, transmission of diseases, rejection of grafts, donor site pain and morbidity, limitation in volume of donor tissue that can be safely harvested, and the possibility of harmful immune responses to allografts, drive surgeons and engineers to seek alternative methods and materials to repair bone defects [14, 15]. This can be advanced technologies in manufacturing field such as the additive manufacturing, including 3D bioprinting, which will be discussed in later chapters.

https://doi.org/10.1515/9783111136691-002

2.1 Procedures and reasons for bone-grafting

During bone-grafting surgery, the ordinary procedures should follow the listed steps [16–18]:

(1) A general anesthesia is applied to a patient to make patient sleep and to temporarily block sensation (pain and/or discomfort); meanwhile a healthcare provider will carefully monitor the patient's vital signs (like heart rate and blood pressure), during the operation.

(2) After cleaning the affected area, a surgeon will make a cut through the skin and muscle surrounding the bone to access the bone that needs grafting.

(3) Bone-graft is a choice for repairing bones almost anywhere in a patient's body. The surgeon might take bone from hips, legs, or ribs to perform the graft. Sometimes, surgeons also use bone tissue donated from cadavers to perform bone-grafting.

(4) The selected and prepared bone-graft is inserted between the two pieces of bone that need to grow together. In some cases, the surgical assistant might secure the bone-graft with special screws.

(5) The layers of skin and muscle around the treated bone will be closed surgically with (preferably) bioresorbable sutures and, if necessary, around where the bone was harvested. The transplanted graft will be held in place using any of the fixing devices such as pins, plates, screws, wires, or cables. In some cases, a cast or splint may be used to support the bone while it heals, although many times, no casting or splint is necessary.

(6) As post-operation cares, a patient might need to take medicines to prevent blood clots (known as "blood thinner") for a little while after the surgery. A patient is suggested not to take certain over-the-counter medicines for pain, because some of these can interfere with bone healing. The patient might be advised to eat a diet high in calcium and vitamin D as the bone heals. Follow-up appointments are scheduled for removing stitches (if they are not bioresorbable type) or staples, a week or so after the surgery, and a series of X-rays should be taken to monitor how the bone has been healing.

A certain type of bone-grafting is needed to promote bone healing and growth for a number of different medical reasons. Some specific conditions that might require a bone-graft include [16, 17, 19]:

(1) An initial fracture that a physician or a healthcare provider suspects that the fracture won't heal without a graft.

(2) A previous fracture that was not treated with a graft and that did not heal well.

(3) Bone disease like osteonecrosis and cancer.

(4) Bone infection (osteomyelitis).

(5) Congenital anomalies, such as uneven limbs or an abnormally small chin (causing facial disfigurement [20, 21]).

(6) Spinal fusion surgery (which might be needed if an unstable spine is noted).
(7) Jawbone reinforcement (dental bone-graft) before receiving a tooth implant.
(8) Joint replacement surgery, which may require bone growth to secure an artificial joint, including total shoulder arthroplasty, total hip arthroplasty, or total knee arthroplasty.
(9) Trauma, including bad fractures that shatter bones.

It is known that most of the skeleton consists of bone matrix, which is the hard material to provide bones the strength. Inside the matrix, there are living bone cells, which make and maintain the matrix. The cells in this matrix can help repair and the heal bone, when necessary [16]. Upon breaking a bone, the healing process begins immediately. If the break in the bone is not too large, bone cells can repair themselves. Sometimes, a fracture results in a large loss of bone, like when a large chunk of the bone crumbles away. In these cases, the bone might not fully heal, and bone-graft is necessitated. During a bone-graft, a new piece of bone is inserted in the place where a bone needs to heal or join. The cells inside the new bone can then seal themselves to the old bone. Bone-grafts can provide a framework for the growth of new, living bones. Hips, knees, and spine are common locations for bone-grafting as autografts, but some patients might need bone-grafting for a different bone in a body. Normally, if the patient's own bone is transplanted, the patient should have to have extra surgery to remove this bone. On the other hand, if the patient chooses not to use autografts and selects to use donated bone, the donated bone has its own risks [16]. Bone-grafting procedure may be necessary if bones do not heal correctly after a fracture; causing (i) delayed union (bone healing or fusion that occurs more slowly than expected), (ii) malunion (bones that heal in an abnormal position), or (iii) nonunion (when bone fusion does not occur) [16].

Moving to the dental field, natural teeth that are embedded in the jawbone help stimulate bone growth through chewing and biting. When missing teeth are left untreated, the bone no longer receives these biomechanical stimulations, which might cause the bone to resorb. It is generally believed that without a replacement tooth or dental solution, 25% of the surrounding bone is lost within the first year of a tooth extraction and will continue to deteriorate over time. Tooth loss is a serious issue that can be caused by different situations. Gum disease is the most common cause, and when gum disease develops, bacteria cause irritation and inflammation of the gum tissue. As a result, gums swell and pockets form, allowing bacteria to fall below the gumline. Then, the bacteria begin attacking the periodontal ligaments and the jawbone, which can ultimately lead to tooth loss. There could be other causes of tooth loss, including (i) facial trauma, resulting from a car accident, a sports injury, or a fall, (ii) severe tooth decay, or (iii) bruxism, or a condition like clenching and grinding teeth [22]. Accordingly, tooth loss (and bone loss as a resultant of the tooth loss) might cause masticatory disorder and poor aesthetic appearance as well.

Bone-grafting is a surgical treatment that is performed to reverse bone loss or enhance an existing bone in areas that may be deficient. In dentistry, there are several bone and soft tissue grafting treatments [23, 24]:

(1) Ridge augmentation. If the alveolar ridge bone, a special type of bone surrounding and supporting teeth, has deteriorated or lost its density, ridge augmentation may be required to widen or heighten the jaw in preparation for the placement of dental implants. During ridge expansion, the alveolar ridge is surgically split, and bone-grafting material is inserted.

(2) Sinus lift. When there is not enough bone present between the upper jaw and the sinus cavity for a successful dental implant, oral surgeons can perform a sinus lift. In this procedure, the sinus membrane is lifted and a bone-graft is placed below the Schneiderian membrane. The graft will integrate with the jawbone over several months. Once the graft has fused with the natural bone (aka, osteointegration), dental implants can be placed with a much higher rate of success.

(3) Socket preservation. This procedure prevents bone loss and prepares for the success of an eventual dental implant by placing a bone-graft into the empty socket, immediately after a tooth extraction. If a tooth was extracted but is not immediately replaced by a dental implant, socket preservation is ideal to maintain the integrity of the empty tooth socket.

(4) Soft tissue graft. If periodontal disease has advanced to cause gum recession, to the extent that the gum does not line up with an implant crown placement, a soft tissue graft will be recommended. This will ensure that the gumline is uniform around the implant for a natural look. Soft tissue grafts consist of small tissue pieces taken from other areas that are surgically implanted into the desired areas. In addition to the aesthetic benefit of this procedure, it helps to stop bone loss and further recession of the gums, and it even helps reduce painful root sensitivity.

(5) Bone morphogenetic protein (BMP). Bone morphogenetic protein is a type of protein found naturally in human bodies that helps stimulate new bone growth. Cytokines such as BMP, IGF-I (insulin-like growth factor) or FGF (fibroblast growth factor) are important for osteoblast development. This is a general term for proteins with molecular weight of about 10,000 to tens of thousands secreted by cells. Cytokines play an important role in the action of biological defenses such as inflammatory responses and immunity, as well as in the transmission of information between cells. This material can be synthesized and used in bone-grafting and augmentation procedures, to increase the jawbone support required for a stable dental implant placement. When BMP is used, an additional procedure to harvest bone is unnecessary because the sample comes from the same patient's body. BMP has been used in oral surgery to expedite healing and improve the clinical outcome of surgical procedures.

2.2 GTR and GBR

Periodontal disease is a common disease in oral surgery, especially in the periodontal supporting tissues (gums, periodontal ligament, alveolar bone, and cementum). The incidence rate is high, manifesting as gingival swelling and bleeding, which affects the quality of life of patients, which is one of the main causes of adult dentition defects [25–27]. The periodontal intraosseous defect is a serious clinical symptom caused by periodontal diseases such as gingival disease and periodontitis. If it is not treated in time, it can lead to alveolar bone resorption and defect, periodontal attachment loss, and teeth loosening and falling off [28, 29]. Periodontal bone defects affect the masticatory function and aesthetic appearance of patients and increase their physiological and psychological pressure [30].

Basically, there are two methods to treat periodontal bone defects: guided tissue regeneration (GTR) and guided bone regeneration (GBR). The former refers to procedures that attempt to regenerate lost periodontal structures (bone, periodontal ligament, and connective tissue attachment) that support tooth structures. This is accomplished using biocompatible membranes, often in combination with bone-grafts and/or tissue stimulating proteins. The latter (GBR) refers to procedures that attempt to regenerate bone, prior to the placement of bridges or, more commonly, implants. This is accomplished using bone-grafts and biocompatible membranes that keep out tissue and allow the bone to grow. A lack of horizontal and/or vertical bone in implant sites may cause major clinical problems [31] and needs to be corrected prior to implant placement. To regenerate enough bone for successful implant placement, a ridge augmentation technique is often required. To regenerate enough bone for successful implant placement, GBR is often required, which is a surgical procedure that uses barrier membranes with or without particulate bone-grafts or/and bone substitutes. There are two approaches to GBR in implant therapy: GBR at implant placement (simultaneous approach) and GBR before implant placement, to increase the alveolar ridge or improve ridge morphology (staged approach). Angiogenesis and ample blood supply play a critical role in promoting bone regeneration [32]. Yuan et al. [30] investigated the curative effect of guided tissue regeneration (GTR), combined with bone-grafting, and improvement in the aesthetic appearance of patients' gingiva. It was concluded that (i) GTR, combined with bone-grafting, has a good effect in repairing periodontal intraosseous defects and can effectively promote the reconstruction and recovery of the periodontal intraosseous defects in patients, and (ii) at the same time, it can significantly improve the gingival aesthetics of patients, which has good clinical application value.

For combining dental implant placement and GBR technique, there are basically two methods: (1) staged procedure and (2) simultaneous procedure.

(1) Staged procedure: A staged procedure is one in which the required bone mass is regenerated and increased by the GBR procedure in advance, and then implant treatment is performed according to the ordinary method. The period for keeping membranes placed by GBR lasts, in general, 6 ~ 9 months. This period is consid-

ered to vary slightly in the condition of bone defects, but it can generally be understood as half year. When using a nonabsorbable membrane, the primary operation of the implant placement is performed at the same time when it is removed. The advantages of this method include: (i) A choice between one-time and two-time implant treatment. (ii) The implant can be placed in an ideal position. (iii) The complication of GTR and implant does not affect each other. On the other hand, the disadvantage can be that the overall treatment period is longer than that of the simultaneous method.

(2) Simultaneous procedure: In this method, performing GBR and fixture placement are carried out at the same time. It is essential that the fixture is initially fixed. The stepwise method should be chosen if initially not fixed. As an advantage, the overall treatment period can be shortened compared to the former method; whilst the disadvantage is that failure of GBR tends to lead to implant failure (in other words, failure of osteointegration).

In summary, in a clinical case of less quality of bone mass, the above discussed bone-graft, GBR or sinus lift, is not the only technique employed. There should include, besides these techniques, socket lifting, distraction osteogenesis, and split crest techniques [33].

2.2.1 Socket lifting

The socket lift procedure allows an increase in bone quantity in cases where the height of the upper jawbone is greater than 5 mm. The socket is prepared for implant placement, followed by carefully tapping the bottom of the sinus with a cylindrical instrument called an osteotome to push up the sinus membrane. An artificial bone material or autogenous bone is inserted into the elevated area to increase the bone height. Due to the elevation of the planned implant site, the wound site is smaller than in the sinus lift technique [34].

There are essential differences between sinus lift and socket lift [35]. The sinus lift technique approaches from the alveolar lateral side to the maxillary sinus and is used in cases in which the thickness of the upper jaw is less than 5 mm or when multiple teeth are lost. The lateral side of the upper gum is incised to expose the bone surface and form a window of about $10 \sim 30$ mm^2 in the alveolar bone. The window is opened to expose the Schneiderian membrane (maxillary sinus mucosa), and the alveolar bone and the Schneiderian membrane are then carefully exfoliated to pack the grafting bone. The bone is formed over the following $3 \sim 6$ months. Thereafter, as an implant is placed, the treatment takes about 9 months. On the other hand, the socket lift technique approaches from the alveolar crest or the extraction cavity to the maxillary sinus, and is applicable in cases where the bone height to the Schneiderian membrane is greater than 5 mm. It can have a shorter treatment period and less pain/

swelling due to the smaller wound, compared with the sinus lift technique. The cavity (space after tooth extraction) is gradually filled with grafted bone, to push up the Schneiderian membrane to obtain sufficient bone thickness for implant placement. Following implant placement, it takes about 4 months before it is possible to chew normally.

2.2.2 Distraction osteogenesis (bone lengthening)

The distraction osteogenesis (aka callus distraction, callotasis, or osteodistraction) is a process used in orthopedic surgery, podiatric surgery, and oral and maxillofacial surgery to repair skeletal deformities and in reconstructive surgery. Distraction osteogenesis is a technique in orthognathic surgery that expands or stretches the bone in the jaw. Simply stated, distraction osteogenesis means slowly moving apart the two bony segments that allows the new bone to fill in the gap created by the bony segments. This technique was initially used to treat defects of the oral and facial region. Over the past two decades, surgical and technological advances in the field of distraction osteogenesis have provided oral and maxillofacial surgeons with a safe and predictable method to treat select deformities of the oral and facial skeleton [36–38].

2.2.3 Split crest technique

The split crest technique (known as a surgical expansion of the alveolar ridge) consists of horizontal and vertical osteotomies, intended to separate the vestibular and lingual/palatal bone cortices, creating a space for the simultaneous implant placement. The main advantages of such a technique are the fact that it allows the installation of implants in the same surgery, decreasing the number of surgical procedures and the time of treatment for the patients; and especially the fact that it does not need a donor area to remove grafts, which reduces the morbidity and the complication rate for the patients. The split crest technique, in addition to the fact that it does not require a donor area to remove the graft, makes it possible to install dental implants in the same operative procedure, decreasing morbidity, treatment time, number of surgical procedures and even the risk of complications to the patients, and it is therefore more easily accepted [39, 40].

2.3 Requirements for ideal bone-graft

Currently, there are various types of bone-grafting materials. They include [41–43]:
(1) Autograft: An autograft uses a sample of a patient's bone tissue. The tissue typically comes from the top of the hip bone (iliac crest). The benefit of using an auto-

graft tissue is that it increases the chances of successful fusion, but the amount of bone tissue that can be collected is limited.

(2) Allograft: This method uses bone tissue from another person (donor). Public health services have strict regulations on how tissues are handled and treated, and the bone tissue is cleaned and processed (sterilized) to ensure the safety of the recipient. This type of graft is common in spinal fusion surgery. It provides a framework around which healthy bone tissue can grow.

(3) Bone marrow aspirate: Marrow is the spongy substance inside bones and contains stem and progenitor cells that can help bone fractures heal. Using a needle, the surgeon gets a bone marrow sample from the hip bone (iliac crest). This bone marrow aspirate is used alone or mixed with other bone-grafts to enhance bone healing for allograft procedures.

(4) Bone substitutes (or synthetic bone-graft): This type of graft uses artificially produced materials made from a variety of porous substances. Some also contain proteins that support bone development.

No matter what type of bone-graft is selected, the basic principle of bone healing should not be ignored because bone healing is essential. Bone healing is a multilateral process that requires mechanical stability and revascularization reaction, along with osteogenesis, osteoinduction, and osteoconduction [44–50].

2.3.1 Osteogenesis

Osteogenesis is the formation of new bone by osteoblasts within the graft material [44–46].

2.3.2 Osteoinduction

Osteoinduction means that primitive, undifferentiated, and pluripotent cells are stimulated to develop into the bone-forming cell lineage [51]. Osteoinduction involves the stimulation of osteoprogenitor cells to differentiate into osteoblasts that then begin new bone formation. The most widely studied type of osteoinductive cell mediators are bone morphogenetic proteins (BMPs) [47]. A bone-grafting material that is osteoconductive and osteoinductive will not only serve as a scaffold for currently existing osteoblasts but will also trigger the formation of new osteoblasts. Osteointegration is the direct contact of vital bone to the graft material [48].

2.3.3 Osteoconduction

Osteoconduction means that bone grows on a bone surface that permits bone growth on its surface or down into pores, channels, or pipes [51]. Accordingly, the most ideal bone substitute should meet such conditions, have no risk of immunological rejection (biocompatible) or disease infection, and achieve incorporation of graft in the host bone by gradually being substituted by the regenerated bone [51]. It should be well molded into the bony defect within a short time and should be osteoinductive, osteo-conductive, and resorbable [52]. In addition, the ideal bone substitute should be ther-mally nonconductive, sterilizable, and readily available at a reasonable cost [52]. Osteoconduction is the ability to support bone growth on a surgical site, when pores, channels, and blood-vessels are formed within the bone. Osteoblasts, from the margin of the defect that is being grafted, utilize the bone-graft material as a framework to spread and generate new bone [45, 46].

Based on what has been mentioned in the above, several requirements for accept-able and successful bone-grating materials and important factors for successful achieve-ment of bone-graft delivery system can be found [53–56]:

(1) Bone-grafts are used as a filler and scaffold to facilitate bone formation and promote wound healing. These grafts (used as a filler and scaffold to facilitate bone formation and promote wound healing) are bioresorbable and have no antigen–antibody reac-tion. These bone-grafts act as a mineral reservoir that induces new bone formation.
(2) There must be enough osteoblasts present to create new bone, or it will fail.
(3) Sufficient blood supply is crucial for clot formation to serve as an anchor for the osteoblasts and cell viability.
(4) The bone-graft delivery must be stable to avoid any disruption in the blood clot and to prevent any fibrous tissue from filling the area. There also can be no ten-sion on soft tissue.

2.4 Benefits and drawbacks

Benefits and drawbacks of bone-graft are based on the grafting technique [57, 58].

2.4.1 Autograft

Benefits:
– No potential for immune reaction (low chance of infection) or no risk of disease transmission
– Commonly used in surgery
– Well-documented success

– May heal large or small defects by itself
– Transplanting your own bone-forming cells to help heal the defect

Drawbacks:
– Risk of pain and/or infection at the harvest site, which may last for a long time
– Additional surgery and anesthesia are required
– May not be an option for some patients

2.4.2 Allograft

Benefits:
– Commonly used in surgery
– May heal small defects by itself
– Portions of the graft may turn into your own bone
– No additional procedure is necessary to harvest bone tissue
– Low risk of spreading disease because bone tissue is sanitized
– Does not take tissue from other bones

Drawbacks:
– Minimal risk for disease transmission
– Does not stimulate your body's cells to form bone
– Portions of the graft may remain in your body for years to come
– Limited in its ability to heal large defects by itself

2.4.3 Xenograft

Benefits:
– Commonly used in surgery
– Not human-derived
– Readily available
– Well-documented success
– May heal small defects by itself
– Portions of the graft may turn into your own bone

Drawbacks:
– Low risk of disease transmission
– Does not stimulate your body's cells to form bone
– Portions of the graft may remain in your body for years to come
– Limited in its ability to heal large defects by itself

2.4.4 Alloplastic graft

Benefits:
– Commonly used in surgery
– Not human-derived
– Readily available
– Well-documented success
– May heal small defects by itself
– Portions of the graft may turn into your own bone
– No risk for disease transmission
– Sterile, free of germs
– Available in unlimited quantities, so it can repair large sections of bone
– Many options exist, making it easier to meet a wide range of medical needs

Drawbacks:
– Does not stimulate your body's cells to form bone
– Portions of the graft may remain in your body for years to come
– Limited in its ability to heal large defects by itself

2.4.5 Growth factors (a synthetic version of a natural protein to regulate bone healing and growth)

Benefits:
– No need for a second surgery to harvest bone from another place of a body
– It has proven and predictable bone growth results
– It is proven clinically safe and effective for bone formation

Drawbacks:
– May experience short-term mild-to-severe facial swelling (edema) after the surgery
– It has not been studied for use in patients under 18 years of age
– It cannot be used in patients with an active infection at the defect site
– It should not be used in pregnant women
– It should not be used in people with immune deficiencies due to other treatments

Gum disease can lead to tooth loss and jawbone atrophy. GTR (guided tissue regeneration) relies on bioactive growth factors, biological membrane barriers, and tissue-stimulating proteins to promote healthy, strong bone growth. This technique helps a patient's body regenerate tissues and bone that may have been lost because of gum disease. GTR can sometimes be used instead of bone-grafting as a minimally invasive alternative. Several benefits associated with the GTR technique are mentioned [59, 60].

1. GTR can help preserve patient's natural teeth. This is particularly true for those at risk of losing teeth due to bone loss from advanced gum disease. GTR can save natural teeth from failing due to bone loss from gum disease. By regenerating the lost bone and tissues surrounding a tooth, these restored structures will create the protective, strong foundation a tooth needs to remain healthy long-term.
2. GTR can be combined with bone-grafts. According to a study, guided tissue regeneration, specifically with barrier material, provided better success rates, when combined with bone-grafting material, when compared with bone-graft alone. In other words, guided tissue regeneration improved the outcomes of the bone-grafting procedure. GTR involves placing a special membrane between the bone and the soft tissues around a tooth. This membrane acts as a barrier so that the bone can regenerate and strengthen without interference from the faster healing gums.
3. Whether used alone or to complement bone-grafting, GTR helps restore lost bone and create a strong, stable foundation for dental implants. Dense bone ensures the implant posts are secured into healthy bone and are able to support new teeth long-term.
4. GTR encourages the growth of jawbone, indicating that this procedure exhibits an ability to reverse some of these destructive consequences. If jawbone can be fortified, the risk of losing teeth decreases.
5. Gum disease, missing teeth, and jawbone atrophy can impact the appearance of a smile. GTR helps to rebuild these tissues, reduces unsightly periodontal pockets, and improves the overall aesthetics of smile. Gum disease or other issues that cause soft tissue and bone loss around a tooth can harm the natural attractiveness of a smile. GTR rebuilds these structures, reducing unsightly periodontal pockets and improving the overall aesthetics of the teeth and smile.

2.5 Risks associated with bone-graft

It is reported that the following clinical situations can be considered as indications [61–63]:
(1) fenestration and dehiscence,
(2) building up bone around implants placed in tooth sockets after tooth extraction,
(3) socket preservation for future implantation of false teeth or prosthetics,
(4) sinus lift elevation, prior to implant placement,
(5) filling of bone after removing the root of a tooth, cystectomy or the removal of impacted teeth,
(6) repairing bone defects surrounding a dental implant caused by peri-implantitis,
(7) vertical and horizontal augmentation of the upper and lower jaws, and
(8) cystic cavity.

Goldstep [62] and Klein et al. [63] pointed out that bone-graft materials are placed in different locations for various indications: (1) in alveolar sockets post extraction, (2) to refill a local bony defect due to trauma or infection, (3) to refill a peri-implant defect due to peri-implantitis, (4) for vertical augmentation of the mandible and maxilla, and (5) for horizontal augmentation of the mandible and maxilla. As to contraindications, the following concerns are listed [61, 64]:
(1) smoking,
(2) inadequate self-performed oral hygiene,
(3) many sites of bony and tissue defects,
(4) unable to achieve wound closure after surgery due to insufficient soft tissues,
(5) severe furcation involvement (i.e., grade 3), and
(6) systemic diseases (e.g., diabetes).

Bateman et al. [64] mentioned that the following issues should be considered as potential complications: (1) unsuccessful treatment procedure, which can lead to recurrent defect, (2) posttreatment infection, (3) barrier membrane being worn away, caused by traumatic toothbrushing, (4) vitality of tooth being compromised in furcation-involved teeth, (5) unfavorable gingival adaptation, which can be of aesthetic concern, (6) dentine hypersensitivity, and (7) requirement for long-term professional maintenance. Meanwhile, although bone-grafting is generally safe, it does have some rare risks [16, 17, 65, 66], including infection, bleeding, blood clot, nerve damage/injury near the bone-grafting area, complications from anesthesia, infection from the donated bone (very rare), problems with bone healing, chronic pain, fractures, hardware failure (meaning plates and screws fail to hold the graft in place), scarring, pain at the body area where the bone was removed, stiffness of the area, swelling, rejection of the bone-graft, inflammation, and/or reabsorption of the graft.

References

[1]	Sallent I, Capella-Monsonís H, Procter P, Bozo IY, Deev RV, Zubov D, Vasyliev R, Perale G, Pertici G, Baker J, Gingras P, Bayon Y, Zeugolis DI. The few who made it: Commercially and clinically successful innovative bone grafts. Front Bioeng Biotechnol, 2020, 8; https://doi.org/10.3389/fbioe.2020.00952.
[2]	Sage K, Levin LS. Basic principles of bone grafts and bone substitutes. Up-To-Date. 2022; https://www.uptodate.com/contents/basic-principles-of-bone-grafts-and-bone-substitutes.
[3]	Campana V, Milano G, Pagno E, Barba M, Cicione C, Salonna G, Lattanzi W, Logroscino G. Bone substitutes in orthopaedic surgery: From basic science to clinical practice. J Mater Sci Mater Med. 2014, 25, 2445–61.
[4]	Baldwin O, Li DJ, Auston DA, Mir HS, Yoon RS, Koval KJ. Allograft, and bone graft substitutes: Clinical evidence and indications for use in the setting of orthopaedic trauma surgery. J Orthop Trauma. 2019, 33, 203–13.
[5]	Keating JF, Simpson AH, Robinson CM. The management of fractures with bone loss. J Bone Joint Surg Br. 2005, 87, 142–50.

[6]   Amini AR, Laurencin CT, Nukavarapu SP. Bone tissue engineering: Recent advances and challenges. Crit Rev Biomed Eng. 2012, 40, 363–408.

[7]   Bose S, Vahabzadeh S, Bandyopadhyay A. Bone tissue engineering using 3D printing. Mater. Today. 2013, 16, 496–504.

[8]   Kumar G, Narayan B. The biology of fracture healing in long bones. Clin Orthop Relat Res. 2014, 531–33.

[9]   Distler T, Fournier N, Grünewald A, Polley C, Seitz H, Detsch R, Boccaccini AR. Polymer-bioactive glass composite filaments for 3D Scaffold manufacturing by fused deposition modeling: Fabrication and characterization. Front Bioeng Biotechnol, 2020, 8; https://doi.org/10.3389/fbioe.2020.00552.

[10] Schemitsch EH. Size matters: Defining critical in bone defect size!. J Orthop Trauma. 2017, 31, 20–22.

[11] Nauth A, Schemitsch E, Norris B, Nollin Z, Watson JT. Critical-size bone defects: Is there a consensus for diagnosis and treatment? J Orthop Trauma. 2018, 32, 7–11.

[12] Mothersill C, Seymour CB, O'Brien A. Induction of c-myc oncoprotein and of cellular proliferation by radiation in normal human urothelial cultures. Anticancer Res. 1991, 11, 1609–12.

[13] Porter JR, Ruckh TT, Popat KC. Bone tissue engineering: A review in bone biomimetics and drug delivery strategies. Biotechnol Prog. 2009, 25, 1539–60.

[14] Palmer W, Crawford-Sykes A, Rose R. Donor site morbidity following iliac crest bone graft. West Indian Med J. 2008, 57, 490–92.

[15] Garg T, Singh O, Arora SR, Murthy SR. Scaffold: A novel carrier for cell and drug delivery. Crit Rev Ther Drug Carr Syst. 2012, 29, 1–63.

[16] Bone Grafting; https://www.hopkinsmedicine.org/health/treatment-tests-and-therapies/bone-grafting.

[17] Bone Grafting; https://my.clevelandclinic.org/health/treatments/16796-bone-grafting.

[18] Bone graft; https://www.mountsinai.org/health-library/surgery/bone-graft.

[19] Osteomyelitis; https://my.clevelandclinic.org/health/diseases/9495-osteomyelitis.

[20] Talley HL. Saving Face. New York University Press. 2014.

[21] Trust D. Overcoming Disfigurement. Thorsons Pub, NY. 1986.

[22] Tooth Loss and Bone Loss; https://optimumoralsurgery.com/p/oral-surgeon-Moorestown-Voorhees-Mullica-Hill-Tooth-Loss-and-Bone-Loss-p31080.asp.

[23] Bone Grafting & Bone Augmentation; https://www.avonomfs.com/procedure/bone-grafting-augmentation-avon-enfield-glastonbury-ct/.

[24] Dental Bone Graft; https://my.clevelandclinic.org/health/treatments/21727-dental-bone-graft.

[25] Nazir MA. Prevalence of periodontal disease, its association with systemic diseases and prevention. Int J Health Sci. 2017, 11, 72–80.

[26] Carrizales-Sepúlveda EF, Ordaz-Farías A, Vera-Pineda R, Flores-Ramírez R. Periodontal disease, systemic inflammation and the risk of cardiovascular disease. Heart Lung Circ. 2018, 27, 1327–34.

[27] Ridgway H. Bone Grafting & Guided Tissue Regeneration; https://www.coolspringsperio.com/peri odontal-disease/reduction-surgery/guided-bone-and-tissue-regeneration/.

[28] John V, Alqallaf H, De Bedout T. Periodontal disease and systemic diseases: An update for the clinician. J Ind Dent Assoc. 2016, 95, 16–23.

[29] Kinane DF, Stathopoulou PG, Papapanou PN. Periodontal diseases. Nat Rev Dis Prim. 2017, 3, 17038; doi: 10.1038/nrdp.2017.38.

[30] Yuan Y, Zhao J, He N. Observation on the effect of bone grafting alone and guided tissue regeneration combined with bone grafting to repair periodontal intraosseous defects. Evid Complement Alternat Med. 2021, 2021, 1743677; doi: 10.1155/2021/1743677.

[31] Lekholm U, Adell R, Lindhe J, Brånemark PI, Eriksson B, Rockler B, Lindvall AM, Yoneyama T. Marginal tissue reactions at osseointegrated titanium fixtures.(II) A cross-sectional retrospective study. Int J Oral Maxillofac Surg. 1986, 15, 53–61.

[32] Liu J, Kerns DG. Mechanisms of guided bone regeneration: A review. Open Dent J. 2014, 8, 56–65.

[33] Bone augmentation method and bone grafting when there is little bone; http://www.implant-office.com/implant2-05-01.html.

[34] Socket lift; https://www.implants-dental.org/surgery/pretreatment/socket-lift.html

[35] Differences between sinus lift and socket lift; https://www.implants-dental.org/surgery/pretreatment/sinuslift-socketlift.html.

[36] Patel PK. Distraction Osteogenesis. 2020; https://emedicine.medscape.com/article/1280653-overview.

[37] Distraction Osteogenesis; https://www.craniofacialteamtexas.com/distraction-osteogenesis/.

[38] Hariri F, Chin SY, Rengarajoo J, Foo QC, Abidin SNNZ, Badruddin AFA. Distraction osteogenesis in oral and craniomaxillofacial reconstructive surgery. In: Yang H (Ed.)., Osteogenesis and Bone Regeneration. 2018; https://www.intechopen.com/chapters/63547.

[39] Waechter J, Leite FR, Nascimento GG, Carmo Filho LC, Faot F. The split crest technique and dental implants: A systematic review and meta-analysis. Int J Oral Maxillofac Surg. 2017, 46, 116–28.

[40] Ventura de Souza CS, Martins de Sá BC, Goulart D, Guillen GA, Macêdo FGC, Nóia CF. Split crest technique with immediate implant to treat horizontal defects of the alveolar ridge: Analysis of increased thickness and implant survival. J Maxillofac Oral Surg. 2020, 19, 498–505.

[41] Scaglione M, Fabbri L, Dell'Omo D, Gambini F, Guido G. Long bone nonunions treated with autologous concentrated bone marrow-derived cells combined with dried bone allograft. Musculoskelet Surg. 2014, 98, 101–06.

[42] Russell JL, Block JE. Surgical harvesting of bone graft from the ilium: Point of view. Med Hypotheses. 2000, 55, 474–79.

[43] Stevens B, Yang Y, Mohandas A, Stucker B, Nguyen KT. A review of materials, fabrication methods, and strategies used to enhance bone regeneration in engineered bone tissues. J Biomed Mater Res B. 2008, 85, 573–82.

[44] Kheirallah M, Almeshaly H. Bone graft substitutes for bone defect regeneration. A collective review. Int J Dentistry Oral Sci. 2016, 3, 247–57.

[45] Greenwald AS, Boden SD, Goldberg VM, Khan Y, Laurencin CT, Rosier RN. Bone-graft substitutes: Facts, fictions and applications. J Bone Joint Surg Am. 2001, 83, 98–103.

[46] Kneser U, Scherfer DJ, Polykandriotis E, Horch RE. Tissue engineering of bone: The reconstructive surgeon's point of view. J Cell Mol Med. 2006, 10, 7–19.

[47] Klokkevold PR, Jovanovic SA. Advanced Implant Surgery and Bone Grafting Techniques. Carranza's Clinical Periodontology. 9th Edition. WB Saunders Co, Philadelphia. 2002, 907–8.

[48] Boyan BD, Weesner TC, Lohmann CH, Andreacchio D, Carnes DL, Dean DD, Cochran DL, Schwartz Z. Porcine fetal enamel matrix derivative enhances bone formation induced by demineralized freeze-dried bone allograft in vivo. J Periodontol. 2000, 71, 1278–86.

[49] Sohn H-S, Oh JK. Review of bone graft and bone substitutes with an emphasis on fracture surgeries. Biomater Res. 2019, 23, 9; doi: https://doi.org/10.1186/s40824-019-0157-y.

[50] Bone Grafting: Bone Grafts and Substitutes. 2015; https://www.ibji.com/blog/orthopedic-care/bone-grafting-real-bone-graft-and-substitutes/.

[51] Kornberg A, Rao NN, Ault-Riché D. Inorganic polyphosphate: A molecule of many functions. Annu Rev Biochem. 1999, 68, 89–125.

[52] Campana V, Milano G, Pagano E, Barba M, Cicione C, Salonna G. Bone substitutes in orthopaedic surgery: From basic science to clinical practice. J Mater Sci Mater Med. 2014, 25, 2445–61.

[53] Kumar P, Vinitha B, Fathima G. Bone grafts in dentistry. J Pharm Bioallied Sci. 2013, 5, S125–7.

[54] Campion CR, Ball SL, Clarke DL, Hing KA. Microstructure and chemistry affect apatite nucleation on calcium phosphate bone graft substitutes. J Mater Sci Mater Med. 2013, 24, 597–610.

[55] Albrektsson T, Johansson C. Osteoinduction, osteoconduction and osseointegration. Eur Spine J. 2001, 10, S96–S101.

[56] Hing KA. Bone repair in the twenty-first century: Biology, chemistry or engineering? Philos Trans A Math Phys Eng Sci. 2004, 362, 2821–50.

[57] Bone Grafting; https://my.clevelandclinic.org/health/treatments/16796-bone-grafting.

[58] Overview of Bone Grafting Options: Dental Bone Grafting; https://www.medtronic.com/us-en/patients/treatments-therapies/bone-grafting-dental/bone-graft-options/overview.html.

[59] Vahadi A, Mirshams V. Benefits of GTR: 5 Benefits of Guided Tissue Regeneration. 2021; https://www.lonestardentalcare.com/blog/5-benefits-of-guided-tissue-regeneration/.

[60] Benefits of Guided Tissue Regeneration. 2022; https://www.michiganimplantcenter.com/4-benefits-of-guided-tissue-regeneration/.

[61] Guided bone and tissue regeneration; https://en.wikipedia.org/wiki/Guided_bone_and_tissue_regeneration.

[62] Goldstep F. Bone grafts for implant dentistry. Basics Oral Health Group. 2015; https://www.oralhealthgroup.com/features/1003918360/.

[63] Klein MO, Al-Nawas B. For which clinical indications in dental implantology is the use of bone substitute materials scientifically substantiated? Eur J Oral Implantol. 2011, 4, 11–29.

[64] Bateman G, Saha S, Chapple ILC. Contemporary Periodontal Surgery: An Illustrated Guide to the Art behind the Science. Quintessence, London. 2007.

[65] What is bone grafting? https://www.hopkinsmedicine.org/health/treatment-tests-and-therapies/bone-grafting.

[66] Bone graft; https://www.mountsinai.org/health-library/surgery/bone-graft.

3 Bone

Before we discuss bone-grafting materials, it would be worthy to understand basic properties of natural bone and its related essential properties. It is, in general, believed that there are 200 bones in the human skeleton and the skeleton accounts for 18% of the average weight of middle-aged males. Of the skeletal weight, the trunk (head, neck, and torso) accounts for 51.2%, the upper limbs account for 14.2%, and the lower limbs account for 34.6%. As to types of human bones, there are five bone types [1, 2]: (1) Long bones are characterized by a shaft, the diaphysis (i.e., much longer than its width) and by an epiphysis (i.e., a rounded head at each end of the shaft). These are mostly densified bone with little marrow and include most of the bones in the limbs. These bones tend to support weight and help movement; e.g., femur, tibia, fibula, and humerus. Long bones such as the clavicle, which have a differently shaped shaft and its end, are also called modified long bones. (2) Short bones are roughly cube-shaped and have only a thin layer of compact bone surrounding a spongy interior. The bones of the wrist and ankle are short bones. (3) Flat bones are thin and generally curved. They consist of two outer layers of compact bone sandwiching a layer of spongy bone. Flat bones include most of the bones of the skull and the sternum or breastbone. They tend to have a protective role. (4) Sesamoid bones are embedded in tendons, such as the patella, pisiform, or kneecap. They protect tendons from biotribological actions such as wear and stress. Since they act to hold the tendon further away from the joint, the angle of the tendon is increased and thus the leverage of the muscle is increased. (5) Irregular bones do not fit into the above categories. They consist of thin layers of compact bone surrounding a spongy interior. Their shapes are irregular and complicated. Often this irregular shape is due to their many centers of ossification or because they contain bony sinuses. The bones of the spine, pelvis, and some bones of the skull are irregular bones. A bone is a rigid organ that constitutes part of the skeleton in most vertebrate animals, protects the various other organs of the body, produces red and white blood cells, stores minerals, provides structure and support for the body, and enables mobility [3]. The Greek word for bone is ὀστέον ('osteon' – which originated from "ostracon," meaning oyster shell), hence the many terms that use it as a prefix such as osteopathy, osteointegration, or osteoporosis [4, 5].

3.1 Structure, compositions and biofunction

3.1.1 Structure and compositions

Natural human bone is a hierarchical assembly of three major nanoscale and macroscale components: (1) organic phase (25–30%) as a mixture of 97% of intracellular matrix (in which 95% is collagen and 5% of noncollagenous) and 3% of bone cells (made

https://doi.org/10.1515/9783111136691-003

out of osteoblast, osteocytes, and osteoclasts), (2) inorganic phase (65–60%) as a mixture of 95% of hydroxyapatite and amorphous, and 5% of trapped ions including Mg,
Na, K, F, Zn, Sr, and C, and (3) 10% of water [6–8]. A number of characteristics of bone
have been recognized as important aspects of bone quality [9–11], leading to a proliferation of studies seeking to determine how these characteristics change during
aging, disease development, and treatment. Some physical and chemical characteristics of bone that may influence biomechanical bone quality are described, categorized
by physical scale [12]. Hernandez et al. [12] categorized four scale group sizes with
relevant bone characteristics: (1) In a scale group of > mm scale, whole bone morphology (size and shape) and bone density spatial distribution are essential properties. (2)
In a size range of mm-µm, microarchitecture, porosity, cortical shell thickness, lacu-
nar number/morphology, remodeling cavity number, size, and distribution should be
considered. (3) In a size range of µm-nm, mineral and collagen distribution/alignment,
and microdamage type, amount, and distribution are important. Finally, (4) in a scale
group of < nm, collagen structure and cross-linking, mineral type and crystal align-
ment, and collagen-mineral interfaces should become dominant considerations [12].

Referring to Figure 3.1, which is prepared based on several sources [13–15], bone
is a nonuniform solid structure, composed of two principle components of extracellular matrix (ECM) and cells along with additional parts such as nerves, blood vessels,
bone marrow, cartilage, and membranes including endosteum and periosteum.

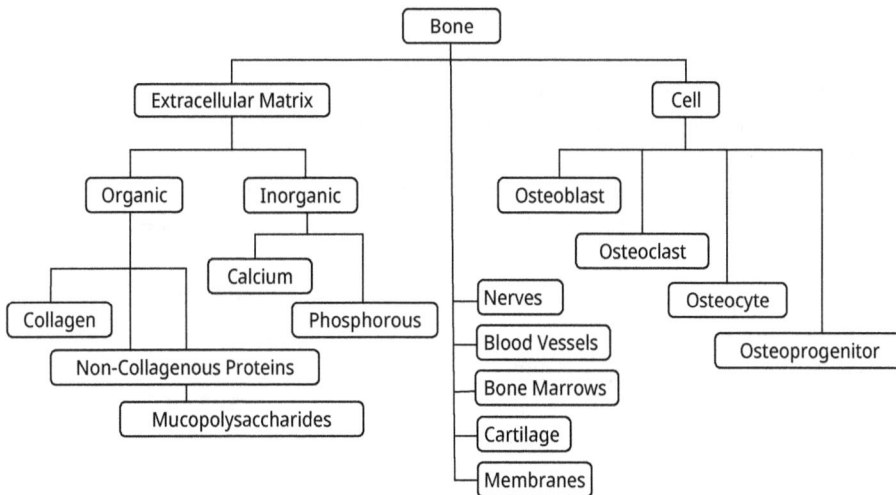

Figure 3.1: Compositions of human bone [13–15].

The composition of a bone can be described in terms of the mineral phase, hydroxyapatite, the organic phase, which consists of collagen type I, noncollagenous proteins,
other components and water. The relative proportions of these various components

vary with age, site, gender, disease and treatment. Any drug therapy could change the composition of a bone [16]. Bone is a heterogeneous composite material consisting, in decreasing order, of a mineral phase, hydroxyapatite, an organic phase (~90% type I collagen, ~5% noncollagenous proteins, ~2% lipids by weight) [17], and water [16]. The amount, proper arrangement, and characteristics of each of these constituents (quantity and quality) define the properties of bone. The relative amount of each of these constituents present in a given bone varies with age, site, gender, ethnicity, and health status [16]. Proteins in the ECM of bone can also be divided as follows with different biofunctions: (a) structural proteins (collagen and fibronectin) and (b) proteins with specialized functions, such as those that: (i) regulate collagen fibril diameter, (ii) serve as signaling molecules, (iii) serve as growth factors, (iv) serve as enzymes, and (v) have other functions [16].

The primary inorganic component of human bone is hydroxyapatite. Hydroxyapatite is a natural mineral form of calcium apatite with formula $Ca_5(PO_4)_3(OH)$ [16–18], but it is usually written as follows; $Ca_{10}(PO_4)_6(OH)_2$ [19] to denote that the crystal unit cell (of hexagonal system) comprises two entities [20].

Bone consists of two types of tissue [21]: (1) compact (cortical) bone: makes up 80% of the total bone in the body and is much stronger than trabecular bone. It is very resistant to various biomechanical movements including bending, torsion, and compression and is much denser with a minimal role in metabolism. It is seen mostly in the shaft of long bones like the femur and the tibia as well as in the outer shell of trabecular bone; and (2) cancellous (trabecular or spongy) bone: makes up only 20% of the total bone but has 10 times the surface/volume ratio of cortical bone. It responds eight times faster to changes in load making it far more dynamic. It occurs in areas that more subject to compression such as the vertebral body, pelvis, and the metaphyses [21].

Osteons represent a single functional unit of bone tissue. It is arranged with concentric lamellae of collagen fiber orientations around a central canal consisting of the osteocyte's arterial, venous, and nerve supply [22].

As mentioned previously, bone is composed of cells and the ECM (see Figure 3.1), which has both organic and inorganic substances and consists of: (1) type I collagen mixed with a matrix of calcium phosphate crystal (which is up to 70% of the dry weight) and (2) proteoglycans and glycoproteins, which are less abundant, but vital for the organization of collagen fibers, mineralization, and resorption of bone [14].

As to cells' components, bone cells make up about 10% of total bone volume. Bones are not a static tissue but need to be constantly maintained and remodeled. There are three main cell types involved in this process; these cells include: (i) osteoblasts, which are involved in the creation and mineralization of bone tissue, (ii) osteocytes, and (iii) osteoclasts, which are involved in the reabsorption of bone tissue. Osteoblasts and osteocytes are derived from osteoprogenitor cells, but osteoclasts are derived from the same cells that differentiate to form macrophages and monocytes [14, 23].

3.1.2 Biofunction

Bone has three major functions; namely, (i) to provide mechanical support for loco-motion, (ii) to protect vital organs, and (iii) to regulate mineral homeostasis [18]. As mechanical biofunctions, bone protects the important and fragile organs (such as heart and the brain) in the body. Bone provides structural integrity, without which human body would have no frame and essentially be an immobile lump of flesh and tissue. Bone facilitates movement; the bones pair up with the joints, ligaments, ten-dons, and muscles to allow the body to move as it does. Bones are also important for conduction of vibrations, which allow us to hear. Hard bones and cartilage have dif-ferent functions. Hard bones are important elements that make up the skeleton due to their hardness, but they are weak against bending force, can hardly deflect, and are easy to break. On the other hand, cartilage is elastic and strong in compressive force. In addition, by nicely arranging the pipe-like structure of the dense part and the cross-linked structure of the spongin part, the strength is increased in the direc-tion where force is easily applied from the outside of the bone. This supports body weight and protects internal organs by placing organs that are vulnerable to impact inside. Furthermore, it also has the function of being a place to store bone marrow having a hematopoietic function [14, 24].

Many metabolic functions of the bone [14, 25–29] can be found. They, at least, in-clude: (1) The bone matrix can store several minerals, chiefly calcium and phospho-rus, as well as iron in the form of ferritin. The process of bone resorption by the osteoclasts releases stored calcium into the systemic circulation and is an important process in regulating calcium balance. Bones act as reserves of minerals important for the body, most notably, calcium and phosphorus. (2) Bone buffers the blood against excessive pH changes by absorbing or releasing alkaline salts, so that the pH balance is regulated as bones may alter the composition of alkaline salts in the serum to maintain the optimal pH level. (3) Mineralized bone matrix stores important growth factors such as insulin-like growth factors (or IGF-1), transforming growth fac-tor, bone morphogenetic proteins, and others. These growth factors are housed in bone and then released periodically. (4) Bone tissues (particularly, osteocytes) can also store toxic heavy metals and other foreign elements, removing them from the blood and reducing their effects on other tissues. These can later be gradually re-leased for excretion as a means of detoxification. (5) Bone controls phosphate metabo-lism by releasing fibroblast growth factor 23 (or FGF-23), which acts on kidneys to reduce phosphate reabsorption. Bone cells also release a hormone called osteocalcin, which contributes to the regulation of blood sugar (glucose) and fat deposition. Chon-droitin sulfate, a carbohydrate moiety, is also a commonly found element in matrices. (6) Marrow adipose tissue (MAT) acts as a storage reserve of fatty acids.

3.2 Formation and remodeling

3.2.1 Bone formation

There are, in general, two modes of bone development [30–33]. (1) Cartilage is replaced by bone (endochondral ossification [34]). Most of mammalian bones are cartilaginous bones, which develop cartilage forming the prototype of the embryonic period and are replaced by bone tissue to form bone. The bones formed in this way are called "replacement bones." (2) Ossification of the membrane (intramembrane ossification [33]). In the part of the skull that covers the brain (cranial vault) and most of the bones and clavicles of the face, the cartilage prototype cannot be formed during the embryonic period, and the cells that form the bone in the connective tissue membrane of that part differentiate and form bone tissue. The bones formed in this way are called "membranous bones."

There are three phases of bone development over time: (I) growth, (II) modeling (or consolidation), and (III) remodeling. During the growth phase, the size of bones increases. Bone growth continues in spurts throughout childhood and adolescence, and eventually ceases in the late teens and early twenties. Bones change shape and thickness and continue accruing mass when stressed during the modeling/consolidation phase bone development. The remodeling phase consists of a constant process of bone resorption (breakdown) and formation that predominates during adulthood and continues throughout life. It was reported that beginning around age 34, the rate of bone resorption exceeds that of bone formation, leading to an inevitable loss of bone mass with age [26, 34]. As mentioned earlier, the cells which are responsible for bone formation and resorption are known as osteoblasts and osteoclasts, respectively. Thus, osteoclasts release calcium and phosphorus from bone in order to restore blood calcium concentrations, and osteoblasts mobilize to replace the resorbed bone. During osteomalacia, however, the deficiency of calcium and phosphorus results in incomplete mineralization of the newly secreted bone matrix [26]. To maintain bone health, it is well known that there are at least essential minerals including Ca, Fe, Na, Mg, Zn, and P. In addition to these elements, there are also important electrolytes in body fluid such as potassium ions (K^+) and chloride ions (Cl^-) as well [35].

3.2.2 Bone remodeling

Once formed, the bone grows and changes shape by modeling, a process in which either bone formation or bone resorption occurs on a given bone surface (bone remodeling is most prominent on cancellous bone surfaces and it is estimated that 80% of bone remodeling activity takes place in cancellous bone, although cancellous bone only comprises 20% of bone [36]). The adult skeleton is renewed by remodeling throughout life. Bone remodeling is a process where osteoclasts (bone resorption) and osteoblasts (bone formation) work sequentially in the same bone remodeling unit. Both processes affect overall bone structure, while remodeling affects material properties such as microdamage, mineralization, and collagen cross-linking [37, 38].

Bone modeling is the process whereby bones are shaped or reshaped by the independent action of osteoblast and osteoclasts. The activities of osteoblasts and osteoclasts are not necessarily coupled anatomically or temporally. Bone modeling defines skeletal development and growth but continues throughout life. Modeling-based bone formation contributes to the periosteal expansion, just as remodeling-based resorption is responsible for the medullary expansion seen at the long bones with aging. The adult skeleton is renewed by remodeling every 10 years [38]. Remodeling persists throughout life. It has been estimated that $3 \sim 4$ million bone remodeling units (BRUs) are initiated each year and that 1 million BRUs are actively engaged in bone turnover at any time [39].

Throughout life, in situ removal and replacement of bone take place without remarkable changing form or density of the bone. Remodeling occurs on both the surface and the interior of the bone. Both processes basically start with osteoclast activation. Internal remodeling initiates with osteoclasts reabsorbing bone by cutting conical spaces through old osteonal systems. Spindle cells, osteoblasts, and blood vessels fill the conical spaces cut by the osteoclasts. Osteoblasts deposit successive lamellae of new osteoid matrix, which will later mineralize. It takes about 50 osteoblasts to fill the cone cut by one osteoclast [40]. Internal remodeling is seen in cortical bone. Surface remodeling occurs on trabecular (which comprises most of the vertebral body), endosteal, and periosteal bones and is very similar to internal remodeling, except that instead of cutting cones, osteoclasts run on the surface of the lamellae excavating a cavity, the so-called Howship lacuna. The rest of the process resembles internal remodeling. Physiologic remodeling serves to repair damaged bone matrix as well as to maintain mineral homeostasis [40].

The aims and results of bone remodeling brings forth various issues, including: (1) replacement of old and damaged bone with new bone, leading to maintain the mechanical strength of bone [41], (2) long-term calcium homeostasis (when our primitive ancestors left the oceans, an environment with a high availability of calcium and ventured on to dry land where calcium is a scarce resource) [38], and (3) bone remodeling also plays a role in the maintenance of acid/base balance, and the release of growth factors embedded in bone. Moreover, it provides a reservoir of labile mineral (short-term homeostasis) and it is the only mechanism by which old, dying, or dead osteocytes can be replaced [42]. It is also reported that there are four characterized phases involved in the remodeling process [38]: (1) the activation phase when the osteoclasts are recruited; (2) the resorption phase, when the osteoclasts resorb bone; (3) the reversal phase, where the osteoclasts undergo apoptosis and the osteoblasts are recruited; and (4) the formation phase, where the osteoblasts lay down new organic bone matrix that subsequently mineralizes.

Ending this section, it should be worth to revisit the so-called Wolff's law. The German anatomist Julius Wolff (1836 ~ 1902) proposed the Wolff's Law (1852), saying, "whether normal or abnormal, bones develop structures that are best suited to resist the forces applied to them." In other words, when an external force is applied to the bone, stress is generated inside the bone and the bone is distorted, but the higher the stress, the more the bone tissue proliferates, becomes thicker and stronger, and resists external forces. Conversely, in areas where stress is low, bone tissue is absorbed and thinned. To put it simply, bones adapt (strength, shape) to external forces (external forces, pressure, stimuli). Namely, if there is no stimulation on the bone, the bone will weaken that much. The theory is supported by the observation that bones atrophy when they are not mechanically stressed and hypertrophy when they are stressed [43, 44]. The law appears to be applied in bone formation, modeling, and remodeling as well [45]. Although Wolff's proposal relates specifically to bone, the law has also been applied to other connective tissues such as ligaments and tendons.

The inverse is true as well: if the loading on a bone decreases, the bone will become less dense and weaker due to the lack of the stimulus required for continued remodeling [45]. This reduction in bone density (osteopenia) is known as stress shielding and can occur as a result of an orthopedic prosthesis such as THA (total hip arthroplasty) or TKA (total knee arthroplasty) [46].

The remodeling of bone in response to loading is achieved via mechanotransduction, a process through which forces or other mechanical signals are converted to biochemical signals in cellular signaling [47]. Mechanotransduction is the process by which physical forces are converted into biochemical signals that are then integrated into cellular responses. It plays a crucial role in bone repair and regeneration, and other bone processes such as physical adaptation, pathological fracture healing, and therapeutic distraction osteogenesis [47].

In the aforementioned manner, the bone morphology and structure are most suitable for external forces. It is thought that when bone is distorted by external force, a piezo voltage is generated there, calcium ions gather at the cathode, and bone tissue is formed by osteocytes. In other words, when an external force is applied to the bone, stress is generated inside the bone and the bone is distorted, but the higher the stress, the more the bone tissue proliferates, becomes thicker and stronger, and resists external forces. Conversely, in areas where stress is low, bone tissue is absorbed and thinned.

3.3 Biomechanics and fracture

3.3.1 Biomechanics

Bone possesses various biomechanical functions [2, 48, 49]. Bones are vital for protecting the important and fragile organs in the body. For example, bones protect the heart and the brain. Bones protect internal organs, such as the skull protecting the brain or the ribs protecting heart and lungs, respectively, from trauma [22]. Due to the way that bone is formed, bone has a high compressive strength of about 170 MPa [50], poor tensile strength of 104 ~ 121 MPa, and a very low shear stress strength (51.6 MPa) [51]. Bones such as the femur are subjected to a bending moment and the stresses (both tensile and compressive) generated by this bending moment account for the structure and distribution of cancellous and cortical bone. In the upper section of the femur, the cancellous bone is composed of two distinct systems of trabeculae. One system follows curved paths from the inner side of the shaft and radiates outwards to the opposite side of the bones, following the lines of maximum compressive stress. The second system forms curved paths from the outer side of the shaft and intersects the first system at right angles. These trabeculae follow the lines of maximum tensile stress, and in general are lighter in structure than those of the compressive system [52]. Hence, stress distribution exhibits a unique directionality as shown in Figure 3.2 [52].

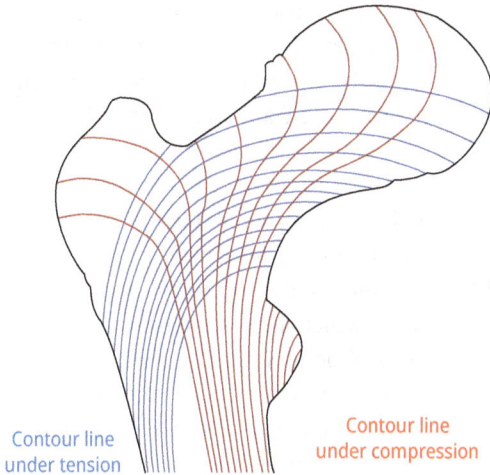

Contour line under tension

Contour line under compression

Figure 3.2: Stress distribution contour lines (computed) under tensile and compressive load [modified after 52].

Under loading material's mechanical property is normally irregular and exhibits directionality. Figure 3.3 demonstrates influences of the tissue anisotropy on stress–strain curves using specimens obtained from a human femoral diaphysis and tested in tensile loading. This direction-dependency of mechanical behavior describes that L indicates the longitudinal direction which is along the long axis of the bone; while T means the transverse direction to the long axis of the bone. The other tested directions (30 and 60 degrees to the long axis of the bone) are also shown [53].

Bones serve a variety of mechanical functions. Together the bones in the body form the skeleton. They provide a frame to keep the body supported, and an attach-

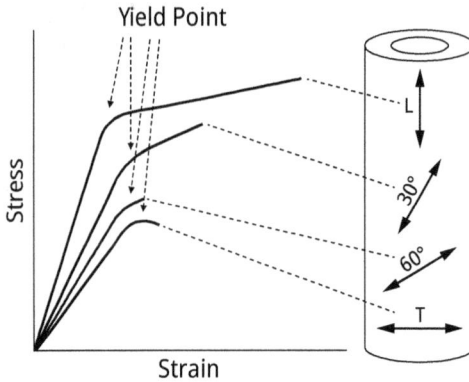

Figure 3.3: Various [stress–strain] curves, depending on the directionality of loading axes [modified after 53].

ment point for skeletal muscles, tendons, ligaments and joints, which function together to generate and transfer forces, so that individual body parts or the whole body can be manipulated in three-dimensional space (the interaction between bone and muscle is studied in biomechanics). Bones provide a frame for other soft tissues of the musculoskeletal system to attach to such as muscles, tendons, and ligaments, to allow support for the body as well as the movement by contracting and relaxing of the muscles which then, in turn, results in flexion, extension, abduction, adduction, and other forms of movement [52]. Mechanically, bones also have a special role in hearing. The ossicles are three small bones in the middle ear, which are involved in sound transduction for conduction of vibrations, which allow us to hear.

Bone strength describes the general integrity of bone. Fragility, stiffness, toughness, ductility, and mechanical strength can be considered as principle biomechanical properties of bone. As described previously, bone strength is ever-changing by age through modeling and remodeling processes, which also directly affects both extrinsic biomechanical properties as well as intrinsic biomechanical properties [54, 55], as shown in Figure 3.4 [54].

Ma et al. [56] mentioned that studies on the biomechanics of the bone matrix can provide a reference for the preparation of more applicable bone substitute implants, bone biomimetic materials and scaffolds for bone tissue repair in humans, as well as for biomimetic applications in other fields. Trabecular bone transfers mechanical loads from the articular surface to the cortical bone.

Cortical (or compact) bone is found mainly in the shafts of long bones and accounts for roughly 80% of bone mass, so that it, in a sense, has low-porosity and less metabolically active tissue. On the other hand, trabecular bone is found in vertebrae and the ends of long bones and is, in contrast, a porous, foam-like (or honeycomb-like) structure with voids filled with bone marrow and possesses a larger remodeling area and higher turnover rate [57]. These intrinsic features go along with different

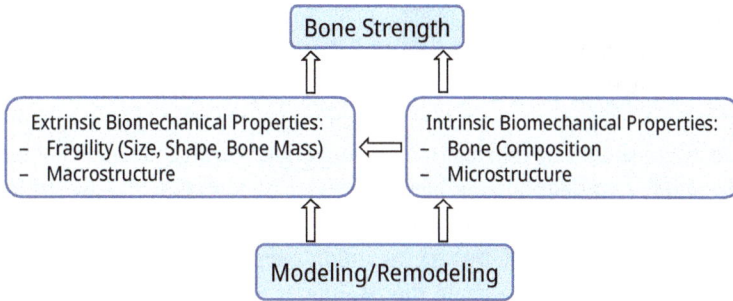

Figure 3.4: Influencing parameters on bone strength [modified after 54].

key roles of each type of bone tissue, e.g., mechanical strength for cortical and mineral homeostasis for trabecular bone [58, 59]. Due to these fundamental different properties between cortical bone and trabecular bone, the essential mechanical properties should apparently differ from each other. Although basic mechanical properties such as elastic behavior, uniaxial strength, and some other density-related and architecture-dependent properties are shared between these, cortical bone is characterized with more mechanical parameters including viscoelastic behavior, strain rate sensitivity, creep behavior, fatigue properties, and fatigue damage accumulation mechanism [59].

3.3.2 Fracture and healing

There are several distinctive terms regarding bone fracture: fracture union, delayed union, and malunion or nonunion [60–63]. Fracture union is a gradual and continuous process whereby the strength of a broken bone is restored by a process of bone regeneration, so that quantitative measures are the most meaningful. However, end point definitions also are useful, but they need empirical validation. The measure that has received the best validation in human fractures is bending stiffness [60]. Delayed union may be defined as a condition in which clinical and roentgenographic examination show that repair is going on slowly although the fracture is still ununited. As the fracture is gently manipulated, varying degrees of solidity and stiffness, and thickening due to the callus may be noted. The roentgenogram often shows a large amount of callus; in fact, the fracture line itself may be partially obscured by this callus [61].

Bone healing is a multilateral process that requires mechanical stability and revascularization along with osteogenesis, osteoinduction, and osteoconduction [64]. Osteoinduction means that primitive, undifferentiated, and pluripotent cells are stimulated to develop into the bone-forming cell lineage. Osteoconduction indicates that bone grows on a bone surface that permits bone growth on its surface or down into pores, channels, or pipes. Accordingly, the most ideal bone substitute should meet such conditions, have no risk of immunological rejection (biocompatible) or disease infection, and achieve incorporation of graft in host bone by gradually being substituted by regenerated bone [65]. It should be molded well into the bony defect within a short time and should be osteoinductive, osteoconductive, and resorbable. In addition, the ideal bone substitute should be thermally nonconductive, sterilizable, and readily available at a reasonable cost [66]. When a bone breaks, the body immediately sets to work trying to heal it. Fractures heal in the following stages: inflammatory phase, repair phase, and remodeling phase [67]. During the inflammatory phase, blood and immune system cells rush to the area to begin the healing process. Immune cells clear out disrupted blood, bone fragments, and other signs of damage. The area around the broken bone is flooded with blood, fluid, and immune cells, resulting in inflammation. This stage peaks a few days after the injury, can last for weeks, and is responsible for most of the pain and swelling after a broken bone. The repair stage follows, overlapping the inflammatory stage. The body builds a callus made of immature bone to stabilize the broken ends of bones. At first, this callus is soft and rubbery and does not contain calcium. This is why the early stages of fracture healing do not show up on an X-ray. Then, finally, the body enters the remodeling stage. That is when calcium starts to enter the callus, and the callus becomes more mature bone tissue. This remodeling stage can last months, but when the fracture is fully healed it may be almost undetectable [67]. The resorption and conversion of the callus progress and the bones return to a form suitable for daily life.

The bone fusion (in other words, osseointegration) does not proceed when the hematoma formed at the beginning flows out. In addition, if a force other than compression force (flexion force, shear force, torsion force, or traction force) is applied to the fracture area, it will hinder bone fusion. Other local causes include missing bone fragments and poor blood circulation. Systemic diseases are adverse factors for bone fusion, and endocrine disorders, especially diabetes, can cause false joints, so that diabetes is one of the contraindications for dental implant treatments. In addition, care should be taken because nutritional disorders can also cause false joints [67].

When the healing process of a fractured bone fails owing to inadequate immobilization, failed surgical intervention, insufficient biological response or infection, the outcome after a prolonged period of no healing is defined as nonunion (or malunion). Nonunion represents a chronic medical condition not only affecting function but also potentially impacting the individual's psychosocial and economic well-being.

Piezoelectricity is the electric charge that accumulates in certain solid materials such as crystals, certain ceramics, and biological matter such as bone, DNA, and various proteins while responding to applied mechanical stress. The word piezoelectricity means electricity resulting from pressure and latent heat [68, 69]. Bone is a complex organ possessing both physico-mechanical and bio-electrochemical properties. As per Wolff's Law, bone can respond to mechanical loading and is subsequently reinforced in the areas of stress. It is well known that when bone is subject to pressure, a negative potential is generated in the bone, and when it is stretched, a positive potential is generated. Negative potentials attract positive calcium ions in the blood, making it easier for calcium in the blood to bind to the bone. Piezoelectricity is one of several mechanical responses of the bone matrix that allows osteocytes, osteoblasts, osteoclasts, and osteoprogenitors to react to changes in their environment [70, 71]. Kao et al. [72] investigated piezoelectric and triboelectric nanogenerators, focusing on their role in the development of wound healing technology. Referring to Figure 3.5, it was reported that: (i) piezoelectricity is one of several mechanical responses of the bone matrix that allows bone cells to react to changes in their environment and (ii) piezoelectric nanogenerators are made by connecting an electrode to a piezoelectric material on a flexible substrate; they generate a current when force is applied and have also been shown to promote the proliferation of human bone-forming cells.

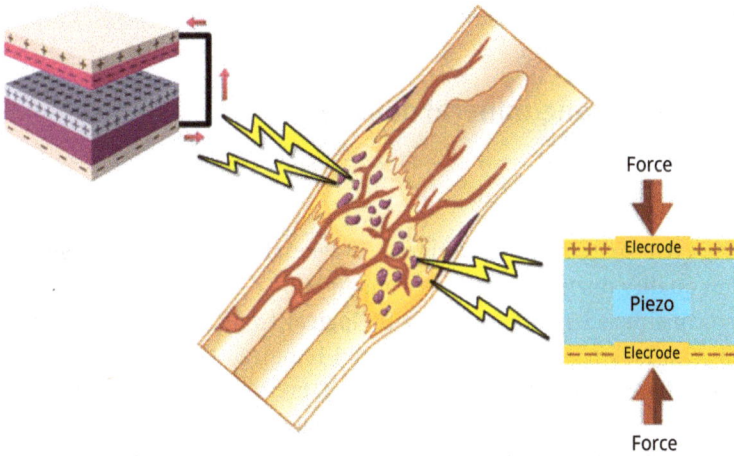

Figure 3.5: Triboelectric nanogenerators (left) and piezoelectric materials (right) to investigate the feasibility of improvement of bone's natural healing properties [modified after 72].

Besides the bone fracture, there are still various clinical significances associated with bone [14, 73–76]. They should, at least, include osteoporosis (a disease that results in a decrease in bone mass and mineral density, causing a decrease in bone strength and an increase in the risk of fracturing), osteopenia (a decrease in bone mineral density below a normal level but not low enough to classify it as osteoporosis), Paget's disease (a condition that affects the bone remodeling process, referring to the action by which the body breaks down old bone tissue and replaces it with new bone tissue), osteogenesis imperfecta (a disorder that causes the bones to fracture easily), osteonecrosis (aka avascular necrosis or aseptic necrosis, occurring when there is a disruption to a bone's blood flow, leading to bone tissue death), osteoarthritis (the most common form of arthritis, is now understood to involve all joint tissues, with active anabolic and catabolic processes), osteomyelitis (describing an infection or inflammation of the bone, with myelitis referring to inflammation of the fatty tissues within the bone, typically occurring when a bacterial or fungal infection enters a bone from the bloodstream or surrounding tissue), fibrous dysplasia (occurring when abnormal fibrous tissue replaces healthy bone tissue and the unusual scar-like tissue makes the bone weaker, causing the bone to change shape and increase the risk of fractures), bone cancer and tumors (an uncommon type of cancer that begins when cells in a bone start to grow out of control), osteomalacia (aka bone softening, referring to a condition where the bone does not harden the way it should after forming), rickets (a childhood bone condition similar to osteomalacia, but it occurs due to imperfect mineralization), and autoimmune conditions (occurring when the immune system attacks the body's own cells, tissue, and organs).

These bone diseases refer to conditions that alter the strength or flexibility of bones. They can result in symptoms such as bone pain, difficulty moving, and a higher risk of bone fractures. These conditions can have many potential causes, including aging, genetics, hormonal changes, and nutritional deficiencies [71–74].

3.4 Bone quality and bone mineral density

3.4.1 Bone quality

The early and long-term success of dental implants depends largely on the alveolar bone quantity and quality during implant placement [47, 77–79]. Poor bone quality and quantity are considered risk factors for biological complications of the implant, associated with lack of primary stability and impaired healing, causing an early implant loss [79]. As Aydin et al. [80] pointed out, bone quality is defined as the amalgamation of all of bone characteristics including bone turnover, bone mineralization, matrix, mineral composition, microarchitecture, and vascularity. Bone strength is determined also by a combination of bone size, shape, and material properties [54, 81].

There are two different sets of bone quality classification: Lekholm and Zarb classification system and Misch classification system. Lekholm and Zarb [82] listed four types of bone quality found in the anterior regions of the jawbone. This classification, widely used in modern implant dentistry, is essentially qualitative and defines bone quality based on the relationship between the compact cortical and the trabecular bones: Type 1 bone (composed mostly of compact bone), Type 2 (mostly a compact bone surrounded by a core of trabecular bone), Type 3 (composed of thin layer of cortical bone surrounded mostly by trabecular bone), and Type 4 (composed of thin layer of cortical bone surrounded by a core of low-density trabecular bone). Misch [83] described bone densities in the edentulous maxilla and mandible on the basis of macroscopic specimens based on cortical and trabecular bones and has classified bone density into four types: D1 is dense cortical bone and mostly found in the anterior mandible, D2 is porous cortical and coarse trabecular bone, and the most common bone type, found most areas of the mandible, D3 is porous cortical bone (thin) and fine trabecular bone, found frequently in the anterior maxilla, and D4 is fine trabecular bone, most often identified in the posterior maxilla. Four bone types are shown in Figure 3.6. Shemtov-Yona [77] mentioned that, despite the wide use of the abovementioned bone classifications, these classifications can be useful during pre-operative or operative stages, particularly during drilling the implant osteotomy [84, 85].

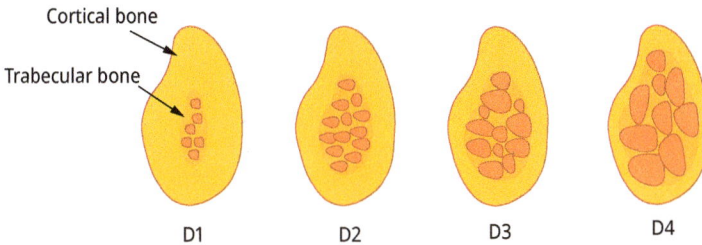

Figure 3.6: Misch bone quality classification [modified after 79].

Since bone quality and quantity are important factors with regard to the survival rate of dental implants, Goiato et al. [86] conducted a systematic review of dental implants inserted in low-density bone to determine the survival rate of dental implants with surface treatments over time, covering the period July 1975 to March 2013. A total of 3,937 patients, who had received a total of 12,465 dental implants, were analyzed. It was reported that (i) the survival rates of dental implants according to the bone density were: type I, 97.6%; type II, 96.2%; type III, 96.5%; and type IV, 88.8%, (ii) the survival rate of treated surface implants inserted in low-density bone was higher (97.1%) than that of machined surface implants (91.6%), and (iii) surface-treated dental implants inserted in low-density bone have a high survival rate and may be indicated for oral rehabilitation.

3.4.2 Bone mineral density

To measure bone mass and density, the dual-energy X-ray absorptiometry (DXA) technique has been widely employed to obtain bone mineral content (BMC in gr) as well as areal bone mineral density (aBMD in gr/cm^2) [12]. DXA results explain a substantial portion of the effects of bone size, shape, and material properties and are strongly correlated with bone mechanical performance and fracture risk [87–90]. DXA method is used to diagnose osteoporosis earlier as a risk for bone fracture and monitor the effectiveness of osteoporosis treatments. The output of a DXA test is a number called a T-score, as seen in Figure 3.2. Normal is zero (0). The more negative the number, the weaker the bones and the more likely they are prone to break. If T-score is −2.5 or below (such as −3.0), then there is a risk for development of osteoporosis, assuming there is no other reason to have such a low T-score [90, 91].

DXA can be used to evaluate bone health in ways that go beyond measuring bone density. Here are some of the other applications of DXA. These tests are available at some but not all DXA facilities [87, 91], including: (1) VFA (vertebral fracture assessment) is a sideways image of the spine that can detect fractures or crushed bones, in the spine, (2) TBS (trabecular bone score) gives a number representing the internal structure of bones in the spine at the microscopic level and the higher the number, the better, (3) FFI (full-length femur imaging) is a technique for using DXA to get an image of the entire femur (thigh bone), instead of just the area around the hip that is seen with standard DXA, and (4) HAS (hip structural analysis, along with DXA, is a method that provides a way to look at size, shape, and configuration of the hip bone to accommodate treatment decisions.

Besides DXA technique, there are still other methods to assess bone density and bone health, including QCT (quantitative computed tomography), MDCT (multidetector computed tomography), BCT (biomechanical computed tomography), and REMS (radiofrequency echographia multi spectrometry) [92].

Osteoporosis is associated with an imbalance in bone remodeling, in which there is relatively greater bone resorption than bone formation. However, the actual rate of bone resorption or bone formation could be above normal (accelerated bone remodeling), normal, or below normal (reduced bone remodeling). In each case, the result of the bone remodeling process leads to a net loss of bone material because of an imbalance in the process, independent of the rate of bone remodeling [93]. Deterioration of skeletal microstructure and bone strength, both associated with loss of bone material, leads to increased susceptibility to fracture [94].

Using dentulous and edentulous mandibular alveolar bones, Oshida prepared four segments which were subjected to the fractal dimension (DF) analysis, using the counting box method [95, 96] (see Figure 3.7). It was obtained that, for the edentulous case A, (i) DF value of sample 1 was 1.47, 1.55 for sample 2, 1.54 for sample 3, and 1.63 for sample 4; (ii) average was 1.55 ± 0.07 and there was no significant difference among the sample locations (from anterior to posterior); for the dentulous case B;, (iii) DF value of sample was 1.78, 1.78 for sample 12, 1.83 for sample 13, and 1.83 for sample 14; (iv) average was 1.81 ± 0.03 and there was no significant difference among sample locations. However, (v) there was a clear difference between the edentulous and dentulous cases and the dentulous case exhibits more higher DF value, suggesting that it has more complicated surface texture.

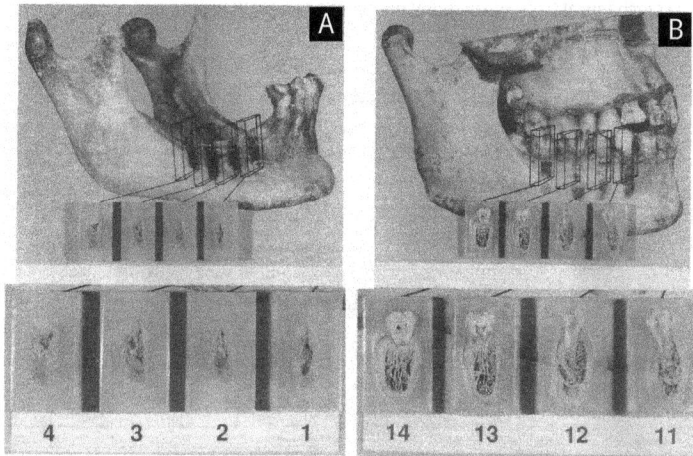

Figure 3.7: Mandibular alveolar bones of edentulous case (A) and dentulous case (B).

References

[1]   Bertazzo S, Bertran CA, Camilli JA. Morphological characterization of femur and parietal bone mineral of rats at different ages. Key Eng Mater. 2006, 309/311, 11–14.

[2]   Bone; https://www.physio-pedia.com/Bone.

[3]   Bone; https://en.wikipedia.org/wiki/Bone.

[4]   Bone; https://www.etymonline.com/word/bone.

[5]   Osteo-; https://www.etymonline.com/word/osteo-.

[6]   Kumar P, Saini M, Dehiya BS, Sindhu A, Kumar V, Kumar R, Lamberti L, Pruncu CI, Thakur R. Comprehensive survey on nanobiomaterials for bone tissue engineering applications. Adv Nanomater Biomed. 2020, 10, 2019; https://doi.org/10.3390/nano10102019.

[7]   Eliaz N, Metoki N. Calcium phosphate bioceramics: A review of their history, structure, properties, coating technologies and biomedical applications. Materials. 2017, 10, 334; https://doi.org/10.3390/ma10040334.

[8]   McMahon R, Wang L, Skoracki R, Mathur A. Development of nanomaterials for bone repair and regeneration. J Biomed Mater Res Part B Appl Biomater. 2013, 101, 387–97.

[9]   Chesnut CH, Rosen CJ. Reconsidering the effects of antiresorptive therapies in reducing osteoporotic fracture. J Bone Miner Res. 2001, 16, 2163–72.

[10] Bouxsein ML. Bone quality: Where do we go from here? Osteoporos Int. 2003, 14, 118–27.

[11] Felsenberg D, Boonen S. The bone quality framework: Determinants of bone strength and their interrelationships, and implications for osteoporosis management. Clin Ther. 2005, 27, 1–11.

[12] Hernandez CJ, Keaveny TM. A biomechanical perspective on bone quality. Bone. 2006, 39, 1173–81.

[13] Composition of bone.png; https://commons.wikimedia.org/wiki/File:Composition_of_bone.png.

[14] Bone; https://www.physio-pedia.com/Bone#cite_note-13.

[15] Bone; https://en.wikipedia.org/wiki/Bone#cite_note-FOOTNOTEDeakin2006190-23.

[16] Boskey AL. Bone composition: Relationship to bone fragility and antiosteoporotic drug effects. Bonekey Rep. 2013, 2, 447; doi: 10.1038/bonekey.2013.181.

[17] Young MF. Bone matrix proteins: Their function, regulation, and relationship to osteoporosis. Osteoporos Int. 2003, 14, S35–S42.

[18] Feng X. Chemical and biochemical basis of cell-bone matrix interaction in health and disease. Curr Chem Biol. 2009, 3, 189–96.

[19] Singh A, Tiwari A, Bajpai J, Bajpai AK, Tiwari A, et al. (Ed.)., Handbook of Antimicrobial Coatings, Polymer-Based Antimicrobial Coatings as Potential Biomaterials: From Action to Application. Elsevier. 2018, 27–61; https://www.semanticscholar.org/paper/Polymer-Based-Antimicrobial-Coatings-as-Potential-Singh-Tiwari/15400846f41cbfd4710aaf544282b4dc69c45439.

[20] Oshida Y. Hydroxyapatite: Synthesis and Applications. Momentum Press. 2015.

[21] Boskey A, Mendelsohn R. Infrared analysis of bone in health and disease. J Biomed Opt. 2005, 10, 031102–08.

[22] Baig MA, Bacha D. Histology, Bone. 2020; https://www.ncbi.nlm.nih.gov/books/NBK541132/.

[23] Marenzana M, Arnett TB. The key role of the blood supply to bone. Bone Res. 2013, 1, 203–15.

[24] Bone; https://ja.wikipedia.org/wiki/%E9%AA%A8.

[25] Doyle ME, Jan de Beur SM. The skeleton: Endocrine regulator of phosphate homeostasis. Curr Osteoporosis Rep. 2008, 6, 134–41.

[26] Bone Health In Depth. Linus Pauling Institute. 2022; https://lpi.oregonstate.edu/mic/health-disease/bone-health.

[27] Hauschka PV, Chen TL, Mavrakos AE. Polypeptide growth factors in bone matrix. Ciba foundation symposium. Novartis Found Symp. 1988, 136, 207–25; doi: 10.1002/9780470513637.ch13.

[28] Styner M, Pagnotti GM, McGrath C, Wu X, Sen B, Uzer G, Xie Z, Zong X, Styner MA. Exercise decreases marrow adipose tissue through β-oxidation in obese running mice. J Bone Miner Res. 2017, 32, 1692–702.

[29] Lee NK, Sowa H, Hinoi E, Ferron M, Ahn JD, Confavreux C, Dacquin R, Mee PJ, McKee MD, Jung DY, Zhang Z, Kim JK, Mauvais-Jarvis F, Ducy P, Karsenty G. Endocrine regulation of energy metabolism by the skeleton. Cell. 2007, 130, 456–69.

[30] Hayashi Y. Bone Health Science. Iwanami Pub. 1999.

[31] Kamiya T. The Story of Bones and Skeletons. Iwanami Junior Series. 2001.

[32] Anatomy & Physiology. OpenStax CNX. Feb 26, 2016; https://openstax.org/books/anatomy-and-physiology/pages/1-introduction.

[33] Bone Growth and Development. 2020; https://courses.lumenlearning.com/wm-biology2/chapter/bone-growth-and-development/.

[34] Nordin BE, Need AG, Chatterton BE, Horowitz M, Morris HA. The relative contributions of age and years since menopause to postmenopausal bone loss. J Clin Endocrinol Metab. 1990, 70, 83–88.

[35] Heaney RP. Sodium, potassium, phosphorus, and magnesium. In: Holick MF, et al (Ed.)., Nutrition and Bone Health. Nutrition and Health. Humana Press, Totowa, NJ. 2004.

[36] Seeman E. Age- and menopause-related bone loss compromise cortical and trabecular microstructure. J Gerontol A Biol Sci Med Sci. 2013, 68, 1218–25.

[37] Allen R, Burr DB. Bone modeling and remodeling. In: Burr DB, et al (Ed.)., Basic and Applied Bone Biology. Academic Press. 2014, 75–90; https://www.sciencedirect.com/science/article/pii/B9780124160156000046.

[38] Langdahl B, Ferrari S, Dempster DW. Bone modeling and remodeling: Potential as therapeutic targets for the treatment of osteoporosis. Ther Adv Musculoskelet Dis. 2016, 8, 225–35.

[39] Manolagas S. Birth and death of bone cells: Basic regulatory mechanisms and implications for the pathogenesis and treatment of osteoporosis. Endocr Rev. 2000, 21, 115–37.

[40] Bone Modeling and Remodeling. 2015; https://clinicalgate.com/bone-modeling-and-remodeling/.

[41] Dempster D. Exploiting and bypassing the bone remodeling cycle to optimize the treatment of osteoporosis. J Bone Miner Res. 1997, 12, 1152–54.

[42] Dempster D. Anatomy and functions of the adult skeleton. In: Favus M (Ed.)., Primer on the Metabolic Bone Diseases and Disorders of Mineral Metabolism. American Society for Bone and Mineral Research, Washington, DC. 2006, 7–11.

[43] Modeling-remodeling; https://teambone.com/education-basic/modelingremodeling/.

[44] Wolff's law; https://en.wikipedia.org/wiki/Wolff%27s_law.

[45] Wolff J. The Law of Bone Remodeling. Springer, Berlin Heidelberg New York. 1986.

[46] Oshida Y, Miyazaki T Biomaterials and Engineering for Implantology. De Gruyter. 2022.

[47] Chenyu H, Ogawa R. Mechanotransduction in bone repair and regeneration. FASEB J. 2010, 24, 3625–32.

[48] Bone; https://en.wikipedia.org/wiki/Bone#cite_note-Schmidt-Nielsen-8.

[49] Bones Functions You Need to Know. 2022; http://www.med-health.net/Functions-Of-Bones.html.

[50] Schmidt-Nielsen K. Scaling: Why is Animal Size so Important? Cambridge University Press, Cambridge. 1984.

[51] Saladin K. Anatomy and Physiology: The Unity of Form and Function. McGraw-Hill, New York. 2012.

[52] Structure of Bones; https://courses.lumenlearning.com/wm-biology2/chapter/structure-of-bones/.

[53] Basic Biomechanical Concepts: The Foundation for Interpreting Local Load History Using Bone Histomorphology; https://teambone.com/education-basic/biomechanics-of-bone/.

[54] Forestier-Zhang L, Bishop N. Bone strength in children: Understanding basic bone biomechanics. Arch Dis Child Educ Pract Ed. 2016, 101, 2–7.

[55] Turner CH. Biomechanics of Bone: Determinants of Skeletal Fragility and Bone Quality. Osteoporosis Int. 2002, 12, 97–104.

[56] Ma C, Du T, Niu X, Fan Y. Biomechanics and mechanobiology of the bone matrix. Bone Res. 2022, 10, 59; https://doi.org/10.1038/s41413-022-00223-y.

[57] Clark B. Normal bone anatomy and physiology. Clin J Am Soc Nephrol. 2008, 3, S131–9.

[58] Leonard MB. A structural approach to skeletal fragility in chronic kidney disease. Semin Nephrol. 2009, 29, 133–43.

[59] Costa LR, Carvalho AB, Bittencourt AL, Rochitte CE, Canziani MEF. Cortical unlike trabecular bone loss is not associated with vascular calcification progression in CKD patients. BMC Nephrol. 2020, 21, 121; https://doi.org/10.1186/s12882-020-01756-2.

[60] Marsh D. Concepts of fracture union, delayed union, and nonunion. Clin Orthop Relat Res. 1998, 355, S22–S30.

[61] Henderson MS. Ununited fractures. J Am Med Assoc. 1926, 86, 81–86.

[62] Orthopaedic Surgery; https://www.pennmedicine.org/departments-and-centers/orthopaedic-surgery/about-us/excellence-in-motion-newsletter/archive/2019-newsletter/treatment-of-nonunion-and-malunion#:~:text=Nonunion%20and%20malunion%20fractures%20are,or%20rotation%20of%20the%20limb.

[63] Wildemann B, Ignatius A, Leung F, Taitsman LA, Smith RM, Pesántez R, Stoddart MJ, Richards RG, Jupiter JB. Non-union bone fractures. Nat Rev Dis Primer. 2021, 7, 57; https://doi.org/10.1038/s41572-021-00289-8.

[64] Sohn H-S, Oh J-K. Review of bone graft and bone substitutes with an emphasis on fracture surgeries. Biomater Res. 2019, 23, 9; https://doi.org/10.1186/s40824-019-0157-y.

[65] Kornberg A, Rao NN, Ault-Riché D. Inorganic polyphosphate: A molecule of many functions. Ann Rev Biochem. 1999, 68, 89–125.

[66] Campana V, Milano G, Pagano E, Barba M, Cicione C, Salonna G. Bone substitutes in orthopaedic surgery: From basic science to clinical practice. J. Mater Sci Mater Med. 2014, 25, 2445–61.

[67] Surgical Options for Fracture Nonunions; https://uoanj.com/trauma-fracture-care/nonunion-repair/.

[68] Piezoelectricity; https://en.wikipedia.org/wiki/Piezoelectricity.

[69] Polland BD. Principles of instrumental analysis. J Chem Educ. 1981, 58, A314.

[70] Kao F-C, Chiu P-Y, Tsai T-T, Lin Z-H. The application of nanogenerators and piezoelectricity in osteogenesis. Sci Technol Adv Mater. 2019, 20, 1103–17.

[71] Yang C, Ji J, Lv Y, Li Z, Luo D. Application of piezoelectric material and devices in bone regeneration. Nanomater (Basel). 2022, 12, 4386; doi: 10.3390/nano12244386.

[72] Kao F-C, Ho H-H, Chiu P-Y, Hsieh M-K, Liao J-C, Lai P-L, Huang Y-F, Dong M-Y, Tsai T-T, Lin Z-H. The application of nanogenerators and piezoelectricity in osteogenesis. Sci Techol Adv Mater. 2019, 20, 1103–17; doi: 10.1080/14686996.2019.1693880.

[73] Neogi T. Clinical significance of bone changes in osteoarthritis. Ther Adv Musculoskelet Dis. 2012, 4, 259–67.

[74] Bone; https://www.kenhub.com/en/library/anatomy/bones.

[75] Lin JS, Santiago JE, Mayerson JL, Schaeschmidt TJ. Clinical significance of bone morphogenetic protein in osteosarcoma: A systematic review. Curr Orthop Prac. 2019, 30, 548–54.

[76] Santhakumar S. What to know about bone diseases. 2022; https://www.medicalnewstoday.com/articles/bone-diseases.

[77] Shemtov-Yona K. Quantitative assessment of the jawbone quality classification: A meta-analysis study. PLoS One. 2021, 16, e0253283; doi: 10.1371/journal.pone.0253283.

[78] Esposito M, Hirsch JM, Lekholm U, Thomsen P. Biological factors contributing to failures of osseointegrated oral implants. (II). Etiopathogenesis Eur J Oral Sci. 1998, 106, 721–64.

[79] Lekholm & Zarb classification; https://www.researchgate.net/figure/Lekholm-Zarb-classification-Type-I-the-entire-bone-is-composed-of-very-thick-cortical_fig3_322512723.

[80] Aydin U, Bulut A, Bulut ÖE. Assessment of maxillary and mandibular bone quality. Conference: ECR 2017 (European Congress of Radiology 2017). 2017; doi: 10.1594/ecr2017/C-219.

[81] van der Meulen MC, Jepsen KJ, Mikic B. Understanding bone strength: Size isn't every-thing. Bone. 2001, 29, 101–04.

[82] Lekholm U, Zarb GA. Patient selection and preparation. In: Brånemrk P-I, et al. (Ed.)., Tissue Integrated Prostheses: Osseointegration in Clinical Dentistry. Quintessence Pub, Chicago. 1985, 199–209.

[83] Misch CE. Bone density: A key determinant for clinical success. In: Misch CE (Ed.)., Dental Implant Prosthetics. Mosby, Maryland Heights. 2005, 130–41.

[84] Resnik RR, Misch CE. Bone density: A key determinant for treatment planning. In: Resnik RR (Ed.)., Misch's Contemporary Implant Dentistry. Elsevier Inc. 2020, 450–66.

[85] Norton MR, Gamble C. Bone classification: An objective scale of bone density using the computerized tomography scan. Clin Oral Implants Res. 2001, 12, 79–84.

[86] Goiato MC, Dos Santos DM, Santiago JFJ, Moreno A, Pellizzer EP. Longevity of dental implants in type IV bone: A systematic review. Int J Oral Maxillofac Surg. 2014, 43, 1108–16.

[87] Marshall D, Johnell O, Wedel H. Meta-Analysis of how well measures of bone mineral density predict occurrence of osteoporotic fractures. BMJ. 1996, 312, 1254–59.

[88] Cummings SR, Bates D, Black DM. Clinical use of bone densitometry – Scientific review. Jama. 2002, 288, 1889–97.

[89] Dhiman P, Andersen S, Vestergaard P, Masud T. Does bone mineral density improve the predictive accuracy of fracture risk assessment? A prospective cohort study in Northern Denmark. BMJ Open. 2018, 8, e018898; doi: 10.1136/bmjopen-2017-018898.

[90] Kim HS, Jeong ES, Yang MH, Yang S-O. Bone mineral density assessment for research purpose using dual energy X-ray absorptiometry. Osteoporos Sarcopenia. 2018, 4, 79–85.

[91] Evaluation of Bone Health/Bone Density Testing. 2022; https://www.bonehealthandosteoporosis.org/patients/diagnosis-information/bone-density-examtesting/.

[92] Lee T-H, Jeong M-A, Kim T-H. Feasibility of assessing maxillary and mandibular bone mineral density for dental implantation by using multidetector computed tomography. Implant Dent. 2019, 28, 367–71.

[93] Haseltine KN, Chukir T, Smith PJ, Jacob JT, Bilezikian JP, Farooki A. Bone mineral density: Clinical relevance and quantitative assessment. J Nucl Med. 2021, 62, 446–54.

[94] Borah B, Gross GJ, Dufresne TE, Smith TS, Cockman MD, Chmielewski PA, Lundy MW, Hartke JR, Sod EW. Three-dimensional microimaging (MRμI and μCT), finite element modeling, and rapid prototyping provide unique insights into bone architecture in osteoporosis. Anat Rec. 2002, 265, 101–10.

[95] Oshida Y, Hashem A, Nishihara T, Yapchulay MV. Fractal dimension analysis of mandibular bones: Toward a morphological compatibility of implants. Bio-Med Mater Eng. 1994, 4, 397–407.

[96] Oshida Y, Munoz CA, Winkler MM, Hashem A, Itoh M. Fractal dimension analysis of aluminum oxide particle for sand blasting dental use. Bio-Med Mater Eng. 1993, 3, 117–26.

4 Natural bone-grafting materials

Bone resorption due to tooth loss leaves the area without sufficient enough bone volume to accept implant placement, which is expected to exhibit successful osseointegration and fixation. Hence, bone-grafting is recognized as the only solution to provide enough bone volume. Bone-grafting is possible because bone tissue has the ability to regenerate completely if space is provided into which it has to grow. As natural bone grows, it generally replaces the graft material completely, resulting in a fully integrated region of a new bone. There are various types of bone-grafting materials developed and available in both dental and orthopedic applications. Although it is normal to classify bone-grafting materials into natural materials group and synthetic material groups, Kumar et al. [1] classified bone-grafting materials based on material group as follows: (1) allograft-based bone-graft involves allograft bone, used alone or in combination with other materials (e.g., Grafton and OrthoBlast), (2) factor-based bone-graft, which are natural and recombinant growth factors, used alone or in combination with other materials such as transforming growth factor-beta (TGF-beta), platelet-derived growth factor (PDGF), fibroblast growth factors (FGF), and bone morphogenic protein (BMP), (3) cell-based bone-grafts that use cells to generate new tissue alone or are added onto a support matrix, for example, mesenchymal stem cells, (4) ceramic-based bone-graft substitutes that include calcium phosphate, calcium sulphate, and bio-glass used alone or in combination; for example, OsteoGraf, ProOsteon, OsteoSet, and (5) polymer-based bone-graft that uses degradable and nondegradable polymers alone or in combination with other materials, for example, open porosity polylactic acid polymer.

In this chapter, natural bone-grafting materials are discussed.

4.1 Requirements for suitable bone-grafting materials

Hudnall [2] mentioned that, to choose dental bone-grafting materials, various factors should be considered as follows: (1) Type of surgery. While the best bone-grafting material for dental implants is simultaneously osteoconductive, osteoinductive, and osteogenic, a material that has all three properties is not always practical or abundantly available. Therefore, the great majority of bone-grafting procedures are performed using the second-best option. Whether the surgery involves ridge or socket preservation, lateral ridge augmentation, a sinus lift, or stabilization of an immediate placed implant, most surgeons will select an allograft, a mixture of cortical and cancellous bone, because it is readily available for purchase in sufficient quality to fill almost any defect and it provides predictable results. (2) Patient concerns. While generally considered safe, patients sometimes decline a procedure involving the use of allografts or xenografts, citing concerns that they could be a possible vector for disease transmission. This forces the surgeon to look for other treatment alternatives such as

https://doi.org/10.1515/9783111136691-004

autographs or alloplasts. (3) Health history. It is true that the patient's systemic health condition has a tremendous effect on the success rate of bone-graft placement for dental implants. After all, the success of the surgery depends on stimulating the body's own response mechanisms. If the patient has a suppressed immune system, altered healing (diabetes, autoimmune diseases), thyroid issues, nutritional deficiency or malabsorption (bariatric surgery patients), or other chronic health issues, the possibility of graft or implant failure is a real concern, and bone-grafting is contraindicated until such time the health issue can be controlled or resolved. (4) Bone quantity and quality. Patients who lack sufficient bone to stabilize and support a dental implant in all dimensions can benefit from bone-grafts. Even those patients with spongy bone can benefit from grafts that promote new and denser bone growth, allowing the implant to be placed partly in solid bone in order to gain stability. Besides these concerns, Oryan et al. [3] and Brydone et al. [4] mentioned that the selection of an ideal bone-graft relies on several factors such as tissue viability, defect size, graft size, shape and volume, biomechanical characteristics, graft handling, cost, ethical issues, biological properties, and associated complications. Furthermore, Kolk et al. [5] added biocompatibility, bioresorbability, sterility, structural integrity, adequate porosity for vascular ingrowth, plasticity, ease of handling, cost, and compressive strength as influencing parameters for the success rate of a bone-graft.

Since the main function associated with bone-grafting materials is to provide mechanical support and stimulate osteoregeneration, the required characteristics of an ideal bone-grafting material are four fundamental biological properties; i.e., osseointegration, osteogenesis, osteoconduction, and osteoinduction [5–10]. Osseointegration refers to an ability of a grafting material to chemically bond to the surface of the bone in the absence of an intervening fibrous tissue layer; hence the connection is a direct-structural and functional connection between the bone and the implant surface. Osseointegration is critical for implant stability and is considered a prerequisite for implant loading and long-term clinical success of endosseous dental implants. Osteogenesis refers to the formation of new bone via osteoblasts or progenitor cells present within the grafting material. Osteogenesis can assist the formation and development of bone, even in the absence of local undifferentiated mesenchymal stem cells. Osteoconduction refers to the ability of a bone-grafting material to generate a bioactive scaffold on which host cells can grow. This structure enables vessels, osteoblasts, and host progenitor cells to migrate into the interconnected osteomatrix; hence osteoinduction is the transformation of undifferentiated mesenchymal stem cells into osteoblasts or chondroblasts through growth factors that exist only in a living bone. Osteoconduction is the process that provides a bio-inert scaffold or a physical matrix, suitable for the deposition of new bone from the surrounding bone, or encourage differentiated mesenchymal cells to grow along the graft surface, and the recruitment of host stem cells into the grafting site where local proteins and other factors induce the differentiation of stem cells into osteoblasts [5–10].

4.2 Types of natural bone-grafting materials

Basically, there are two types of bone-grafting materials: i.e., they are natural bone materials and synthetic bone substitutes. Figure 4.1 illustrates all these bone-grafting materials. Referring to Figure 4.1 [3], a patient possesses variety of bone sources from own body (autograft), from a human donor (allograft), or from an animal model (xenograft), or various types of synthetic and biologically based, tissue-engineered biomaterials and combinations of these substitutes [11]. Besides the above three natural bone-grafting materials (autografts, allografts, and xenografts), there are still phytogenic materials such as algae-based or coral-based materials [5, 6, 12, 13]. All grafting materials have one or more of these three mechanisms of action (osteogenesis, osteoinduction, and osteoconduction). The mechanisms by which the grafts act are normally determined by their origin and composition. An autogenous bone harvested from the patient forms a new bone by osteogenesis, osteoinduction, and osteoconduction. Allografts harvested from cadavers have osteoconductive and possibly osteoinductive properties, but they are not osteogenic. Xenografts and alloplasts are typically only osteoconductive [7]. In this section, natural bone-grafting materials are discussed.

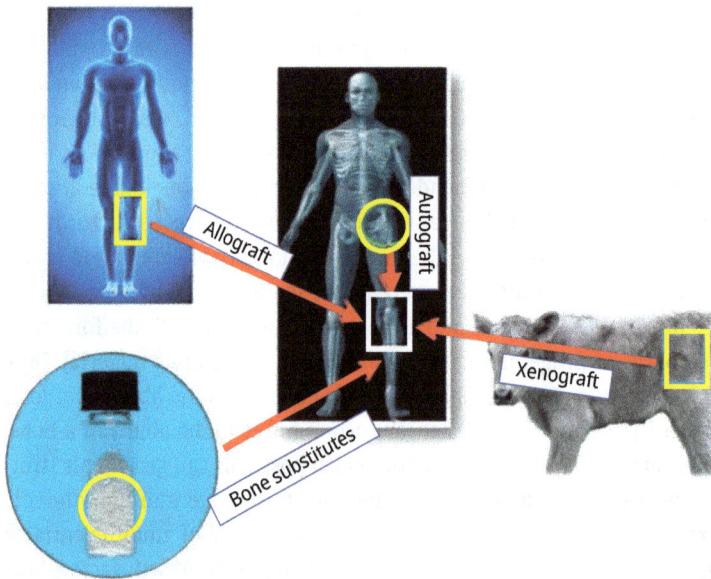

Figure 4.1: Types of bone-grafting materials and their recipient [3].

4.2.1 Autografts

Autografts can be defined as bone-grafts that are harvested from one site and implanted into another site within the same individual. They are also termed autologous or autogenous bone-grafts [14]. Autografts are considered as the gold standard in reconstructing small bone defects due to having three properties (osteogenesis, osteoinduction and oesteocondiction [15]) and provide osteogenic cells, osteoinductive growth factors, and an osteoconductive scaffold, all relevant to bone healing, modeling, and remodeling [3, 16]. Autografts are commonly obtained from intraoral and extraoral sites from the same individual, such as the mandibular symphysis, mandibular ramus, external oblique ridge, iliac crest (tope of hip bone), proximal ulna, or distal radius, due to being good sources of cortical and cancellous bone [6, 17–20].

The grafted material maintains the characteristics of embryological site of origin: bone density, which matures on the site, reflects that principle [21]. Fresh autografts contain surviving cells and osteoinductive proteins such as BMP-2, BMP-7, FGF, IGF, and PDGF [4]. From a biological point of view, they are the best material available, since they totally lack immunogenicity [3]. They retain their viability immediately after transplantation, and the lack of immunogenicity enhances the chances of graft incorporation into the host site [22]. Furthermore, as mentioned previously, the osteogenic, osteoinductive, and osteoconductive properties of fresh autografts are optimal, given the presence of MSCs (mesenchymal stem cells), osteoprogenitor cells, osteogenic cells, and GF (growth factors) [23].

The greatest chance for successful transplantation of a live bone is associated with a cancellous autograft or pedicled, vascularized cortical autograft [24]. In general, the success of grafting the autogenic bone depends on the survival and proliferation of the osteogenic cells, the type of graft chosen, handling of the graft, and shaping of the graft during the operative procedure, to adapt it into the host's bone [20]. While fresh autologous graft has the capability of supporting new bone growth by all four means (induction, genesis, conduction, and integration), it may not be necessary for a bone-graft replacement to inherently have all four properties in order to be clinically effective [3]. When inductive molecules are locally delivered on a scaffold, the mesenchymal stem cells are ultimately attracted to the site and are capable of reproducibly inducing new bone formation, provided that minimal concentration and dose thresholds are met [3, 25]. Although many benefits are recognized with an autograft transplantation, there are still drawbacks [3, 26, 27], including that (i) the amount of bone tissue that can be collected is limited, (ii) a patient may have pain at the site where the bone-graft is collected, (iii) autografts carry the limitations of morbidity at the harvesting site as also limited availability, and (iv) pain and donor site morbidity as well as other risks such as major vessel or visceral injuries during harvesting.

4.2.2 Allografts

Allograft (homologous) is provided by the tissue banks (from a human donor) in various formulations as sticks, granules or paste. This, along with xenografts, is an osteoconductive and osteoinductive (but lacks osteogenic properties) material that provides mechanical properties even in large defects [28, 29]. Allografts and xenografts have osteoinductive and osteoconductive characteristics but lack the osteogenic properties of autografts [4, 31, 32]. There are three types of bone allografts: fresh or fresh-frozen bone, FDBA (freeze-dried bone allograft) and DFDBA (demineralized freeze-dried bone allograft) [1]. The use of allografts for bone repair often requires sterilization and deactivation of proteins normally found in healthy bone. Contained in the extracellular matrix of bone tissue is the full cocktail of bone growth factors, proteins, and other bioactive materials, necessary for osteoinduction and successful bone healing; the desired factors and proteins are removed from the mineralized tissue using a demineralizing agent such as hydrochloric acid. The mineral content of the bone is degraded, and the osteoinductive agents remain in a demineralized bone matrix (DBM) [1]. Allografts are used in both morselized and structural forms [27], and are provided as cortical, cancellous, or cortico-cancellous grafts [30] and in various shapes such as powder, cortical chips, and cancellous cubes [3]. They also can be processed as mineralized or demineralized, fresh, fresh-frozen, or freeze-dried forms [14, 31]. Fresh-frozen and freeze-dried bone allografts induce more prompt graft vascularization, incorporation, and bone regeneration than fresh allograft [32]. Freeze drying produces a safer graft in terms of reducing the risk of immunologic responses in the donor, and in transmission of viral diseases. However, despite modern sterilization and storage methods, processing of allografts using freeze-drying techniques and treating the graft by hypotonic solutions, acetone, ethylene oxide, or gamma irradiation, which can eliminate cellular and viral particles, therefore reduces the risk of infectious and transmissible diseases [33]. However, the use of allografts is not completely safe [14, 32]. These processes may destroy the bone cells and denature proteins present in the graft and alter osteoconductive and osteoinductive characteristics, essentially eliminating the osteogenic properties [27]. Therefore, freeze-dried allografts, in comparison to autografts, take longer to become revascularized and incorporated than autografts [34]. Freeze-drying procedure also reduces the mechanical strength of the graft, and the cost of processed and ready-to-use allografts is high [32, 35]. The mineral component of the allogeneic bone can be removed by demineralization to obtain demineralized bone matrix (DBM) which has osteoinductive and partly osteoconductive properties [14]. DBM revascularizes quickly, and its biological activity is attributed to proteins and various growth factors present in the extracellular matrix [30]. Given these major disadvantages, allografts are not the perfect substitute for autograft [3].

There are several advantages associated with allografts. One of the most obvious one is that no additional surgical site or procedure is necessitated. Allograft materials can be prepared in three primary forms – fresh, frozen, or freeze-dried. Fresh and frozen allograft materials possess superior osteoinductive properties but are rarely used nowadays due to the higher risk of a host immunogenic response, limited shelf life, and increased risk of disease transmission [6]. Further processing of allograft material through freeze-drying can increase the shelf life of the material and decrease the immunogenicity, though at the cost of decreased osteoinductive potential, decreased structural strength, and osseointegration [13]. Allografts have been successfully used in combination with xenografts for guided bone regeneration (GBR) in bone augmentation procedures [6]. Allografts exhibit good histocompatibility and are found available in various forms, from whole bone segments, cortico-cancellous, and cortical pieces to chips, wedges, pegs, powder, and DBM (demineralized bone matrix). Allograft materials can also be produced in custom shapes to satisfy the requirements of the recipient sites [6].

At the same time, there are drawbacks recognized with allografts. Allografts and xenografts carry the risk of disease transmission and rejection. Tissue engineering is a new and developing option that has been introduced to reduce the limitations of bone-grafts and improve the healing processes of bone fractures and defects [3]. Since the allograft method uses bone tissue from another person (donor), public health services have strict regulations on how tissues are handled and treated. The bone tissue is cleaned and processed (sterilized) to ensure the safety of the recipient. This type of graft is common in spinal fusion surgery. It provides a framework around which healthy bone tissue can grow [26]. Allografts are an alternative option with major limitations associated with rejection, transmission of diseases, and cost. In recent decades, the use of allograft materials has often been preferred [31]. This is largely due to the alleviation of many of the major concerns associated with autografting procedures described previously, especially in large bony defects. However, limitations persist relating to the risk of infectious disease transmission, such as for human immunodeficiency virus (HIV) and Hepatitis B and C. Studies have found that there is ~8% prevalence of unknown diseases in osteoarthritic femoral heads removed during hip arthroplasty [6, 35]. Bone allografts carry the risk of transmitting bacterial contamination and viral diseases, such as HIV and Hepatitis B and C, and they may also induce immunological reactions that interfere with the bone healing process and can lead to rejection of the graft [30, 34, 36–38]. In addition, the rate of healing, using allografts, is generally lower than with autografts [3]. These concerns can generally be alleviated through tissue processing such as sterilization, mechanical debridement, ultrasonic washing and cleaning, and gamma irradiation [6, 15].

4.2.3 Bone bank

Bone allografts are distributed through regional tissue banks and by most major orthopedic and spinal companies [3, 27]. Bone Bank Allografts (BBA) is the leading provider of regenerative medicine technologies to the surgical community by maximizing the gift of tissue donation. BBA provides a wide range of surgical options from traditional bone and soft tissues to specialty custom products for transplant, and continues to innovate tissue processing and distribution [39]. A bone bank harvests bone and tissue from a cadaver to provide them to surgeons who perform transplants. Bone banks perform rigorous infection control testing on all the grafts prior to releasing for transplantation. If the grafts do not pass infection control standards, the graft will not be released to the surgeon.

Bone banks are necessary for providing biological material for a series of orthopedic procedures. To increase the safety of transplanted tissues, standards for bone bank operation have been defined by the government, which has limited the number of authorized institutions. The good performance in a bone bank depends on strict control over all stages, including formation of well-trained harvesting teams; donor selection; conducting various tests on the tissues obtained; and strict control over the processing techniques used. A combination of these factors enables greater scope of use and of number of recipient patients, while the incidence of tissue contamination becomes statistically insignificant, and there is traceability between donors and recipients [40]. For details, there are several technical considerations.

(1) Donor selection should be done carefully by the coordinator of the harvesting team and by the medical director of the tissue bank, taking into consideration data such as the donor's age and sex, cause of death, previous medical history, physical examination, and numerous laboratory tests [40]. The biggest complication to be avoided is transmission of diseases from the donor to the recipient, which might be of viral nature (HIV or Hepatitis B and C) or bacterial nature, caused by an organism present in the donor or by contamination at the time of harvesting the skeletal tissues [41]. To avoid this complication, sterilization methods can be used (autoclaving or use of ethylene oxide or irradiation). Each has its advantages and disadvantages. Sterile processes applied from the time of harvesting to the final storage, with serial tests performed, alternating with periods of quarantine [40]. There are two types of donors of homologous tissues: (i) live donors, consisting mainly of donations of femoral heads after total hip arthroplasty procedures, which have the advantage that the donor patients can be called back for new tests in suspected cases; and (ii) cadaver donors, from which much greater quantities of tissues can be harvested from practically any segment of the skeleton. Additionally, generally being young, donors have better quality bone tissue than seen in live donors [40].

(2) Tissue harvesting should be carried out very cautiously, and generally by a team formed by four members (two surgeons, one packer, and one auxiliary), following all the guidance for antisepsis and asepsis that would be observed in a large-sized orthopedic surgical procedure [40]. After the material has been removed, material for testing is collected. The specimens are packed individually, identified and transported in thermally insulated boxes, and packed in dry ice or ordinary ice. There is a legal re-

quirement, as a matter of respect for the families of donors, that the body structure of the cadaver should be reconstructed: this is done using PVC pieces that were assembled beforehand [40].

(3) It was also pointed out that storage should be carefully done. The tissues are generally kept in freezers at a temperature of −85 °C. They also have an alarm and a supply of liquid CO_2 for additional security. Under ideal conditions and at constant temperature, the tissues can be stored for a period of five years [40].

4.2.4 Xenografts

Autografts and allografts possess inherent limitations despite their excellent success rates in bone-grafting practice [6]. Therefore, natural bone substitutes have been developed to encourage improved osteogenic, osteoconductive, and osteoinductive potentials by creating a favorable bone growth microenvironment [42]. Xenografts are grafting materials that are derived from a genetically unrelated species from the host [43]. The most common source of xenograft materials in the dental field is deproteinized bovine bone, which is commercially available as Bio-Oss [6]. The heterologous (xenograft) material that has bovine or equine origin is a non-stoichiometric less resorbable apatite, which does not resist traction forces and masticatory load [44, 45]. Bovine bone is treated with a stepwise annealing process, followed by chemical treatment with NaOH, to produce a porous hydroxyapatite (HA) material containing only the inorganic components of bovine bone. The resulting porous structure highly resembles that of human bone and can provide good mechanical support and stimulate bone healing through osteoconduction [6]. The porous structure exhibits a vast surface area and promotes the growth of new blood vessels via angiogenesis, which enhances bone growth, confirming the morphological compatibility, which was firstly proposed by Oshida [46] as one of the requirements for a successful implant system. Bovine bone substitutes have been used extensively in maxillary sinus lifting and in implant procedures due to their superior stability and low immunogenicity [43, 47, 48].

A xenograft (bone from animals with similar bone structures as humans) possesses several advantages, including (i) there is no additional surgical site or procedure necessary, (ii) animal bones available are extremely similar in structure to human bones, and (iii) animal bones may reduce the risk of infection [54]. In contrast, it was recognized that some patients may feel uncomfortable using animal bone and xenografts. In addition to the disadvantages of allografts, they carry the risks of transmission of zoonotic diseases, and rejection of the graft is more likely and aggressive [37, 38].

Oshida [46] proposed a morphological compatibility between surface configuration and receiving vital bone surface as one of three major compatibilities for successful implant treatment; the other two are biochemical compatibility and biomechanical compatibility. Surface micro- and macro-structure are also related to mechanical properties in the surface zone. Accordingly, morphological compatibility and biomechanical compatibility influence each other. A similar argument can be applied to the case of natural bone structure and bone structure of the receiving area. Bone-grafts may be cortical, cancellous, or cortico-cancellous [3]. Figure 4.2 shows SEM images of three distinctive bone structures [3]. Referring to the figure, the cortical bone has higher mineral content than the trabecular or cancellous bone [4]. In addition, given the presence of spaces within the structure of the cancellous bone, the latter is more osteogenic than cortical bone [3]. Compressive stiffness and strength of the cortical bone are much higher than those of the cancellous bone. In selecting a graft or combination of grafts, the surgeon must be aware of these fundamental differences in bony structures [3, 17].

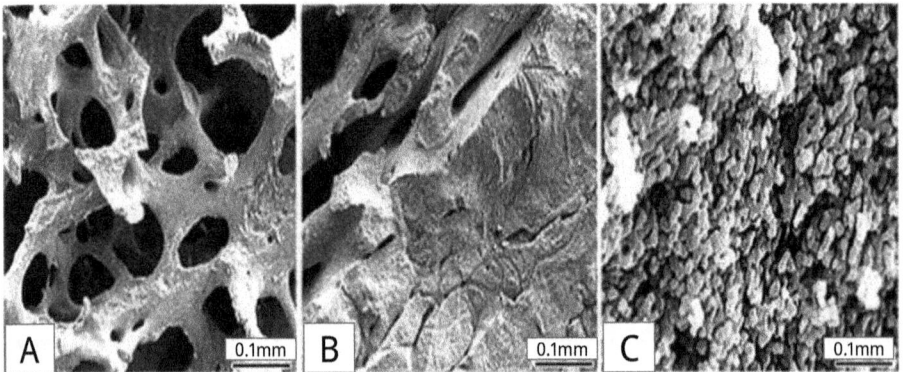

Figure 4.2: SEM images of (A) trabecular or cancellous bone-graft, showing the porous honeycomb microstructure of cancellous bone-graft, (B) cortico-cancellous bone-graft, and (C) cortical or compact bone-graft [3].

4.2.5 Phytogenic material

Phytogenic materials are bone substitute materials obtained from a plant-based origin, such as Gusuibu, coral-based bone substitutes, and marine algae [6]. Wong et al. [49] compared the amount of new bone produced by Gusuibu (a traditional Chinese herbal medicine), integrated with a collagen grafts as a structural scaffold, to that produced by bone-grafts and collagen grafts, and found that (i) a total of 24% and 90% more new bone were present in defects grafted with Gusuibu in collagen grafts than those grafted with bone and collagen, respectively, and (ii) no bone was formed in the passive control group; suggesting that (iii) the efficacy of Gusuibu in promoting new bone formation was comparable to that of the autograft material, and (iv) Gusuibu may present a promising viable bone substitute material when used with a collagen carrier. In dental applications, Gusuibu has been shown to accelerate bone remodeling, following orthodontic tooth movement through promotion of osteoblastic activity, and cell culture studies have revealed that Gusuibu is able to mediate bone remodeling through regulation of osteoclast and osteoblast activity [50, 51].

Coral-based bone substitutes are predominantly composed of calcium carbonate used either in its naturally occurring form or processed by heat treatment with ammonium phosphate and converted into crystalline hydroxyapatite (HA), which subsequently possesses minimal residual carbonate [52–54]. HA is a natural polymer of calcium phosphate obtained from bone or natural materials, such as coral, and widely used to promote bone healing due to its ability to act as a structural scaffold. The main issue associated with naturally occurring coralline HA is its brittleness and high capability of resorption (or resorbability); hence coral-based materials are most commonly used as crystalline HA in granule or block form to provide a structural framework, very similar to that of trabecular bone [53].

The material (known as AlgiPore) is a naturally occurring HA, derived from marine algae, and possesses desirable properties such as good resorbability over time, a large surface area for protein adhesion, and low immunogenicity [6, 55]. The material can act as a carrier for GFs (growth factors) and MSCs (mesenchymal stem cells) [56, 57]. There have been recent developments to combine the material with β-TCP, which claims to decrease resorption times while maintaining the volume support required for bone healing. Zhou et al. [58] compared three calcium phosphate ceramics, two of which have similar structures but different composition: 100% HA (algae-derived) and HA/β-tricalcium phosphate (β-TCP) 20/80 (algae-derived), and two with different structures but similar composition: HA/β-TCP 20/80 (algae-derived) and HA/β-TCP 15/85 (synthetic). It was reported that all three calcium phosphate ceramics demonstrated osteoconductivity and performed similarly in supporting new bone formation, suggesting that the differences in their composition, structure, or degradation did not significantly affect their ability to promote bone healing in this application.

There are still natural materials such as chitosan or silk and their composites, which will be reviewed in Chapter 5.

References

[1]   Kumar P, Vinitha B, Fathima G. Bone grafts in dentistry. J Pharm Bioallied Sci. 2013, 5, S125–7.

[2]   Hudnall D. Selecting the Best Bone Graft Material for a Dental Implant; https://stomadentlab.com/best-bone-graft-material-dental-implant/.

[3]   Oryan A, Alidadi S, Moshiri A, Maffulli N. Bone regenerative medicine: Classic options, novel strategies, and future directions. J Orthop Surg Res. 2014, 9; https://josr-online.biomedcentral.com/articles/10.1186/1749-799X-9-18.

[4]   Brydone AS, Meek D, Maclaine S. Bone grafting, orthopaedic biomaterials, and the clinical need for bone engineering. Proc Inst Mech Eng H. 2010, 224, 1329–43.

[5]   Kolk A, Handschel J, Drescher W, Rothamel D, Kloss F, Blessmann M, Heiland M, Wolff K-D, Smeets R. Current trends and future perspectives of bone substitute materials – From space holders to innovative biomaterials. J Cranio Maxillofac Surg. 2012, 40, 706–18.

[6]   Zhao R, Yang R, Cooper PR, Khurshid Z, Shavandi A, Ratnayake J. Bone grafts and substitutes in dentistry: A review of current trends and developments. Molecules. 2021, 26, 3007; doi: 10.3390/molecules26103007.

[7]   Liu J, Kerns DG. Mechanisms of guided. Bone regeneration: A review. Open Dent J. 2014, 8, 56–65.

[8]   Misch CE, Dietsh F. Bone-grafting materials in implant dentistry. Implant Dent. 1993, 2, 158–67.

[9]   Horch H-H, Sader R, Pautke C, Neff A, Deppe H, Kolk A. Synthetic, pure-phase β-tricalcium phosphate ceramic granules (Cerasorb®) for bone regeneration in the reconstructive surgery of the jaws. Int J Oral Maxillofac Surg. 2006, 35, 708–13.

[10]   Buser D, Hoffmann B, Bernard J-P, Lussi A, Mettler D, Schenk RK. Evaluation of filling materials in membrane-protected bone defects. A comparative histomorphometric study in the mandible of miniature pigs. Clin Oral Implant Res. 1998, 9, 137–50.

[11]   Dimitriou R, Jones E, McGonagle D, Giannoudis PV. Bone regeneration: Current concepts and future directions. BMC Med. 2011, 9, 66; doi: 10.1186/1741-7015-9-66.

[12]   Wang W, Yeung KW. Bone grafts and biomaterials substitutes for bone defect repair: A review. Bioact Mater. 2017, 2, 224–47.

[13]   Roberts TT, Rosenbaum AJ. Bone grafts, bone substitutes and orthobiologics: The bridge between basic science and clinical advancements in fracture healing. Organogenesis. 2012, 8, 114–24.

[14]   Zimmermann G, Moghaddam A. Allograft bone matrix versus synthetic bone graft substitutes. Injury. 2011, 42, S16–S21.

[15]   Tonelli P, Duvina M, Barbato L, Biondi E, Nuti N, Brancato L, Rose GD. Bone regeneration in dentistry. Clin Cases Miner Bone Metab. 2011, 8, 24–28.

[16]   Athanasiou VT, Papachristou DJ, Panagopoulos A, Saridis A, Scopa CD, Megas P. Histological comparison of autograft, allograft-DBM, xenograft, and synthetic grafts in a trabecular bone defect: An experimental study in rabbits. Med Sci Monit. 2010, 16, BR24–BR31.

[17]   Elsalanty ME, Genecov DG. Bone grafts in craniofacial surgery. Craniomax Traum Rec. 2009, 2, 125–34.

[18]   Lee M, Song HK, Yang KH. Clinical outcomes of autogenous cancellous bone grafts obtained through the portal for tibial nailing. Injury. 2012, 43, 1118–23.

[19]   Mauffrey C, Madsen M, Bowles RJ, Seligson D. Bone graft harvest site options in orthopaedic trauma: A prospective in vivo quantification study. Injury. 2012, 43, 323–26.

[20]   Vittayakittipong P, Nurit W, Kirirat P. Proximal tibial bone graft: The volume of cancellous bone, and strength of decancellated tibias by the medial approach. Int J Oral Maxillofac Surg. 2012, 41, 531–36.

[21]   Schlegel KA, Schultze-Mosgau S, Wiltfang J, Neukam FW, Rupprecht S, Thorwarth M. Changes of mineralization of free autogenous bone grafts used for sinus floor elevation. Clin Oral Impl Res. 2006, 17, 673–78.

[22]   Janicki P, Schmidmaier G. What should be the characteristics of the ideal bone graft substitute? Combining scaffolds with growth factors and/or stem cells. Injury. 2011, 42, S77–S81.

[23]   Oryan A, Meimandi-Parizi AH, Shafiei-Sarvestani Z, Bigham AS. Effects of combined hydroxyapatite and human platelet rich plasma on bone healing in rabbit model: Radiological, macroscopical, histopathological, ultrastructural and biomechanical studies. Cell Tissue Bank. 2012, 13, 639–51.

[24]   Yazar S. Onlay bone grafts in head and neck reconstruction. Semin Plast Surg. 2010, 24, 255–61.

[25]   Greenwald AS, Boden SD, Goldberg VM, Yusuf K, Laurencin CT, Rosier RN. Bone-graft substitutes: Facts, fictions, and applications. J Bone Joint Surg. 2001, 83, S98–S103.

[26]   Bone Grafting; https://my.clevelandclinic.org/health/treatments/16796-bone-grafting.

[27]   Ehrler DM, Vaccaro AR. The use of allograft bone in lumbar spine surgery. Clin Orthop Relat Res. 2000, 1, 38–45.

[28] Boyan BD, Ranly DM, McMillan J, Sunwoo M, Roche K, Schwartz Z. Osteoinductive ability of human allograft formulations. J Periodontol. 2006, 77, 1555–63.

[29] Keskin D, Gundogdu C, Atac AC. Experimental comparison of bovine-derived xenograft, xenograft-autologous bone marrow and autogenous bone graft for the treatment of bony defects in the rabbit ulna. Med Princ Pract. 2007, 16, 299–305.

[30] Parikh SN. Bone graft substitutes: Past, present, future. J Postgrad Med. 2002, 48, 142–48.

[31] Bostrom MP, Seigerman DA. The clinical use of allografts, demineralized bone matrices, synthetic bone graft substitutes and osteoinductive growth factors: A survey study. HSS J. 2005, 1, 9–18.

[32] Malinin T, Temple HT. Comparison of frozen and freeze-dried particulate bone allografts. Cryobiology. 2007, 55, 167–70.

[33] Muller MA, Frank A, Briel M, Valderrabano V, Vavken P, Entezari V, Mehrkens A. Substitutes of structural and non-structural autologous bone grafts in hind foot arthrodeses and osteotomies: A systematic review. BMC Musculoskelet Disord. 2012, 14, 59; doi: 10.1186/1471-2474-14-59.

[34] Gomes KU, Carlini JL, Biron C, Rapoport A, Dedivitis RA. Use of allogeneic bone graft in maxillary reconstruction for installation of dental implants. J Oral Maxillofac Surg. 2008, 66, 2335–38.

[35] Folsch C, Mittelmeier W, Bilderbeek U, Timmesfeld N, von Garrel T, Peter Matter H. Effect of storage temperature on allograft bone. Transfus Med Hemother. 2012, 39, 36–40.

[36] Palmer SH, Gibbons CLMH, Athanasou NA. The pathology of bone allograft. J Bone Jt Surgery Br. 1999, 81, 333–35.

[37] Moshiri A, Oryan A. Role of tissue engineering in tendon reconstructive surgery and regenerative medicine: Current concepts, approaches and concerns. Hard Tissue. 2012, 1, 11; doi: 10.13172/2050-2303-1-2-291.

[38] Oryan A, Alidadi S, Moshiri A. Current concerns regarding healing of bone defects. Hard Tissue. 2013, 2, 13; doi: 10.13172/2050-2303-2-2-374.

[39] Bone Bank Allografts; https://www.bonebank.com/.

[40] De Alencar PGC, Vieira IFV. Bone Banks. Rev Bras Ortop. 2010, 45, 524–28.

[41] Tomford WW. Transmission of disease through transplantation of musculoskeletal allografts. J Bone Joint Surg Am. 1995, 77, 1742–54.

[42] Kozusko SD, Riccio C, Goulart M, Bumgardner J, Jing XL, Konofaos P. Chitosan as a bone Scaffold biomaterial. J Craniofacial Surg. 2018, 29, 1788–93.

[43] Kao ST, Scott DD. A review of bone substitutes. Oral Maxillofac Surg Clin North Am. 2007, 19, 513–21.

[44] Berglundh T, Lindhe J. Healing around implants placed in bone defects treated with Bio-Oss. Clin Oral Impl Res. 1997, 8, 117–24.

[45] Artzi Z, Dayan D, Alpern Y, Nemcovsky CE. Vertical ridge augmentation using xenogenic material supported by a configured titanium mesh: Clinicohistopathologic and histochemical study. Int J Oral Maxollofac Implants. 2003, 18, 440–46.

[46] Oshida Y. Bioscience and Bioengineering of Titanium Materials. Elsevier. 2007.

[47] Oliveira G, Pignaton TB, de Almeida Ferreira CE, Peruzzo LC, Marcantonio E. New bone formation comparison in sinuses grafted with anorganic bovine bone and β-TCP. Clin Oral Implants Res. 2019, 30, 483; doi: 10.1111/clr.438_13509.

[48] https://aiceducation.com/what-you-need-to-know-about-dental-bone-regeneration/#:~:text=What%20Is%20Bone%20Regeneration%3F,bone%20will%20begin%20to%20deteriorate.

[49] Wong RW, Rabie ABM. Effect of Gusuibu graft on bone formation. J Oral Maxillofac Surg. 2006, 64, 770–77.

[50] Li Y. Local use of iontophoresis with traditional Chinese herbal medicine, e.g., Gu-Sui-Bu (Rhizoma Drynariae) may accelerate orthodontic tooth movement. Dent Hypotheses. 2013, 4, 50; doi: 10.4103/2155-8213.113008.

[51] Sun J-S, Lin C-Y, Dong G-C, Sheu S-Y, Lin F-H, Chen L-T, Wang Y-J. The effect of Gu-Sui-Bu (Drynaria fortunei J. Sm) on bone cell activities. Biomaterials. 2002, 23, 3377–85.

[52] Fernandez de GG, Keller L, Idoux-Gillet Y, Wagner Q, Musset AM, Benkirane-Jessel N, Bornert F, Offner D. Bone substitutes: A review of their characteristics, clinical use, and perspectives for large bone defects management. J Tissue Eng. 2018, 9, 2041731418776819; doi: 10.1177/2041731418776819.
[53] Bhatt RA, Rozental TD. Bone graft substitutes. Hand Clin. 2012, 28, 457–68.
[54] McPherson R. Bone grafting with coralline hydroxyapatite. EC Dent Sci. 2019, 18, 2413–23.
[55] Galindo-Moreno P, Padial-Molina M, Lopez-Chaichio L, Gutiérrez-Garrido L, Martín-Morales N, O'Valle F. Algae-derived hydroxyapatite behavior as bone biomaterial in comparison with anorganic bovine bone: A split-mouth clinical, radiological, and histologic randomized study in humans. Clin Oral Implant Res. 2020, 31, 536–48.
[56] Smiler D, Soltan M, Lee JW. A histomorphogenic analysis of bone grafts augmented with adult stem cells. Implant Dent. 2007, 16, 42–53.
[57] Wanschitz F, Nell A, Patruta S, Wagner A, Ewers R. Influence of three currently used bone replacing materials on the in vitro proliferation of human peripheral blood mononuclear cells. Clin Oral Implants Res. 2005, 16, 570–74.
[58] Zhou A-J-J, Clokie CML, Peel SAF. Bone formation in algae-derived and synthetic calcium phosphates with or without poloxamer. J Craniofacial Surg. 2013, 24, 354–59.

5 Synthetic bone-grafting materials

Bone-grafting material comes in many forms and from different sources. In the previous chapter, natural sources for bone substitutes (including the patient's own body, bone tissue from another human-body-like (usually cadavers), or tissue from an animal such as bovine bone) have been reviewed. In this chapter, synthetic bone-graft materials are covered. In general, synthetic bone substitutes include polymers, ceramics, metals, and their hybrid and composites. The most successful synthetic bone-graft materials are those substances that mimic and closely resemble human bone scaffolding [1, 2]. These graft materials are sometimes called alloplastic grafting material, which is synthetically derived or made from natural materials. The major advantages of alloplastic bone-grafts include zero risk of disease transmission and low antigenicity. Alloplastic grafting materials mainly include hydroxyapatite, dicalcium phosphates, and bioactive ceramics. Alloplastics are osteointegrative materials with a different degree of resorption; they have biomechanical properties; they are partially replaced in bone remodeling, based on their size and porosity [3].

Synthetic bone-graft products first became available to surgeons in the early 1990s [2]. Sandberg [2] classified bone-grafting materials into passive type and active type. Many of such first-generation products were based on porous calcium phosphate material systems [4]. They solely functioned as an osteoconductive scaffold that supported bone growth on the surface of the material. These materials are typically porous and allow bone formation on the material's surface, through its porosity. The graft resorbs over time and, as a result, eventually the graft is fully replaced by the bone. Products in this category have been used for years with clinically successful outcomes. However, due to their limited role in the bone regeneration process, passive bone-graft products may take longer to fully fill with new bone or may result in a reduced bone formation response in challenging bone-grafting applications, such as large voids or patients with multiple comorbidities. As a result, these bone-grafting materials functioned as bone formation and were not actively involved in the cellular healing process. These grafting materials can be considered as passive bone-grafts. On the contrary, the limitations associated with passive grafts can be overcome with the use of products based on bone-graft materials. Bone-graft materials are considered biologically active when they can promote or enhance the cellular regeneration process: active type. This can be accomplished through multiple mechanisms of action. For example, a growth factor product that contains bone morphogenetic protein (specifically BMP-2) has been shown to increase osteoblast proliferation and function, and promote stem cell differentiation due to the interaction of the BMP protein with bone-forming cells. Bioglass products accomplish the same cellular effect, but through a different mechanism. Once implanted, bioglass undergoes a slow dissolution process that releases ions that stimulate the surrounding cells. Hence, bioactive bone-graft

https://doi.org/10.1515/9783111136691-005

materials are osteoconductive and also have the ability to positively influence the cellular bone formation process [4].

Various types of synthetic bone-grafting materials discussed in this chapter possess different extents of biological properties and biological features. As to biological properties [4–7], there should be osteoconductivity, osteoinductivity, osteogenicity, osteostimulation, and bioactivity. Biological features [4, 7] should include porosity, origin, resorption rate, handling properties and physical features, ability to mix, immunogenicity, and composition. To overcome potential immunogenicity and morbidity at donor sites, artificial synthetic bone substitute materials are generated to closely mimic the biological properties of the natural bone. Despite these characteristics, currently available synthetic materials display only osteointegrative and osteoconductive properties [8, 9]. Materials that fall into this category include calcium phosphate ceramics, such as hydroxyapatite (HA), tricalcium phosphate (TCP) and bioglass; metals, such as nickel-titanium; polymers, such as polymethylmethacrylate (PMMA), and polyglycolides and calcium phosphate cements [10, 11].

5.1 Calcium phosphate material systems

The majority of the artificially prepared calcium orthophosphates of high purity appear to be well tolerated by human tissues in vivo and possess excellent biocompatibility, osteoconductivity and bioresorbability; these biomedical properties of calcium orthophosphates are widely used to construct bone-grafts [12]. Briefly, by definition, all calcium orthophosphates consist of three major chemical elements: calcium, phosphorus, and oxygen, as a part of the orthophosphate anions [12].

Basically, the Ca-P compound group is a big family of MCPA (monocalcium phosphate anhydrous), MCPM (monocalcium phosphate monohydrate), DCPA (dicalcium phosphate anhydrous), DCPD (dicalcium phosphate dihydrate), OCP (octacalcium phosphate), α- and β-TCP (tricalcium phosphate), HA (hydroxyapatite), ACP (amorphous calcium phosphate), and TTCP (tetracalcium phosphate). Each of the Ca-P family members possesses different Ca/P ratio, solubility, bioactivity, and biostability [13]. Ca-P compound exhibits advantages; (i) its chemistry mimics normal biological tissue (C, P, O, H), (ii) it has excellent biocompatibility, (iii) attaches with hard and soft tissues, (iv) shows minimal thermal and electrical conductivities, (v) its modulus of elasticity is close to that of the bone than many other implantable materials, and (vi) its color is similar to hard tissues. On the other hand, there are some disadvantages associated with Ca-P compounds; namely, (i) variable chemical and structural characteristics (technology and chemistry related), (ii) low mechanical, tensile, and shear strengths under fatigue (or repeated cyclic) loading, (iii) low attachment with coating and with substrate, (iv) variable solubility, and (v) variable mechanical stability of coatings under load-bearing conditions [14].

In living organisms, the biological HA is in constant contact with body fluids such as blood serum and saliva. Thus, dissolution, solubility, and precipitation take place as part of the interaction of the material with biological fluids in tissues [15]. It might be worth making a clear difference for relevant terms. The term dissolution refers to a kinetic process of disintegration of solid crystalline structure as separate ions, atoms, and molecules form, and is quantified by its rate [16]. The solubility describes one stage of dissolution and quantifies the dynamic equilibrium state achieved when the rate of dissolution equals the rate of precipitation. Solubility is the property of a solid, liquid, or gaseous chemical substance called solute to dissolve in a solid, liquid, or gaseous solvent to form a homogeneous solution of the solute in the solvent. The solubility of a substance fundamentally depends on the physical and chemical properties of the solute and solvent as well as on temperature, pressure, and the pH value of the solution [17]. The extent of the solubility of a substance in a specific solvent is measured as the saturation concentration, where adding more solute does not increase the concentration of the solution and it begins to precipitate the excess amount of solute. The order from the most soluble to the least soluble (or the most chemically stable) is as follows: MCPM > MCPA > TTCP > α-TCP > DCPD > DCPA > β-TCP > ACP > HA in 25 °C water at pH of 7.4 [18–20], where MCPM: monocalcium phosphate monohydrate (Ca/P: 0.50), MCPA: monocalcium phosphate anhydrous (Ca/P: 0.50), TTCP: tetracalcium phosphate (Ca/P: 2.00), TCP: tricalcium phosphate (Ca/P: 1.50), DCPD: dicalcium phosphate dihydrate (Ca/P: 1.00), DCPA: dicalcium phosphate anhydrous (Ca/P: 1.00), ACP: amorphous calcium phosphate (Ca/P: 1.50), and HA: hydroxyapatite (Ca/P: 1.67). Even though not listed in the above, there is another type of OCP (octacalcium phosphate: $Ca_8H_2(PO_4)_6.5H_2O$) (Ca/P:1.33), which is considered to be a precursor for HA production [20, 21]. Here, one must stress that the prefixes "mono", "di", "tri" and "tetra" are related to the amount of hydrogen ions replaced by calcium.

5.1.1 Hydroxyapatite

Hydroxyapatite can be found in teeth and bones within the human body. Many modern implants, for example, hip or knee replacements, dental implants, and osteoconductive implants, are coated with hydroxyapatite (HA), due to the facts that (i) HA possesses similarities with the mineral part of the bone, and (ii) HA promotes osseointegration, which can be defined as the direct structural and functional connection between the ordered, living bone and the surface of a load-carrying implant [22]. Although a chemical similarity of HA to that of inorganic component of the natural bone enables it to be used as a bone-grafting material [23], synthetic HA does not contain trace amounts of Na^+, Mg^{2+}, K^+, and Sr^+, which are found in naturally derived HA, such as bovine bone, which influences various biomechanical reactions [8]. Synthetic HA does not possess a microporous structure, as seen in bovine-derived HA [24]. Synthetic HA has a delayed resorption rate due to its relatively high Ca/P ratio and crystallinity [8, 13]. Another major concern associated with HA is its relatively low mechanical strength, preventing it from being used at high load-bearing sites. Previous studies have found that the quality and quantity of new bone formed, following grafting with synthetic HA alone or in combination with a polymer, was insufficient for preservation of alveolar ridge heights for the placement of endosseous implants, maxillary sinus lifting, and management of periodontal osseous defects [25], providing some limitation in application in dentistry for the coating on implants, external fixator pins, or in sites with low loading stress [23, 26].

HA materials' application is not limited as coating materials but are also used in bone-grafting substitutes. Among the many types of bone-graft substitute materials, a continued interest in avoiding donor sites and utilizing the convenience of off-the-shelf bone-substitute products has stimulated the development of several synthetic, yet biocompatible, bone substitutes. To date, the calcium phosphate apatites, including HA cements, have been the most useful synthetic bone-graft substitutes [17]. Synthetic HA cement has excellent biocompatibility when used to repair bone defects and is capable of osseointegration and substitution by bone, when placed in direct contact with a viable host bone due to a well-operated balance between osteoclasts and osteoblasts cells to control the amounts of Ca and P in blood, during the bone remodeling process [17].

The original composition of HA may be altered by two reasons, during (1) deposition operation, and (2) resorption upon placing the HA-coated implant by the biological actions. After plasma spray coating, both the purity and crystallinity of the HA decreases because of the decomposition of HA at high temperature and the rapid cooling rate (or quenching). New phases appear in the HA coating, including an amorphous phase, tricalcium phosphate, tetracalcium phosphate, and calcium oxide. The calcium oxide phase is not biocompatible, and should be avoided [22]. The general agreement is that the chemical purity of HA should be as high as possible (>90%), with a Ca/P ratio of 1.67 [27]. Plasma spraying has an ability to produce crystalline coatings from 30% to 70%; while under a normal deposition process, the crystallinity is approximately 65% [28]. Resorption means the destruction, disappearance, or dissolution of a tissue or a part of the tissue by biochemical activity, as the loss of bone or tooth dentin; in other words, it describes a remodeling process, since the human body is in a constant state of bone remodeling [29–31]. Bone is resorbed by osteoclasts, and is deposited by osteoblasts in a process called ossification. Similarly, the bone resorption can be defined as a process by which osteoclasts break down the bone and release minerals, resulting in a transfer of calcium from bone fluid to the blood, leading to the calcium paradox [32–34].

Recent advances in HA-based bone substitute materials have looked into producing nano-sized HA, which displays enhanced biomechanical properties that more closely mimics the composition of natural bone [22, 35, 36]. The rationale for the development of these nano-sized materials includes its much closer resemblance to bone extracellular matrix; a faster response to external environmental stimuli; enhanced delivery, and controlled release of bioactive molecules, such as growth factors, allowing for enhanced osteoregenerative properties [37, 38]. Nanocrystalline HA exhibits improved biological performance and dissolution compared with its conventional forms of HA [39]. The nanostructure allows for a larger surface-to-volume ratio, promoting more effective adhesion, proliferation, and differentiation of osteogenic progenitor cells. It also improves sinterability and enhances densification, resulting in improved fracture toughness and other mechanical properties [23, 40, 41]. Despite significant improvements in performance across all domains compared with conventional forms of HA, there remains insufficient evidence to support nanocrystalline HA's widespread use [23, 37].

5.1.2 Complex hydroxyapatite, incorporated with substitutional element(s)

The name "apatite" comes from the Greek word απατω, which means to deceive. The reason for this was that the mineral appeared in different colors with a variety of crystals and was thus often mistaken for precious minerals such as aquamarine or amethyst [42]. What actually happens inside the apatite crystal structure is dependent on the unique characteristics of ion exchangeability in M-site, B-site and A-site in M-$_{10}B_6A_2$, where M is bivalent cations, B is trivalent anions representative of XO_4, and A is monovalent anions that can be substituted by fluoride, chloride or carbide [13], as listed in Table 5.1, which shows Oshida's summarization of the elemental cation and anion substitutes into hydroxyapatite composition and their primary influences.

5.1.3 Tricalcium phosphates

Tricalcium phosphates (TCP) possess two crystallographic forms, α-TCP and β-TCP [43, 44]. β-TCP is a type of calcium phosphate material widely used as a bone substitute material over many years, exhibiting a faster biodegradation and absorption, compared with HA, due to its lower Ca/P ratio [26]. Pure phasic β-TCP possesses many desirable properties, such as its ease of handling, radiopacity (allowing monitoring of healing), good osteoconductivity due to macroporosity (promoting fibrovascular ingrowth and osteogenic cell adhesion), good resorbability (compared with bovine bone-grafts), and low immunogenicity and risk of disease transmission [45, 46]. Whilst the interconnected porous structure of β-TCP allows for improved vascularization, it also results in the material's poor mechanical strength under compression [26, 45, 47]. This results in β-TCP being unsuitable as a bone substitute; however, it is suitable for use as filler in bony defects, repairing at morphological sites [48]. It is commonly used to repair marginal periodontal and periapical defects and as a partially resorbable filler in alveolar bony defects [49]. Nakajima et al. found that the bone regenerative potential of β-TCP is comparable to that of freeze-dried bone allo-

Table 5.1: Various types of cation- and anion-substitutes into hydroxyapatite structure.

	Ag	Zn	Cu	Fe	Sr	Mg	La	Ti	Y	W	Pb	Ba	Na	K	SiC	TiO$_2$	SiO$_2$	ZrO$_2$	Al$_2$O$_3$	C	Si	F
Antimicrobial activity	•	•	•		•																	•
Bioactivity		•							•			•						•	•	•		•
Promoting bone growth		•	•	•	•							•		•							•	•
Improve biocompatibility																				•		
Anticancer therapy			•																			
Assisting drug delivery system						•																
Improve radiopacity					•					•												
Control transform to HA					•	•							•									
Improve mechanical strengths						•	•	•					•						•			
Surface energy control																					•	
Enhance adhesive strength					•	•										•						•
Improve corrosion resistance				•												•						•
Control environment										•												

graft, and deproteinized freeze-dried bone allograft and autograft materials [50]. However, the mechanical properties of this material limit its wider application [51].

5.1.4 Comparison between HA and TCP

As we have seen in the above, HA and TCP are two major members in Ca-P family, and each of these possesses different characters in its composition (Ca/P ratio), solubility behavior, crystallinity, preparation method, bioactivity, etc. Highly dense sintered HA is hard to be dissolved by osteoclast. At the same time, it is stable inside the body, which is saturated by HA. Therefore, due to bone formation by bone modeling (which happens due to HA resorption by the osteoclast), the major role of HA inside the body is just as a space-making material. On the other hand, TCP is not a rapid self-hardening material, and is very slowly resorbed into body, which is already saturated by HA, so bone reformation is very difficult. TCP has a stoichiometry similar to amorphous bone precursors, whereas HA has a stoichiometry similar to bone mineral [52].

Liu et al. [53] investigated the influences of Ca/P stoichiometry of calcium phosphates on biological performances, including osteoblast (bone-forming cell) viability, collagen production, alkaline phosphatase activity, and nitric oxide (NO) production. A group of calcium phosphates with Ca/P ratios between 0.5 and 2.5 was obtained by intentionally adjusting the Ca/P stoichiometry of the initial reactants necessary for calcium phosphate precipitation. TCP and $Ca_2P_2O_7$ phases were observed for samples with Ca/P ratios of 0.5 and 0.75.. In contrast, for samples with Ca/P ratios of 1.0 and 1.33, the only stable phase was TCP. For samples with a Ca/P ratio of 1.5, the TCP phase was dominant; however, small amounts of the HA phase started to appear. For samples with a 1.6 Ca/P ratio, the HA phase was dominant. Lastly, for samples with 2.0 and 2.5 Ca/P ratios, the CaO phase started to appear in the HA phase, which was the dominant phase. It was furthermore reported that (i) the average grain size and the average pore size decreased with increasing Ca/P ratios, (ii) the porosity of calcium phosphate substrates also decreased with increasing Ca/P ratios, (iii) the collagen production by osteoblasts was similar between all the calcium phosphates, but slightly lower with a 1.6 Ca/P ratio, and (iv) greater alkaline phosphatase activity by osteoblasts was observed in all the cultures with various calcium phosphates (0.5–2.5 Ca/P ratios) than in the control (only cells in culture); suggesting that the Ca/P ratio of calcium phosphate is a very important factor that should be considered when selecting nano-to-micron particulate calcium phosphates for various orthopedic applications [53].

There are several studies to compare osteoblast cell attachment and growth behaviors between HA and TCP. An essential property of bone substitute materials is that they are integrated into the natural bone remodeling process, which involves the resorption by osteoclast cells and the formation by osteoblast cells. When the monocyte cells adhere to a calcium phosphate surface (bone or bone substitute material), they can bind together and form multi-nucleated osteoclast cells. Detsch et al. [54] showed that osteoclast-like cells derived from a human leukoma monocytic lineage responded in a different way to TCP than to HA ceramics. It was reported that (i) both bioceramics were degraded by resorbing cells; however, HA enhanced the formation of giant cells, (ii) the osteoclast-like cells on HA formed a more pronounced actin ring, and larger lacunas could be observed, (iii) TCP ceramics are medically used as bone substitute materials because of their high dissolution rate; on the other hand, highly soluble calcium phosphate ceramics like TCP seem to be inappropriate for osteoclast resorption because they produce a high calcium concentration in the osteoclast interface and in the environment [54]. Kasten et al. [55] compared three resorbable biomaterials (CDHA: calcium-deficient hydroxyapatite, TCP, and DBM: demineralized bone matrix) in terms of seeding efficacy with human bone marrow stromal cells (BMSCs), cell penetration into the matrix, cell proliferation, and osteogenic differentiation. The three biomaterials were seeded with human BMSCs and kept in human serum and osteogenic supplements for 3 weeks. Morphologic and biochemical evaluations were performed on day 1, 7, 14, and 21. It was reported that (i) the allograft DBM and CDHA exhibited both an excellent seeding efficacy while the performance of β-TCP was lower, when compared, (ii) the total protein content and the values for specific alkaline phosphatase (ALP) increased on all matrices, and no significant difference was found for these two markers, (iii) BMSCs in monolayer had a significant increase of protein, but not of ALP, (iv) osteocalcin (OC) values increased significantly higher for BMSC in cultures on DBM, when compared to CDHA and β-TCP, (v) all three matrices promoted BMSC proliferation and differentiation to osteogenic cells, and (vi) DBM allografts seem to be more favorable with respect to cell ingrowth, tested by histology, and osteogenic differentiation, ascertained by an increase of OC. Based on the these findings, it was concluded that CDHA, with its high specific surface area, showed more favorable properties than β-TCP, regarding reproducibility of the seeding efficacy [55].

HA and TCP are also compared in terms of bone formation in sinus augmentations. Hadi et al. [54] compared two different graft materials (bovine HA and β-TCP) associated with platelet-rich plasma (PRP) in one patient referred for bilateral sinus augmentations. It was reported that (i) the bovine HA and TCP resulted in bone formation after the sinus lift procedures, and (ii) histomorphometric analysis showed some differences, but both biomaterials allowed for stable clinical results [56]. Similarly, Kurkcu et al. [57] compared the biological performance of the bovine HA graft material and the synthetic β-TCP material in sinus augmentation procedure. The study consisted of 23 patients (12 male and 11 female) who were either edentulous or partially edentulous in the posterior maxilla, and required implant placement. After an average of 6.5 months of healing, bone biopsies were taken from the grafted areas. It was found that (i) the mean new bone formation was 30.13% in bovine HA group and 21.09% in β-TCP group, (ii) the mean percentage of residual graft particle area was 31.88% for bovine HA group and 34.05% for β-TCP group, and (iii) the mean percentage of soft tissue area was 37.99% in bovine HA group and 44.86% in β-TCP group. Based on these clinical data, it was concluded that both graft materials demonstrated successful biocompatibility and osteoconductivity in sinus augmentation procedure; although the bovine HA appears to be more efficient in osteoconduction, when compared to β-TCP [58]. Yang et al. [58] compared the bone formation on titanium implant surfaces coated with biomimetically deposited calcium phosphate or electrochemically deposited hydroxyapatite and found that the electrochemically deposited HA coating has good bone formation properties, while the biomimetically deposited CaP coating has weaker bone formation properties.

5.1.5 Calcium phosphate biocomposites

There are materials under development utilizing the aforementioned differences between HA and TCP to fabricate biphasic bioceramics. Solubility and degradation profile of HA and TCP varies between very stable and almost insoluble appearance (100% HA) to fast degradation and loss of structure (100% α-TCP). For optimization of the properties of the composite materials for cell seeding and tissue engineering applications, several studies of different ratios of HA/TCP biphase biocomposites have been conducted and reported.

Calcium phosphate cements (CPCs) are generally two- or three-component systems consisting of an aqueous component and a powder component, commonly containing sintered CP material, such as α-TCP and HA. Mixing of the components results in a workable paste that hardens in situ in a self-setting manner, to form HA nanocrystals at room temperature [11, 59]. The main advantages of CPCs include their self-setting ability, the ability to shape the paste into the defect site, their ability to replicate the structure and composition of bone in a repeatable manner, their high biocompatibility and availability in different forms for different types of bony defects, and their osteoconductive properties [26, 51, 59]. However, CPC lacks a macroporous structure, which limits the speed of cell adhesion, fluid exchange, and restorability [45, 59]. Additionally, the potential of an incomplete setting reaction, resulting in an inflammatory reaction, presents a major drawback associated with CPCs. Recent research has aimed to address these shortcomings and strategies used. These include the development of prefabricated 3D-printed CPC scaffolds and improved CPC injectability through various mechanisms, including the addition of viscous binders such as chitosan, gelatin, and hyaluronic acid; optimizing the particle size, distribution, shape, and interparticle interactions of the CPC powder; regulation of the setting reaction and modification of external factors like syringe and needle sizes [60–62]. Furthermore, research has focused on improving the material properties of CPC products by doping CPCs with various ions, such as silicon and strontium, to improve osteoconductivity; incorporating bioactive glass to improve bioactivity; and infusion with growth factors and stem cells to improve osteoinductivity [63]. Lyu et al. [64] demonstrated that using CPCs, in addition to a collagen membrane, delayed new alveolar bone formation at extraction sites. In addition, CPCs are generally brittle when subjected to tensile and shear forces, and thus they are indicated for use only in non-load-bearing sites [51, 65]. Another major concern associated with CPCs is the potential extrusion of the material into the surrounding tissues, causing potential damage to adjacent tissues [47]. In clinical dentistry, CPCs have been used as filler for bony defects, for reconstruction of bony fractures, and in dental implantology.

Arinzeh et al. [66] prepared 100% HA, 76%HA/24%TCP, 63/37, 56/44, 20/80, and 100% β-TCP (by weight) and seeded human mesenchymal stem cells (hMSCs). The hMSC-loaded implants and cell-free implants were implanted subcutaneously in the backs of 24 SCID mice. Total porosity of the scaffolds was 60–70%, the pore size ranged between 300 and 600 μm, and the grain size ranged between 0.5 and 1.5 mm.

Implants were harvested at 6 and 12 weeks and processed for routine decalcified histology. The hMSC-loaded ceramics demonstrated osteogenesis at 6 weeks. It was reported that (i) the composition significantly influenced bone growth, and 100% HA and 100% TCP promoted the least bone growth, (ii) in the compositions of 100% HA, 76/34 HA/TCP, 62/37, 56/44, and 100% TCP, bone was only present in the peripheral pores of the implant, (iii) for hMSC-loaded 20/80 HA/TCP implants, bone was detected to be lining all of the pores throughout the ceramic structure, although bone was also only present in the peripheral pores of the implant, and (iv) the culture media and the donor of cells effected on the amount of bone growth, but it was also shown that the different compositions of HA and TCP were working better than pure 100% HA and 100% TCP alone [66]. Similarly, Jensen et al. [67] prepared three biphasic calcium phosphate bone substitute materials with HA/TCP ratios of 20/80, 60/40, and 80/20 and compared them in terms of reaction to coagulum, particulated autogenous bone, and deproteinized bovine bone mineral (DBBM) in membrane-protected bone defects. It was found that (i) 20/80 biphase biocomposite showed bone formation and degradation of the filler material, similar to autografts, whereas 60/40 and 80/20 biocomposites rather equaled DBBM, (ii) among the three biphase biocomposites, the amount of bone formation and degradation of filler material seemed to be inversely proportional to the HA/TCP ratios, and (iii) the fraction of filler surface covered with bone was highest for autografts at all time points and was higher for DBBM than 80/20 and 60/40 biocomposites at the early healing phase [67].

Suzuki et al. [68] prepared 100 HA, 80/20, 60/40, 30/70, and 100 β-TCP, and seeded osteoblast-like cells of calvarial cells from neonatal rats and the well-characterized mouse-fibroblast cell line, L-929 (NCTC Clone 929) onto ZrO_2 and LUX plates and the ceramic disk plates (30 mm OD x 2 mm thick) consisting of the aforementioned biphase biocomposites of HA and β-TCP with five different ratios. It was reported that (i) the total number of cells on LUX plates at 2 d of culture was higher than that on any TCP-HA plates, (ii) the total number of cells on ZrO_2 plates at 2 d of culture was higher than that on TCP-HA plates, except on the 100% HA plate, (iii) the initial number of cells adhering to the TCP-HA plates, after seeding, was lower than that found on LUX and ZrO_2 plates, but thereafter, the total number of cells increased rapidly on the TCP-HA plates during the rapid growth phase between 4 d and 9 d of culture, and (iv) at 9 d of culture, the numbers of cells on 100% β-TCP 100% HA and composition containing 30/70 (HA/TCP) were higher than those on the reference LUX and ZrO_2 plates. Based on these results, it was suggested that the HA/TCP biphase biocomposites are able to stimulate differentiation of osteoblasts and promote osteogenesis in vivo [68]. Alam et al. [69] prepared five biphasic calcium phosphate ceramics (100 HA, 75/25, 50/50, 25/75, and 100 β-TCP) and evaluated these ceramics as carriers for rhBMP-2. These bioceramics, impregnated with different doses of recombinant human bone morphogenetic protein 2 (rhBMP-2) (1, 5, and 10 µg), were used for experimental purpose and ceramics without rhBMP-2 were used as control. The pellets were placed into subcutaneous pockets on the dorsum of 4-week-old male Wistar rats. The animals were sacrificed 2 and 4 weeks after implantation. It was found that (i) all experimental pellets exhibited new bone formation whereas the control pellets produced only fibrous connective tissue, and (ii) 100% HAP ceramic showed most amount of bone formation, whereas 25HA/75TCP ceramic produced bone least in amount among different biphasic ceramics at the end of 4 weeks; indicating that formation of new bone depends on the ceramic content with high HA/TCP ratio and high dose of rhBMP-2 [69].

The effects of HA coating and biphasic HA/TCP (50/50) coating on the osseointegration of grit-blasted Ti-6Al-4V implants were compared [70]. Each coated implant was compared with uncoated grit-blasted implants as well. The implants were press-fit into the medullary canal of rabbit femora, and their osseointegration was evaluated 3 to 24 weeks after surgery. It was reported that (i) the coated implants had significantly greater new bone ongrowth than the uncoated implants, (ii) unmineralized tissue (cartilage and osteoid) was seen on the uncoated implants but never on the coated implants, (iii) the coated implants had significantly greater interfacial shear strength than the uncoated implants, and (iv) there was no difference between HA and HA/TCP coating with regard to new bone growth or interfacial shear strength [70]. De Kok et al. [71] studied MSC (marrow stromal cell)-based alveolar bone regeneration in a canine alveolar saddle defect model using biphasic HA/TCP (60/40) material with a mean pore size of 300–500 μm. It was reported that (i) histomorphometrical analysis showed that equivalent amounts of new bone were formed within the pores of the matrices loaded with autologous MSCs and MSCs from an unrelated donor, and (ii) bone formation in the cell-free HA/TCP matrices was less extensive [69].

Miao et al. [72] investigated the penetration and the attachment of bone marrow stromal stem cells (BMSC) into HA/TCP scaffolds (average pore size: 500 μm; porosity: 87%). The scaffold contained biphasic HA/TCP (80%/20%) and some of them were further processed and coated with polylactideglycolide copolymer (PLGA) for the improvement of mechanical properties. It was observed that (i) the bone marrow stromal cells attach well to the surfaces of the PLGA-coated HA/TCP scaffolds, and (ii) the cells attach well to the exposed ceramic HA/TCP surface and the PLGA coating surface, indicating the biocompatibility of both the ceramic phases and the PLGA phase [72]. Yuan et al. [73] conducted a 2.5-year study in dorsal muscles of dog with porous HA, porous biphasic calcium phosphate ceramic (HA/TCP), porous α-TCP, and porous β-TCP. It was reported that (i) normal compact bone with bone marrow was found in HA implants and in HA/TCP implants, 48% pore area was filled with bone in HA implants, and 41% in HA/TCP implants, (ii) bone-like tissue, which was a mineralized bone matrix with osteocytes but lacked osteoblasts and bone marrow, was found in β-TCP implants and in one of the four α-TCP implants, and (iii) both normal bone and bone-like tissues were confined inside the pores of the implants; suggesting that calcium phosphate biphasic HA/TCP ceramics are osteoinductive in muscles of dogs [73].

Dorozhkin [74] described the state-of-the-art calcium orthophosphate ($CaPO_4$)-containing biocomposites and hybrid biomaterials suitable for biomedical applications. It was mentioned that (i) various types of CaPO4-based biocomposites and hybrid biomaterials are either already in use or are being investigated for biomedical applications with many different formulations in terms of the material constituents, fabrication technologies, structural and bioactive properties, as well as both in vitro and in vivo characteristics, (ii) among others, the nano-structurally controlled biocomposites, those containing nanodimensional compounds, biomimetically fabricated formulations with collagen, chitin and/or gelatin, as well as various functionally graded structures seem to be the most promising candidates for clinical applications, and (iii) among them, porous bioceramics made of $CaPO_4$ appear to be very prominent due to both the excellent biocompatibility and bonding ability to living bone in the body; directly relating to the fact that the inorganic material of mammalian calcified tissues, i.e., of bone and teeth, consists of $CaPO_4$ [75, 76]. In general, it has been desirable for the bone-graft material to be a material having the property of producing HA and binding to the bone and curing, while retaining the shape at the time of transplantation. The reasons should include (1) since the body fluid is supersaturated with the HA of the phosphate calcium, HA is chemically stable in the body, and does not apparently dissolve, so there should not be any volume loss, (2) since hard tissues have HA as its main component, the body inevitably shows excellent biocompatibility for HA, so it does not induce inflammatory rejection, and (3) the bone-graft has the effect of activating osteoblasts, and it is absorbed by osteoclasts itself in a short time and replaced by bone. Eventually, the graft needs to move into a cycle of bone remodeling. Research and development has been progressing to meet these requirements, and Sugawara et al. [77–79] have developed a bioresorbable self-curing phosphate calcium.

5.2 Ceramic-based bone substitute materials

Composite bone substitute materials aim to improve the mechanical properties of the resulting combination of different materials, such as bioglass and polymers, through combining their osteoconductive properties. They are commonly used to expand the utility of autograft products and are often combined with bone marrow or act as carriers for BMPs to improve their osteoconductive and osteoinductive properties [8, 11, 47]. Composite bone substitutes combining two or more materials have been used to exploit the advantages of various materials [80].

Majority of bone-grafts available involve ceramics, either alone or in combination with another material (e.g., calcium sulfate, bioactive glass, and calcium phosphate). The use of ceramics, like calcium phosphates, is calcium hydroxyapatite, which is osteoconductive and osteointegrative; and in some cases, osteoinductive. They require high temperatures for scaffold formation and have brittle properties. Most clinical applications of bioceramics relate to the repair of the skeletal system, comprising bone, joints, and teeth, and to augment both hard and soft tissue. The most common bioceramics used in bone tissue applications is calcium phosphate; its is widely used for the regeneration of bone tissue because of its ability to induce osteoblastic differentiation in progenitor cells [81]. To restore skeleton function in the field of orthopedic and oral-maxillofacial surgery, bone tissue regeneration remains an important challenge. Bone Tissue Engineering (BTE) involves generating new bone for skeletal use, which helps humans in bone transplantation. In the past two decades, the advent of tissue engineering has brought new ideas, leading to the discovery and/or development of innovative biomaterials for bone tissue engineering purposes. The inorganic component of bone is hydroxyapatite (HA). For this reason, synthetic calcium phosphate materials, including HA, are FDA-approved and are among the most investigated materials for scaffold composition for over three decades [81].

Bioactive glasses (BAG) are a group of synthetic silicate-based ceramics composed of silicates coupled to other minerals, such as Ca, Na_2O, H, and P [26, 51]. The original composition of bioglass consists silicon dioxide (SiO_2), sodium oxide (Na_2O), calcium oxide (CaO), and phosphorus pentoxide (P_2O_5), though this has been recently modified to a more stable composition through the addition of potassium oxide (K_2O), magnesium oxide (MgO), and boric oxide (B_2O) [26]. Upon exposure to body fluids during implantation, silicon ions can leach out and accumulate, forming a layer of HA on the surface of the material, which allows for adherence of osteogenic progenitor cells. The desirable properties of bioglass include good biocompatibility, osteoconductivity, antimicrobial activity, and a porous structure, promoting vascularization [26, 51, 82]. Bioactive glass (bioglass) is a biologically active silicate-based glass [83], having high modulus and brittle nature; it has been used in combination with polymethylmethacrylate (PMMA) to form bioactive bone cement and with metal implants as a coating to form a calcium-deficient carbonated calcium phosphate layer, which facilitates the chemical bonding of implants to the surrounding bone. Different types of calcium phosphates can include tricalcium phosphate, synthetic hydroxyapatite, and coralline hydroxyapatite, available in pastes, putties, solid matrices, and granules [83].

Recent research has shown that incorporation of various ions with BAG is able to enhance the material's properties. Esfahanizadeh et al. [84] found that zinc-doped BAG resulted in reduced biofilm formation for microbes associated with periodontal disease. Additionally, another research group, Lovelace et al. [85] found that silver-doped BAG showed controlled release of ions from the material, enhancing its antibacterial properties against Porphyromonas gingivalis (P.g), Prevotella intermedia (P.i), and Aggregatibacter actinomycetemcomitas (A.a). These microbes play a central role in the destruction of periodontal tissues, leading to gum disease and peri-implantitis [85]. However, BAG can be brittle and possess low mechanical strengths and poor fracture resistance. Thus, their use in dentistry is limited to low-stress environments or in combination with other grafting materials [82, 86]. Bioglass materials have been successfully used to augment the unilateral cleft alveolar bone, manage periodontal osseous defects, and preserve alveolar bone, following tooth extractions in orthodontic patients [87–89].

Bone healing differs from any other soft tissue since it heals through the generation of new bone rather than by forming fibrotic tissue. Osteogenesis, osteoinduction, osteoconduction, and adequate blood and nutrient supply are the four critical elements for bone regeneration, along with the final bonding between the host bone and the grafting material, which is called osseointegration [90, 91]. The ideal bone composite material with composition and mechanical properties equivalent to that of bone should have adequate biocompatibility, tailorable biodegradability, and ability to initiate osteogenesis; in short, the graft should closely mimic the natural bone. Biodegradability, together with biocompatibility and suitable mechanical properties, is found only in a small group of materials [85]. Bioactive glasses exhibit osteoinductive and osteoconductive properties [92], and can be manufactured into microspheres, fibers, and porous implants. Since silica is a majority composition of BAGs, the bioactivity depends upon its content [93, 94]. Valimaki et al. [95] reported that the bonding between bone and glass is most excellent if the bioactive glass contains 45% ~ 52% SiO_2.

5.2.1 BAG composites incorporated with HA

The combination of hydroxyapatite with BAG results in better composite bioactivity and biocompatibility compared to hydroxyapatite alone [96]. Its preparations have significantly greater mechanical strength when compared to calcium phosphate preparations. After contact with body fluids, a silicate-rich layer is formed, leading to mechanical strong graft/bone bonding. Above this, a hydroxyapatite layer forms, which directs new bone formation, together with protein absorption. The extracellular proteins magnetize macrophages, mesenchymal stem cells, and osteoprogenitor cells. Consequently, the osteoprogenitor cells proliferate into matrix-producing osteoblasts [95]. Bellucci et al. [97] developed bioactive glass/hydroxyapatite composites for bone tissue repair and regeneration, with its low tendency to crystallize, so it enables to sinter samples at a relatively low temperature. It was reported that the mechanical properties of the composites with 80 wt% of BAG + Ca mixture, incorporated with 20 wt% of HA, are sensibly higher than those of Bioglass® 45S5 reference samples due to the presence of HA (mechanically stronger than the 45S5 glass).

Different compositions of BAG and hydroxyapatite composite coatings were fabricated by electrophoretic deposition on titanium substrate and their various properties were investigated [98]. Stability and characteristics of the prepared suspensions using isopropanol and triethanolamine (TEA) were also studied by zeta potential, FTIR (Fourier transform infrared), and conductivity measurements. It was obtained that the optimum amount of TEA to make stable suspension is influenced by the particle size, and consequently this affects suspension conductivity and deposition kinetics. Moreover, the porosity of the coating microstructure was affected by the particles size, as the incorporation of finer HA particles among coarser BG particles led to more compact coatings. It was also indicated that (i) BAG coating includes 50 wt% HA had the highest bioactivity response as well as the maximum bonding strength within one day of immersion in SBF (simulated body fluid) solution, and (ii) HA particles acted as the pre-available seeds for the apatite nucleation while the dissolution of BAG particles provided the required ions for their growth, resulting in a synergistic effect.

5.2.2 BAG composites incorporated with synthetic polymers

Several drawbacks associated with traditional bone-grafting materials and methods (including the limited supply of autografts, transmission of diseases, rejection of grafts, donor site pain and morbidity, limitation in volume of donor tissue that can be safely harvested, and the possibility of harmful immune responses to allografts, etc.) cause surgeons, material scientists, and engineers to seek alternative methods and materials to repair bone defects [99, 100]. Forming hybrid compounds or composites is one of the remarkable solutions as promising bone-grafting materials. BAG has been incorporated with either synthetic polymers or natural polymers [e.g., [101–103]]; the latter will be discussed later in this section.

Combined with biopolymers, BAG has been used to develop composite scaffolds for bone engineering, e.g., [104–107]. Direct solvent-assisted printing has been demonstrated to successfully process polymer-BAG composites. Russias et al. [108] demonstrated that hybrid organic/inorganic scaffolds with controlled microstructures can be built using robotic assisted deposition at room temperature. Polylactide (PLA) or polycaprolactone (PCL) scaffolds, with pore sizes ranging between 200 and 500 μm and hydroxyapatite contents up to 70 wt%, were fabricated. It was mentioned that (i) compressive tests revealed an anisotropic behavior of the scaffolds, which is strongly dependent on their chemical composition, (ii) the inclusion of an inorganic component increased their stiffness but they were not brittle, and could be easily machined even for ceramic contents up to 70 wt%, (iii) the mechanical properties of hybrid scaffolds did not degrade significantly after 20 days in simulated body fluid; however, the stiffness of pure polylactide scaffolds increased drastically due to polymer densification, and (iv) scaffolds containing bioactive glasses were also printed and after 20 days in simulated body fluid, they developed an apatite layer on their surface. Similar studies were carried out on PCL-borate BAG composites [109, 110] and PLA/PEG/BAG composite (PEG: polyethyleneglycol) [107].

In the above, new technology for BAG-based composite with polymers has been reviewed. In particular, it emphasized the fabrication technology of additive manufacturing (AM) such as 3D bioprinting. There are also other ways to improve BAG-based composite – manipulating micro- or nano-structural approaches. Nanoscale bioactive glasses have been recognized for their superior osteoconductivity when compared to conventional (micron-sized) bioactive glass materials. The combination of bioactive glass nanoparticles or nanofibers with polymeric systems enables the production of nanocomposites with the potential to be used in a series of orthopedic applications, including scaffolds for tissue engineering and regenerative medicine. Boccaccini et al. [102] reviewed the state of art of the preparation of nanoscale bioactive glasses and the corresponding composites with biocompatible polymers and the recent developments in the preparation methods of nano-sized bioactive glasses, covering sol-gel routes, microemulsion techniques, gas phase synthesis method (flame spray synthesis), laser spinning, and electro-spinning.

5.2.3 BAG composites incorporated with natural polymers

Including synthetic polymers that have been discussed in the previous section, all synthetic polymers can be classified as (i) hydrophobic, non-water-absorbing materials such as PE: polyethylene, PP: polypropylene, PMMMA: poly(methylmethacrylate); (ii) more polar systems like PLGA: copoly(lactic-glycolic acid); (iii) water-swelling materials as PHEMA: poly (hydroxyethyl methacrylate); and (iv) water-soluble materials such as PEG: poly (ethylene glycol). On the contrary, natural polymers can be broadly categorized as (i) proteins such as collagen, gelatin, and silk fibroin; (ii) polysaccharides such as chitosan, hyaluronic acid, alginate, and cellulose; and (iii) polynucleotides (DNA, RNA) [111, 112]. These proteins and polysaccharides are considered to be biocompatible, non-cytotoxic, and attractive natural polymers for medical devices for both soft and hard tissues [113]. Furthermore, natural polymers show more similarity to the extracellular matrix (ECM, network of biomacromolecules, including glycosaminoglycans, which are polysaccharides, and fibrous proteins such as collagen, laminin, elastin, and fibronectin) [111], being readily recognizable by the body, compared to the synthetic ones. Such similarity to ECM could be summarized as a suspension of macromolecules that support everything from local tissue growth to the maintenance of an entire organ. In this context, since ECM provides structural support and modulates the activity of growth factors, it represents one of the main footprints for designing biomaterials [114].

Despite these positive aspects, natural polymers show lower stability in terms of physical and mechanical properties, and lower bioactivity, compared to synthetic ones [113]. Moreover, natural polymers suffer from some limitations due to their solubility and industrially acceptable processability: (i) variation in the final properties of the polymer due to their source, (ii) some contamination, caused by the presence of microbes, (iii) uncontrolled water uptake, and (iv) an unpredictable degradation route. Furthermore, since most natural polymers are water-soluble, various crosslinking methods to control their structure, water uptake, and degradation in aqueous environment have been developed [115, 116]. To overcome these drawbacks associated with natural polymers, these have been combined with one another to improve workability; in addition, natural polymers are typically combined with ceramics fillers (i.e., ceramics, glass-ceramics, and bioactive glasses) to reinforce the structure of the final system and, thus, to produce composites with better mechanical performance [117, 118]. These natural polymers are incorporated with BAG nanoparticles and microparticles to produce advanced composites, improving the mechanical properties of the final system as well as its bioactivity and regenerative potential, as illustrated in Figure 5.1 [113].

hyaluronic acid

collagen

silk fibroin

phosphates

gelatin

Natural polymers extracted from animals waste and plants

+

Bioactive glasses

Possibility to design advanced composites

Scaffolds, membranes, hydrogels, films, coatings, injectable systems, wound dressings for both hard and soft tissues

chitosan

borates

silicates

cellulose

alginate

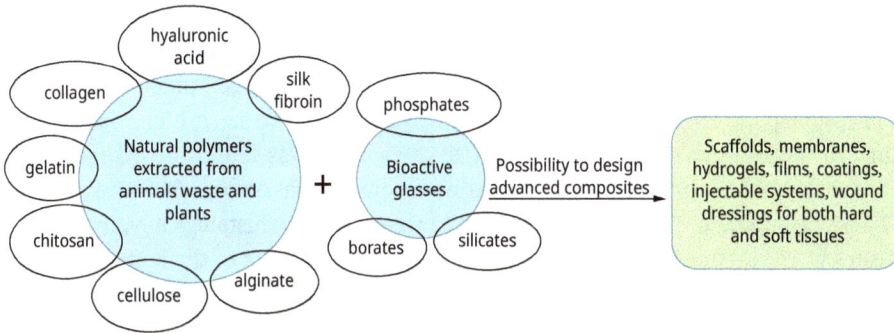

Figure 5.1: Natural polymeric materials, incorporated with BAGs to fabricate advanced composites [modified after 113].

= Silk =

Silk is a natural biopolymer obtained from the silkworm, Bombyx mori [8]. Silk fibroin from various silkworms and spider species, and sericin from various silkworm species, have been researched for their potential applications in the healthcare industry such as tissue-engineered grafts, cancer therapeutics, high-throughput tissue-on-chip models, food preservatives, biomedical imaging, biosensing, biomedical textiles, implants, and cosmetics and bioremediation products [119]. It is predominantly composed of the proteins, fibroin, and sericin. After removal of sericin through degumming, silk fibroin (SF) is commonly used as a bone scaffold in sponge, fibers, film, and hydrogel forms; it exhibits excellent biocompatibility, degradability, tissue integration, and oxygen and water permeability [119–122]. Mandal et al. [123] mentioned that studies using silk particles (fabricated by milling) reinforced into a silk matrix resulted in improved scaffolds for bone applications, with compressive properties in hydrated state of approximately 3 MPa, improving the ingrowth of human-bone-marrow -derived mesenchymal stem cells (hMSCs), in vitro, toward forming bone-like tissues, based on several studies [123–125].

A good bone implant material should have porous structure, mechanical strength, and close resemblance to the framework of extracellular matrix of bone [126, 127]. It should also have temporary architecture for the attachment and proliferation of bone-forming cells for bone regeneration [128, 129]. Silk fibroin (SF) and hydroxyapatite (HA) have been used in bone tissue regeneration as biomaterials due to the mechanical properties of SF and due to the biocompatibility of HA. There has been growing interest in developing SF/HA composites to reduce bone defects. In this regard, several attempts have been made to study the biocompatibility and osteoconductive properties of this material [130].

Koh et al. [131] fabricated hybrid composites of silk and hydroxyapatite to simulate natural bone tissue to overcome the softness and brittleness of the individual components. Three treatment groups were tested: an empty defect group (group I), a silk fibrin scaffold group (group II), and a hydroxyapatite-conjugated silk fibrin scaffold group (group III). New bone formation was assessed using computed tomography and histology at 4, 8, and 12 weeks, and semiquantitative western blot analysis was done to confirm bone protein formation at 12 weeks. It was found that (i) radiomorphometric volume analysis revealed that new bone formation was 64.5% in group I, 77.4% in group II, and 84.8% in group III at 12 weeks, and (ii) histologically, the osteoid tissues were surrounded by osteoblasts, not only at the border of the bone defect but in the center of the scaffold implanted area, in group III from week 8 on, concluding that (iii) new bone formation was higher in hybrid scaffolds and (iv) both osteoconduction at the defect margin and osteoinduction at the center of the defect were confirmed.

Recent research has shown that silk's favorable biological properties enable its use as a membrane for guided bone regeneration, despite its poor mechanical properties [8]. In several clinical trials performed in 2016, patients who received a silk mat membrane, following extraction of impacted mandibular third molars, displayed a significant gain of new bone of approximately 4 mm, six months following the grafting procedures [122, 132, 133]. SF has been indicated for use as a GBR membrane, following tooth extraction, cyst/tumor excision, and deficient alveolar bone, for implant placement [134]. SF-based membranes have been shown to exhibit excellent tensile strength, good osteogenic potential, and good mechanical properties in several in vivo studies [133–135]. HA can also be blended with an integrated membrane of SF and chitosan to improve the membrane's mechanical strength and stability [124]. Despite the promising outlook for the many xenograft materials described, there are still some limitations associated with the use of xenograft bone substitutes. These include the variable resorption rates, lack of viable cells and biological components, and the need for tissue treatment processes, which enable the retention of osteoinductive cells [136].

Peng et al. [137] prepared silk fibroin-octacalcium phosphate composite with silk fibroin by freeze-drying technology, and the freeze-dried scaffolds were further surface-modified with self-polymerization by using dopamine. The silk fibroin/octacalcium phosphate/polydopamine composite bone scaffold was constructed. The effects of different concentrations of polydopamine coatings on the porosity, swelling property, hydrophilicity, compressive strength, mineralization ability, and biocompatibility of scaffolds were measured. It was obtained that (i) the scaffolds had interconnected 3D network structure, and polydopamine coating increased the compressive strength of SF scaffolds, delayed the degradation rate of scaffolds, and enhanced the apatite deposition rate on the surface of scaffolds, (ii) the relative viability of cells increased with the rising of the concentration of coated polydopamine after co-culture with MC3T3-E1 cells, and (iii) the polydopamine coating conferred the scaffolds excellent bioactivity. Based on these findings, it was indicated that silk fibroin/octacalcium phosphate/polydopamine composite scaffold may provide a new idea for bone tissue engineering.

Moses et al. [138] fabricated copper-doped bioactive glass-functionalized silk micro-fiber reinforcements to improve the physicochemical and osteoinductive properties of two silk scaffolding matrices (mulberry *Bombyx mori* and non-mulberry *Antheraea assama*). It was mentioned that (i) the reinforced composite matrices increase the surface area and present an open porous biomimetic micromillieu, favoring stem cell and endothelial cell migration within the matrix, (ii) biochemical results indicate the stabilization of hypoxia-inducible factor-1α and expression of C-X-C chemokine receptor type-4 in adipose-derived human mesenchymal stem cells, which regulate the downstream proangiogenic signaling and endothelial cell homing, respectively, (iii) osteoinduction, matrix turnover, and resorption effectiveness are favored better in the non-mulberry silk matrices, and (iv) the composite matrices significantly promote neo-osseous tissue formation in volumetric femur defect in rabbits, with periosteal restoration seen in the non-mulberry silk composite matrices [138].

Silk sericin is an active ingredient in bone-grafts. However, the optimal scaffold for silk sericin has yet to be identified. Hence, Lee at el. [139] studies and compared collagen and gelatin sponges as silk sericin scaffolds. A critical-sized bone defect model in rat calvaria was used to evaluate bone regeneration. Silk sericin from *Yeonnokjam, Bombyx mori*, was incorporated into gelatin (group G) and collagen (group C). Bone regeneration was evaluated using micro-computed tomography (mCT) and histology. It was obtained that (i) group C showed a larger bone volume than group G in the mCT analysis, (ii) histological analysis showed a larger area of bony defects in group G than in group C, and (iii) the bone regeneration area in group C was significantly larger than that in group G; concluding that compared with gelatin, collagen shows better bone regeneration in silk sericin-based bone-grafts.

Wu et al. [140] evaluated a weakly alkaline, biomimetic and osteogenic, three-dimensional composite scaffold with HA and nano magnesium oxide (MgO), embedded in the fiber of silkworm cocoon and silk fibroin (SF), for its bone repair potential in vivo and in vitro experiments, particularly focusing on the combined effect between HA and MgO. It was found that (i) magnesium ions has long been proven to promote bone tissue regeneration, and HA is provided with osteoconductive properties, (ii) the weak alkaline microenvironment from MgO may also be crucial to promote rat bone mesenchymal stem cells BMSCs proliferation, osteogenic differentiation, and alkaline phosphatase activities, and (iii) the fabricated SF/HA/nano MgO scaffold composite with superior biocompatibility and biodegradability has better mechanical properties, BMSCs proliferation ability, osteogenic activity, and differentiation potential, compared with scaffolds that have HA or MgO alone or neither; suggesting that hybrid composite scaffold promoting bone regeneration in bone tissue engineering.

= Chitosan =
Chitin and its deacetylated derivative, chitosan, are a family of linear polysaccharides, composed of varying amounts of (β1→4) linked residues of *N*-acetyl-2 amino-2-deoxy

-D-glucose (glucosamine, GlcN) and 2-amino-2-deoxy-D-glucose (*N*-acetyl-glucosamine, GlcNAc) residues. Chitosan is soluble in aqueous acidic media via primary amine protonation. In contrast, in chitin, the number of acetylated residues is high enough to prevent the polymer from dissolving in aqueous acidic media [141]. Chitin is a very abundant biopolymer that can be found in the exoskeleton of crustacea, insect's cuticles, algae, and in the cell wall of fungi. Chitosan is less frequent in nature, occurring in some fungi (*Mucoraceae*). Historically, commercial chitosan samples were mainly produced from chemical deacetylation of chitin from crustacean sources. More recently, chitosan from fungi is gaining interest in the market, driven by vegan demands. Moreover, these samples are better controlled in terms of low viscosity, and exhibit a very high deacetylation degree [142]. Production from insect cuticles is also gaining interest, driven by the increased interest in protein production from these sources. Chitosan is the only polycation in nature and its charge density depends on the degree of acetylation and pH of the media. The solubility of the polymer depends on the acetylation degree and molecular weight. Chitosan oligomers are soluble over a wide pH range, from acidic to basic ones (i.e., physiological pH 7.4). On the contrary, chitosan samples with higher molecular weight are only soluble in acidic aqueous media, even at high deacetylation degrees. This lack of solubility at neutral and basic pH has hindered the use of chitosan in some applications under neutral physiological conditions (i.e., pH 7.4). This is the reason why a great number of chitosan derivatives, with enhanced solubility, have been synthetized [141].

Chitosan is a polymeric biomaterial obtained by the deacetylation of chitin. With the properties of easy acquisition, antibacterial and hemostatic activity, and the ability to promote skin regeneration, hydrogel-like functional wound dressings (represented by chitosan and its derivatives) have received extensive attention for their effectiveness and mechanisms in promoting skin wound repair [143].

Chitosan is being used as a wound-healing accelerator in veterinary medicine. Chitosan enhances the functions of inflammatory cells such as polymorphonuclear leukocytes (PMN) (phagocytosis, production of osteopontin and leukotriene B4), macrophages (phagocytosis, production of interleukin (IL)-1, transforming growth factor beta 1, and platelet derived growth factor), and fibroblasts (production of IL-8). As a result, chitosan promotes granulation and organization, therefore chitosan is beneficial for the large open wounds of animals [144].

Naturally derived polymers have been extensively used in scaffold production for cartilage tissue engineering. Alves da Silva et al. [145] evaluated and characterized extracellular matrix (ECM) formation in two types of chitosan-based scaffolds using bovine articular chondrocytes (BACs); the influence of these scaffolds' porosity, as well as their pore size and geometry on the formation of cartilagineous tissue was also investigated. Chitosan-poly(butylene succinate) (CPBS) scaffolds were produced by compression molding and salt leaching, using a blend of 50% of each material. Different porosities and pore size structures were obtained. Bovine articular chondrocytes were seeded onto CPBS scaffolds using spinner flasks. Constructs were then transferred to the incubator, where one half was cultured under stirred conditions, and the other half under static conditions for 4 weeks. It was reported that (i) both materials showed good affinity for cell attachment, (ii) cells colonized the entire scaffolds and were able to produce ECM (extracellular matrix), (iii) large pores with random geometry improved proteoglycans and collagen type II production; however, that structure has the opposite effect on glycosaminoglycan (GAG) production, and (iv) stirred culture conditions indicate enhancement of GAG production in both types of scaffold [145].

A promising xenograft material currently being researched is chitosan, a naturally occurring polymer derived from the exoskeletons of crustaceans composed of glucosamine and N-acetylglucosamine [146]. Chitosan is able to stimulate bone regeneration by providing a structural scaffold that supports osteoblastic activity, the formation of mineralized bone matrix and inducing differentiation of MSCs into osteoblasts in various in vitro environments [147].

Chitosan is available in a variety of forms, including beads, films, hydrogels, and more complex structures, such as porous scaffolds [8], leading to a variety of application potentials. Composites of water-soluble chitosan and HA, the paste form, were investigated for their the use as synthetic bone-grafting material [148]. HA powder was added to 5% aqueous solution of chitosan in different HA-to-chitosan ratios. Different techniques were developed to measure the viscosity, setting time, compressive moduli, and push-out resistance through cadaver bone holes of the HA-chitosan pastes. An optimum formulation was selected based on the results of these measurements. A system was developed to deliver the pastes to the focal defects in the bone by modifying an arthroscopic syringe [148].

Bone-graft substitutes composed of calcium phosphate (CP) – a component of natural bone, and chitosan (CS) – a biocompatible biopolymer were fabricated and characterized [149]. CP-CS composites were synthetized, molded, dried, and characterized. The effect of drying temperatures (38 and 60 °C) on the morphology, porosity, and chemical composition of the composites was evaluated. The effects of drying temperature and period of drying (3, 24, 48 and 72 h) on the mechanical properties – compressive strength, modulus of elasticity and relative deformation-of the demolded samples – were as well investigated. Results indicated that (i) scanning electron microscopy and gas adsorption-desorption analyses of the CS-CP composites showed interconnected pores, indicating that the drying temperature played an important role on pores size and distribution, and (ii) drying temperature altered the color (brownish at 60 °C due to Maillard reaction: an organic chemical reaction in which reducing sugars react with amino acids to form a complex mixture of compounds.) and the chemical composition of the samples, confirmed by FTIR. It was, therefore, concluded that the prolonged period of drying improved the mechanical properties of the CS-CP composites dried at 38 °C, which can be designed according to the mechanical needs of the replaceable bone [149].

Kjalarsdóttir et al. [150] assessed bone regenerative properties of an injectable chitosan and calcium phosphate-based composite, and identified the optimal degree of deacetylation (%DDA) of the chitosan polymer. Holes were drilled on the left side of a mandible in Sprague-Dawley rats, and the hole was either left empty or filled with the implant. The animals were sacrificed at several time points after surgery (7 ~ 22 days) and the bone was investigated using micro-CT and histology. No significant new bone formation was observed in the implants themselves at any time point. However, substantial new bone formation was observed in the rat mandible, further away from the drill hole. Morphological changes indicating bone formation were found in specimens explanted on Day 7 in animals that received implant. Similar bone formation pattern was seen in control animals with an empty drill hole, at later time points but not to the same extent. A second experiment was performed to examine if the %DDA of the chitosan polymer influenced the bone remodeling response. Results suggest that chitosan polymers with %DDA between 50% and 70% enhance the natural bone remodeling mechanism.

Huang et al. [151] developed a biocomposite scaffold using chitosan (CS) and bovine-derived hydroxyapatite (BHA). The prepared CS-BHA biocomposite scaffold was characterized for its physiochemical and biological properties, and compared against control BHA scaffolds to evaluate the effects of CS. Energy-dispersive X-ray analysis confirmed the elemental composition of the CS-BHA scaffold, which presented peaks for C and O from CS and Ca and P along with trace elements in the bovine bone such as Na, Mg, and Cl. Fourier transform infrared spectroscopy confirmed the presence of phosphate, hydroxyl, carbonate, and amide functional groups attributed to the CS and BHA present in the biocomposite scaffolds. It was also mentioned that (i) the CS-BHA scaffolds demonstrated an interconnected porous structure, with pore sizes ranging from 60 to 600 µm and a total porosity of 64% ~ 75%, as revealed by scanning electron microscopy and micro-CT analyses, respectively, (ii) thermogravimetric analysis revealed that the CS-BHA scaffold lost 70% of its weight when heated up to 1,000 °C, which is characteristic of CS phase decomposition in the biocomposite, and (iii) the in vitro studies demonstrated that the CS-BHA scaffolds were biocompatible toward Saos-2 osteoblast-like cells, showing high cell viability and a significant increase in cell proliferation across the measured time points, compared to the controls [151].

Kowalczyk et al. [152] described a method of obtaining chitosan-bioceramic granulate with better properties and bone regeneration potential for small or non-load-bearing voids. Both commercially available β-TCP and genuine human bone were employed to determine the effect of the natural-origin material presence on cell growth. This developed composite, chitosan-calcium phosphate-human bone material, could augment guided bone regeneration with higher efficiency than fully synthetic granulates, and could provide an alternative to purely ceramic bone filler, xenografts, etc. This material is also sterilizable with an autoclave, a cheaper and more commonly available sterilization method than ethylene oxide or gamma radiation methods. It was reported that (i) composite material retained its physicochemical properties after thermal sterilization, unlike pure chitosan granules, and (ii) confocal microscopy revealed stable MG63 cell growth on the surface of obtained materials; β-TCP/human bone composites performed significantly better than β-TCP composites, and induced increased alkaline phosphatase activity in cells, suggesting that the obtained material can be easily prepared and thermally sterilized with an autoclave, showing potentiality of bone regeneration of the composites [152].

Alternatives to autograft and allograft bone substitutes currently being researched are synthetic and natural graft materials that are able to guide bone regeneration. One promising material currently being researched is chitosan, a highly versatile, naturally occurring polysaccharide, derived from the exoskeleton of arthropods that is comprised of glucosamine and N-acetylglucosamine [147]. Chitosan is efficacious in bone regeneration due to its lack of immunogenicity, biodegradability, and physiologic features. Chitosan, combined with growth factors and/or other scaffold materials, has proven to be an effective alternative to autologous bone-grafts. Additionally, current studies have shown that it can provide the additional benefit of a local drug delivery system. As research in the area of bone scaffolding continues to grow, further clinical research on chitosan, in conjunction with growth factors, proteins, and alloplastic materials, will likely be at the forefront [147].

Due to the poor mechanical properties exhibited by chitosan, it is often combined with other materials such as gelatin, calcium phosphates and bioglass to provide more desirable properties [153–156]. Wattanutchariya et al. [157] found that mixing chitosan with gelatin and HA produces a porous scaffold with more desirable properties, including decreased degradability and an open pore structure conducive to cell attachment and vascularization [157]. Chitosan-based bone substitute materials also possess other beneficial properties, such as low immunogenicity, fibrous encapsulation, structural versatility and a hydrophilic surface, which promotes cell adhesion and proliferation [147]. The versatility associated with chitosan-based bone substitute materials indicates it is as such a promising alternative to the conventional gold-standard of autografts. Recent studies in the dental field have reported the successful use of chitosan-based materials as a membrane for GBR, GTR (guided tissue regeneration), coating implant surfaces, periodontal regeneration, and for restoring alveolar bone height [158, 159].

Jayash et al. [160] assessed and compared the quantity and the quality of the new bone generated when using chitosan-based gel scaffold and osteoprotegerin-chitosan gel scaffold. A total of 18 critical-sized defects on New Zealand white rabbit craniums were created. In 12 defects, either chitosan gel or osteoprotegerin-chitosan gel was implanted and the last six defects were kept unfilled as a control. Bone formation was examined at 6 and 12 weeks. The bone's specimens were scanned using the high-resolution peripheral quantitative computed tomography. Histological and histomorphometric analysis were carried out to compare the volume and area of the regenerated bone. It was reported that (i) the results of the HR-pQCT (high resolution peripheral quantitative computed tomography) showed that bone volume and densities in the osteoprotegerin-chitosan gel group were significantly higher than the chitosan gel and control groups, whereas the bone volume density in the chitosan gel group was significantly higher than the control group in both time intervals, and (ii) no significant difference in bone volume between the chitosan gel and control groups was observed; however, similar findings were shown in the histomorphometric analysis, with the highest new bone formation being observed in the OPG (osteoprotegerin)-chitosan gel group followed by the chitosan group, indicating that the mean percentage of new bone was greater at 12 weeks, compared to 6 weeks in all groups. Based on these findings, it was concluded that the chitosan-based gel demonstrated a significant bone quantity and quality compared to unfilled surgical defects, and consistently, osteoprotegerin enhanced the chitosan gel in bone regeneration [161].

5.3 Polymer-based bone substitute materials

Polymer-based bone-graft substitutes can be divided into natural polymers and synthetic polymers [10], and synthetic polymers can be further classified into degradable and nondegradable subtypes [8, 10]. The most widely used polymers for bone regeneration include polylactic acid, polyglycolic acid, poly-caprolactone, and their copolymers and derivatives; these polymers are generally known as aliphatic polyesters [11, 136]. These materials possess several advantages including low immunogenicity, controllable bioresorbability, and various degrees of porosity and physiochemical structure as well [136, 161].

Pilipchuk et al. [162] reviewed recent advances on regenerative technologies (scaffolding matrices, cell/gene therapy, and biologic drug delivery) to promote reconstruction of tooth and dental implant-associated bone defects. Growth factors and other biologics with clinical potential for osteogenesis were examined, with a comprehensive assessment of preclinical and clinical studies. It was mentioned that an analysis of the existing scaffold materials, their strategic design for tissue regeneration, and the use of growth factors for improved bone formation in oral regenerative therapies results in the identification of current limitations and required improvements to continue moving the field of bone tissue engineering forward into the clinical arena, suggesting that the introduction of growth factor biologics and cells has the potential to improve the biomimetic properties and regenerative potential of scaffold-based delivery platforms for next-generation patient-specific treatments with greater clinical outcome predictability. Along with these concerns, there should be other issues relating to the release of acidic degradation products, resulting in the alteration of local pH, osteoconductivity, and poor cell adhesion capacity, thus restricting their use in the dental field [8].

It is a common method to modify synthetic polymer-based scaffolds, such as (i) addition of Ca-P materials like HA or TCP to improve the bone regeneration potential [163, 164], or (ii) coating silk fibroin to achieve controlled release delivery of bioactive molecules to enhance angiogenic properties and improve osseointegration [165, 166].

Guduric et al. [167] evaluated the proliferation and differentiation of human bone marrow stromal cells (HBMSCs) and endothelial progenitor cells (EPCs) in two (2D) and three dimension (3D) using an LBL assembly of polylactic acid (PLA) scaffolds, fabricated by 3D printing. It was reported that 3D-printed polylactic acid-based biopolymer possessing 200 μm pore diameters exhibited improved cell proliferation and differentiation. Cristache et al. [168] described computer aided design and manufacturing (CAD/CAM) additive techniques and synthetic polymers for bone reconstruction in the maxillofacial region. It was mentioned that (i) additive manufacturing (AM) represents a promising field for future research in bone replacement/regeneration; however, standard guidelines for mimicking clinical environment with the different bone characteristics are strongly required, and (ii) the rapid prototyping techniques, particularly, bioprinting allows the construction of 3D living functional tissues to be able to replace, in the near future, large defects caused by tumor excision, trauma, clefts or infections, which can limit the autogenous bone-graft requirement.

Alloplasts are synthetic, inorganic, biocompatible bone substitutes that function as defect fillers to repair skeletal defects. The acceptance of these substitutes by host tissues is determined by the pore diameter, and the porosity and interconnectivity [169]. Unlike permanent autogenous dentine, alloplasts are synthetic, inorganic, biocompatible bone substitutes that primarily function as defect fillers. In this case, the material used functions as bone filler, with the potential to upregulate host bone regeneration, but possibly lacks the quality of allografts and xenografts. In such instances, the bony defect may not form the bone in its entirety. Besides biocompatibility, the acceptance of these substitutes by host tissues is determined by three important features – pore diameter, porosity, and interconnectivity. In addition, Palma et al. [170] reported the influence of different formulations of bone-grafts in providing an adequate scaffold, thus emphasizing the importance of the type of carrier in the three-dimensional distribution of particles and space provision in new bone formation.

Compared to natural ones, synthetic polymers can be synthesized under more controllable conditions. Therefore, their physicochemical and biological properties, e.g., mechanical strength, degradation rate, and microstructure, are more predictable and reproducible, and the desired properties can be conveniently obtained by wisely designing the segments and functional groups of the polymers [171]. Aliphatic polyesters, including PLA, PGA, and their copolymer PLGA, are the most widely used polymers for bone tissue engineering. These polymers degrade in vivo by non-enzymatic hydrolysis and result in nontoxic degradation products. Their degradation rate can be readily tailored by altering the chemical composition, crystallinity, molecular weight, and distribution. However, these pure polymers have drawbacks such as insufficient cell recognition sites and osteoconductivity. To improve their performance, these polymers can be put through PDA coating, grafting with bioactive molecules such as peptide and proteins, or compositing with HA nanoparticles [162, 172, 173]. For example, 3D-printed PLA scaffolds, with PDA coating, were reported to enhance the osteogenic differentiation of adipose-derived stem cells (ADSCs) [174]. Electrospun PLGA scaffolds, incorporated with gelatin/

HA, were also reported to possess excellent biocompatibility and osteogenic activity [175]. Compared to PLA, PGA, and PLGA, PCL has higher crystallinity and hydrophobicity, and much lower degradation rate. Therefore, surface modification, such as plasma treatment and loading of bioactive molecules, is usually employed to overcome these disadvantages. PCL can also be copolymerized with other monomers to bear different functional groups, which can be further modified to enhance its interaction with cells for improved performance [176].

= PLA, PGA, PLGA =

PLA (polylactic acid), PGA (polyglycolic acid), and their co-polymer (PLGA) are available in different shapes, from mesh for orthopedic applications to drug-eluting coatings on vascular stents [169].

PLA is a biodegradable thermoplastic polyester developed by polymerization of chiral semicrystalline molecules, named D- and L-isomer [177]. Currently, most 3D-printed scaffolds use PLA and PLGA (poly(lactic-co-glycolic acid)) to create composites with other inorganic material to produce customized substitutes. In an experimental study, 3D maxillary sinus model was fabricated using the composite material from osteogenic HA-PLA [178].

PGA is an aliphatic polyester and exhibits controlled solubility and a high degradation rate. The degradation product of the PGA, glycolic acid, is excreted in urine. Clinically, it has been used as the first biodegradable suture for many years. Compared to other polyesters such as PCL and PLA, PGA has a higher mechanical strength. Nonetheless, it is not suitable to be used alone for bone repair because of its high degradation rate in vivo. Three-dimensional porous composite scaffolds of PGA/β-TCP (in 1:1 ratio) showed a strong ability to regenerate bone, with a degradation rate of 90 days [179].

PLGA is a linear copolymer of lactic acid and glycolic acid monomers, formed by the ring-opening polymerization of PLA and PGA. The performance, mechanical properties, and its degradation rate can be adjusted by the different ratios of these two polymers. It has been reported that the scaffold of PLGA with the lactic acid and glycolic acid ratios of 75/25, respectively, has approximately half the degradation rate of scaffold, with the ratio of 85/15 [180]. Like other polyesters, PLGA scaffolds have been used as carriers. Recently, various bioactive molecules have been loaded with PLGA/HA scaffolds to aid bone healing. It has been demonstrated that nano-HA could improve the bone repairability of scaffolds [181].

=PLC =

PCL (poly(ε-caprolactone)) is a semicrystalline thermoplastic polymer with a slow degradation rate that maintains its mechanical feature. It has been studied due to its microstructure being similar to the trabecular bone and its activity to encourage vascularization and cell communication [182]. It is considered less costly compared to the other polyesters such as PLA, PGA, and their copolymers; however, it has higher hydrophobicity and crystallinity, and a slower degradation rate than the others. To overcome these disadvantages, surface modification, such as loading of bioactive molecules and plasma treatment, is usually employed. Wang et al., found the addition of pristine graphene to PCL has a positive impact on cell viability and proliferation [183]. PCL polymer membrane contributed to the early biodegradation of β-TCP, without affecting the bone regeneration capacity in a canine mandibular defect [183]. Furthermore, PCL could be copolymerized with other monomers to take on different functional groups, which can be further modified to enhance its bioactivity [184, 185]. It can be combined with other biomaterials such as gelatin to enhance cell adhesion, proliferation, and to accelerate its biodegradation rate [186].

A novel hybrid scaffold composed of a 3D-printed PCL HA/β-TCP scaffold was developed with simultaneous use of implant fixtures [187]. Clinically, another novel 3D-printed PCL-TCP device for ridge preservation has been tested and is in the midst of undergoing further clinical trials. It is shown to have high porosity and bioactivity that promotes osteogenesis and reduces resorption, while leveraging its 3D shape to fit snugly in the tooth socket [188]. In dentistry, PCL alone has been shown to provoke differentiation, colonization, proliferation of odontogenic human dental pulp cells isolated from mature teeth into functional odontoblast-like cells within the PCL cone, and secretion of extracellular matrix similar to the mineralized dentine matrix [189].

= PVA, PE =

Poly(vinyl alcohol) (PVA) is a polyalcohol that is synthesized through the hydroxylation of polyvinyl acetate. The degree of hydroxylation largely determines many characteristics of PVA, with higher degree of hydroxylation leading to lower water solubility and crystallinity [190]. Polyphosphoester (PPE) is a phosphate-containing polymer with potential osteoinductive capacity for bone regeneration. It was reported that the phosphoester groups of PPE promoted the osteogenic differentiation of human MSCs [191]. Moreover, PPE was also used to modify the PLA surface to enhance the proliferation and functions of osteoblastic cells [192]. Polytrimethylene carbonate (PTMC) is an amorphous biocompatible polymer with low elasticity at room temperature. In a study that evaluated the performance of PTMC membrane in guided bone regeneration of rat mandibular defects, similar amount of newly generated bone was observed for PTMC, collagen, and PTFE membranes, two weeks after implantation, predicting the potential use of PTMC in guided bone regeneration [193]. Biodegradable polyurethanes (PUs) are a series of synthetic polymers that have gained increasing attention for bone tissue engineering. The properties of PUs can be tailored in a broad range by varying the chemical composition, segment ratio and structure, and molecular weight. PUs generally possess much better mechanical properties compared to other conventional biodegradable polymers [194]. PUs are usually composited with inorganic materials and modified with bioactive substance to stimulate bone tissue regeneration [194]. Polypeptides used in bone regeneration include acidic poly(amino) acids such as γ-poly(glutamic acid) (γ-PGA) and poly(aspartic acid) and basic poly(amino acids) such as polylysine and polyarginine. As an example, a chitosan matrix was composited with γ-PGA to enhance its hydrophilicity and cytocompatibility [195].

Some treatment options available to repair bone defects are the use of autogenous and allogeneic bone-grafts. The drawback of the first one is the donor site's limitation and the need for a second operation on the same patient. In the allograft method, problems are associated with transmitted diseases and high susceptibility to rejection. As an alternative to biological grafts, polymers can be used in bone repair. Some polymers used in the orthopedic field are poly(methyl methacrylate), poly(ether-ether-ketone), and ultra-high molecular weight polyethylene (UHMWPE). UHMWPE has drawn much attention since it combines low friction coefficient, and high wear and impact resistance. However, UHMWPE is a bioinert material, which means that it does not interact with the bone tissue. Senra et al. [196] further mentioned that UHMWPE composites and nanocomposites with hydroxyapatite (HA) are widely studied in the literature to mitigate these issues. HA is the main component of the inorganic phase in the natural bone, and the addition of this bioactive filler to the polymeric matrix aims to mimic bone composition. This brief review discusses some polymers used in orthopedic applications, focusing on the UHMWPE/HA composites as a potential bone substitute.

= PMMA =

PMMA (ploy(methyl methacrylate)) is a rigid hydrophobic thermoplastic polymer, produced by the polymerization of methyl methacrylate through a mass, emulsion, or solution polymerization process [197]. Despite the disadvantages mentioned above, it is still used clinically with similar success rate as bone cement, which is formed through a mass polymerization reaction of methyl methacrylate (MMA) via free radicals, where their reaction products can induce local inflammation. Effectively, it is possible to make the properties of PMMA-based bone cements more tissue-friendly by adding 10% vitamin E as an antioxidant, which could reduce the number of free radicals formed. In addition, a functional active composite structure of PMMA that is bioactive and porous could be created by slightly changing its content, such as the addition of bioactive material (bioactive glass) to its matrix [197].

The surface properties (e.g., surface chemistry and wettability) of polymer scaffolds significantly affect the adsorption of protein molecules and interactions between the cell and the matrix, which ultimately determine the cellular functions and quality of bone tissue formation. Engineering the surface properties is an effective way to improve the osteoconductive and osteoinductive performance of polymer scaffolds [169]. In general, most polymeric materials are hydrophobic and do not provide an ideal environment for protein adsorption and cell–matrix interactions. Various methods such as plasma treatment, corona discharge, and irradiation are therefore applied to generate functional groups such as hydroxyl and amine groups on the surface to improve their hydrophilicity [198]. Grafting scaffold surface with different chemical functional groups can direct differentiation of human MSCs. For example, charged phosphate groups effectively promoted osteogenesis [199]. This approach provides simple yet powerful technique to control the complex cell behaviors for bone tissue engineering. Surface coating with inorganic materials is a popular way to modify polymers, especially synthetic ones, to enhance their biocompatibility and osteoconductivity. A tough DN hydrogel coated with HAs showed much improved osteointegration by forming a hybrid layer of hydrogel/bone at the interface [199]. Polydopamine (PDA) coating is a novel universal coating technique. PDA-coated PLA scaffolds were proven to enhance cell adhesion, proliferation, and hence osteogenesis [200]. Furthermore, the catechol groups on the PDA coating can react with primary amine and thiol groups to immobilize many other bioactive molecules [200]. Bioactive ligands such as peptides and polysaccharides can also be adsorbed or covalently grafted onto the surface to facilitate cell adhesion and spreading. Adhesion of cells on PLLA and PLGA scaffolds was greatly improved after grafting with an adhesive peptide RGD (Arg-Gly-Asp), and bone regeneration was promoted when combined with other bioactive molecules [201].

Despite biocompatibility, many polymer scaffolds for bone engineering may suffer from implant failure due to bacterial infection. During implantation, bacteria can easily adhere to and colonize the polymer scaffold surfaces, ultimately leading to serious implant infection. Reducing bacterial adhesion to scaffold surface and bactericidal coating are two major methods to address the bacteria-associated concerns [202]. Grafting with inert polymer brush, such as PEG, is a passive way to inhibit bacterial adhesion on the surface, and prevent infection. However, it may also inhibit the adhesion of mammalian cells [203]. The reduced adhesion of osteogenic cells can be addressed by modifying the scaffold with a RGD ligand peptide [204]. In comparison, bactericidal coating is an active approach that can actually kill the bacteria and eliminate the root cause of infection. Silver is widely known to possess a broad spectrum of antimicrobial activities. A highly porous PLGA fibrous composite containing TCP and silver was fabricated using electrospinning [205]. The encapsulated silver showed steady release and provided prolonged antibacterial effects. A cationic polymer, chitosan, is a promising alternative to silver as an antibacterial agent. For example, chitosan and berberine were applied to coat a polyamide66/HA scaffold [206]. The combination of chitosan and the drug berberine provided the scaffold with significant antibacterial efficacy. Hydrogels containing $ZnCl_2$ and $SrCl_2$ were also reported to effectively hamper the growth of bacteria on the polymer scaffold for bone regeneration [207].

5.4 Metal-based bone substitute materials

Bone-grafts have been predominated used to treat bone defects, delayed union or nonunion, and spinal fusion in orthopedics, clinically for a period of time, despite the emergence of synthetic bone-graft substitutes [26]. Nevertheless, the integration of allogeneic grafts and synthetic substitutes with the host bone did not insure long-term survival. Therefore, the enhancement of osseointegration of these grafts and substitutes with the host bone becomes more clinically crucial. To overcome these issues, addition of various growth factors, such as bone morphogenetic proteins (BMPs), parathyroid hormone (PTH), and platelet rich plasma (PRP), into structural allografts and synthetic substitutes have been considered [26]. It was further mentioned that although clinical applications of these factors have exhibited good bone formation, their further application was limited due to high cost and potential adverse side effects; hence, alternatively, bioinorganic ions such as magnesium (Mg), strontium (Sr), and zinc (Zn) are considered alternatives of osteogenic biological factors [26], as indicated in Table 5.1 [13].

In the dental field, Zhao et al. [8] mentioned that the use of Ni-Ti materials for bone regeneration has been explored due to their numerous desirable properties, including good mechanical strength, good biocompatibility, corrosion resistance, and elastic modulus. However, although Ni-Ti (depending on the pre-heat treatment) exhibits unique superelastic capacity as well as shape memory function [208], its nickel element content (normally 50 wt%) might be potential toxic. Accordingly, majority of the studies have been done on commercially pure titanium (CpTi).

Jeng et al. [209] utilized autogenous bone-grafts and titanium mesh-guided alveolar ridge augmentation for patients with alveolar bone deficiency, but requiring dental implantation. Four months after the abovementioned procedures, cone-beam computed tomography showed adequate alveolar bone formation. The titanium mesh was removed and the dental implant was placed in the augmented alveolar ridge at the same time. It was found that (i) the secondary bone-graft, combined with autogenous bone and inorganic bovine bone, was covered by the pseudo-periosteum and suitable for dental implantation in our four patients, and (ii) the implants were submerged for 3–4 months till uncovering, and then the prostheses were delivered one month afterward with successful clinical outcomes; concluding that (iii) the clinical outcomes of four patients indicate that the vital autogenous bone-grafts and the titanium mesh possess the ability to induce and guide new bone formation in four months, and can be successful used for alveolar ridge augmentation and subsequent dental implantation. One of the most often used bone augmentation techniques is the GBR procedure. Di Stefano et al. [210] reported the case of a 75-year-old man with an atrophic right posterior mandible who underwent bone augmentation through guided bone regeneration with a preshaped titanium mesh adapted on a stereolithographic model of the patient's jaw. The graft volume was simulated with a light-curing resin. The actual site was grafted with a mixture of autogenous and equine-derived bone. Five months later, the mesh was retrieved, three cylindrical implants were positioned, and a bone biopsy was collected for histomorphometric analysis. A provisional prosthesis was delivered three-and-a-half months later. Definitive rehabilitation was accomplished after one additional month. It was also mentioned that (i) the graft allowed for effective bone formation (newly formed bone, residual biomaterial, and medullar spaces were, respectively, 39%, 10%, and 51% of the core volume), and (ii) the patient functioned successfully throughout six-and-a-half years of follow-up; indicating that using the preshaped titanium mesh, in association with the enzyme-treated equine bone substitute, provided effective bone regeneration.

Guided bone regeneration (GBR) is an effective and simple method for bone augmentation, which is often used to reconstruct the alveolar ridge when the bone defect occurs in the implant area. Titanium mesh has expanded the indications of GBR technology due to its excellent mechanical properties and biocompatibility, so it can be used to repair alveolar ridges with larger bone defects, and can deliver excellent and stable bone augmentation results. Currently, GBR with titanium mesh has various clinical applications, including different clinical procedures. Bone-graft materials, titanium mesh covering methods, and titanium mesh fixing methods are also optional. Moreover, the research of GBR with titanium mesh has led to multifarious progress in digitalization and material modification [211].

Atef et al. [212] evaluated the quantity and the quality of the bone gained using collagen membrane, with a 1:1 mixture of autogenous and anoraganic bovine bone mineral, compared to titanium mesh with the same mixture of bone for GBR of horizontally deficient maxillary ridges. Two different grafting techniques were evaluated, 10 patients received GBR using a native collagen membrane using 1:1 autogenous and anorganic bovine bone mineral (ABBM) bone mixture, and 10 patients received GBR using titanium mesh with the same mixture of bone. It was reported that (i) statistical analysis showed a significant increase in alveolar bone width in both techniques, with a mean bone gain of 4.0 mm for collagen group and 3.7 mm for titanium mesh group, (ii) bone area percent was almost 28% for both groups, (iii) for Ti-mesh group, healing was uneventful in six soft tissue sites, with no signs of wound dehiscence; however, four cases showed mesh exposure – first, three patients showed this exposure 3 weeks postoperative, while the fourth patient showed exposure 4 months postoperative, and (iv) the mean graft resorption in the collagen and mesh group 6 months postoperative was considered nonsignificant. Based on these findings, it was concluded (v) GBR, with both collagen membrane and titanium mesh, using a 1:1 mixture of autogenous and ABBM is a viable technique for horizontal augmentation of deficient maxillary alveolar ridges, (vi) titanium mesh is a more sensitive technique compared to collagen membrane, and (vii) soft tissue dehiscence and difficulty during the second stage removal should limit its use in augmentation of horizontally deficient maxillary ridges. Zhang et al. [213] introduced a newly designed L-shaped Ti-mesh used for GBR with simultaneous implant placement, and evaluated the bone augmentation effectiveness, resorption, and long-term stability of peri-implant bone with this newly designed L-shaped Ti-mesh. Twelve patients (16 implants) who underwent a GBR procedure using L-shaped Ti-mesh with simultaneous implants placement were reviewed. Complications, implant success, and survival rate were recorded and calculated. Furthermore, the bone gain values, labial bone resorption, and remaining labial bone volume were measured by cone beam CT (CBCT) during the 13–41 months follow-up. It was found that (i) the average bone gain values, after GBR using L-shaped Ti-mesh, were 3.61 ± 1.50 mm vertically and 3.10 ± 2.06 mm horizontally, (ii) the implant success and surviving rates were 93.75% and 100%, respectively during the longest 41 months follow-up, (iii) the vertical labial bone resorption was -0.81 ± 1.00 mm and the horizontal labial bone resorption was -0.13 ± 1.19 mm at the top of the implants, (iv) the remaining labial bone thickness was 2.24 ± 1.29 mm at the top of the implants and 2.86 ± 1.08 mm at 2 mm apically from the implant tops after loading, and (v) there was still 1.13 ± 1.18 mm vertical labial bone remaining above the top of implants for approximately 87.5% sites. It was, therefore, concluded that (vi) GBR using L-shaped Ti-mesh, with simultaneous implant placement (one-step surgery) in the esthetic zone, could achieve predictable results for vertical and horizontal bone augmentation at the same time. Meanwhile, L-shaped Ti-mesh could preferably reconstruct and reduce the labial bone resorption to achieve long-term esthetics, and (vii) the newly designed L-shaped Ti-mesh may offer predictable and excellent outcomes for the implant restoration in the esthetic zone.

Mg and its alloys are considered attractive metallic materials for the development of potential medical implants; biocompatibility is the most essential aspect for consideration, so numerous researches are being conducted for enhancement or improving the biocompatible ability of Mg-based alloys by alloying, coating, or microstructural alterations [214]. In recent years, Liu et al. [215] have developed a magnesium-based bone substitute, fabricated using pure Mg (99.9%) and a Mg-30 wt% Sr alloy, in a high-purity graphite crucible generated in a mixed gas atmosphere. This material utilized the combined properties of degradability, excellent mechanical properties, and biocompatibility of pure Mg and Mg-Sr alloys (e.g., [216, 217]). It was mentioned that when compared with conventional commercial bone-grafts, such as calcium sulfates, HA, and TCP materials, the newly developed Mg-based material displayed improved tensile and compressive strengths, improved biocompatibility, and improved antibacterial properties, indicating its potential use in load-bearing areas as a bone substitute material [215].

References

[1] Robinson JL, Brudnicki P, Lu HH. Polymer-bioactive ceramic composites. Compr Biomater II. 2017, 1, 460–77.

[2] Sandberg J. Types of synthetic bone graft products: Active vs passive materials. 2022; https://orthospinenews.com/2022/10/05/types-of-synthetic-b0ne-graft-products-active-vs-passive-materilas/.

[3] Bauer TW, Muschler GF. Bone graft materials. An overview of the basic science. Clin Orthop. 2000, 371, 10–27.

[4] Sage K, Levin LS. Basic principles of bone grafts and bone substitutes; https://www.medilib.ir/upto date/show/134456.

[5] Campion CR, Ball SL, Clarke DL, Hing KA. Microstructure and chemistry affects apatite nucleation on calcium phosphate bone graft substitutes. J Mater Sci Mater Med. 2013, 24, 597–610.

[6]   Albrektsson T, Johansson C. Osteoinduction, osteoconduction and osseointegration. Eur Spine J. 2001, 2, S96–S101.

[7]   Hing KA. Bone repair in the twenty-first century: Biology, chemistry or engineering? Philos Trans A Math Phys Eng Sci. 2004, 362, 2821–50.

[8]   Zhao R, Yang R, Cooper PR, Khurshid Z, Shavandi A, Ratnayake J. Bone grafts and substitutes in dentistry: a review of current trends and developments. Molecules. 2021, 26, 3007; doi: 10.3390/molecules26103007.

[9]   Moore WR, Graves SE, Bain GI. Synthetic bone graft substitutes. ANZ J Surg. 2001, 71, 354–61.

[10]   Kumar P, Vinitha B, Fathima G. Bone grafts in dentistry. J Pharm Bioallied Sci. 2013, 5, S125–7.

[11]   Kolk A, Handschel J, Drescher W, Rothamel D, Kloss F, Blessmann M, Heiland M, Wolff K-D, Smeets R. Current trends and future perspectives of bone substitute materials – From space holders to innovative biomaterials. J Cranio Maxillofac Surg. 2012, 40, 706–18.

[12]   Dorozhkin SV. Calcium orthophosphates: Occurrence, properties and major applications. Bioceram Dev Appl. 2014, 4, 081; doi: 10.4172/2090-5025.1000081.

[13]   Oshida Y. Hydroxyapatite – Synthesis and Application. Momentum Press. 2015.

[14]   Lemons JE, Misch-Dietsh F, McCracken MS. Biomaterials for dental implants. In: Misch CE (Ed.)., Dental Implant Prosthetics. Dentistry, Mosby. 2015.

[15]   Bertazzo S, Zambuzzi WF, Campos DDP, Ogeda TL, Ferreira CV, Bertran CA. Hydroxyapatite surface solubility and effect on cell adhesion. Colloids Surface B: Biointerfaces. 2010, 78, 177–84.

[16]   Chander S, Fuerstenau DW. On the dissolution and interfacial properties of hydroxyapatite. Coll Surf. 1982, 4, 101–20.

[17]   Atlas E, Pytkowicz RM. Solubility behavior of apatites in seawater. Limnol Oceanog. 1977, 22, 290–300.

[18]   Sugawara A. Technology of Bone Regeneration. Zenith Press, Tokyo Japan. 2008, 36.

[19]   Ozeki K, Fukui Y, Aoki H. Influence of the calcium phosphate content of the target on the phase composition and deposition rate of sputtered films. App Sur Sci. 2007, 253, 5040–44.

[20]   Dorozhkin SV. Calcium orthophosphates as bioceramics: State of the art. J Funct Biomater. 2010, 1, 22–107.

[21]   Cheng PT. Formation of octacalcium phosphate and subsequent transformation to hydroxyapatite at low supersaturation: A model for cartilage calcification. Calcif Tissue Int. 1987, 40, 339–43.

[22]   Sicilia A, Cuesta S, Coma G, Arregui I, Guisasola C, Ruiz E, Maestro A. Titanium allergy in dental implant patients: A clinical study on 1500 consecutive patients. Clin Oral Implants Res. 2008, 19, 823–35.

[23]   Kattimani VS, Kondaka S, Lingamaneni KP. Hydroxyapatite – past, present, and future in bone regeneration. Bone Tissue Regen Insights. 2016, 7, 36138; doi: 10.4137/BTRI.S36138.

[24]   Ratnayake JTB, Mucalo M, Dias GJ. Substituted hydroxyapatites for bone regeneration: A review of current trends. J Biomed Mater Res Part B Appl Biomater. 2017, 105, 1285–99.

[25]   Dewi AH, Ana ID. The use of hydroxyapatite bone substitute grafting for alveolar ridge preservation, sinus augmentation, and periodontal bone defect: A systematic review. Heliyon. 2018, 4, e00884; doi: 10.1016/j.heliyon.2018.e00884.

[26]   Wang W, Yeung KW. Bone grafts and biomaterials substitutes for bone defect repair: A review. Bioact Mater. 2017, 2, 224–47.

[27]   Sun MH, Berdt C, Gross KA, Kucuk A. Material fundamentals and clinical performance of plasma-sprayed hydroxyapatite coatings: A review. J Biomed Marer Res. 2001, 58, 570–92.

[28]   Ong JL, Appleford M, Oh S, Yang Y, Chen WH, Bumgatdner JD, Haggard WO. The characterization and development of bioactive hydroxyapatite coating. J of Metals. 2006, 5, 67–69.

[29]   Goto T, Kojima T, Iijima T, Yokokura S, Kawano H, Yamamoto A, Matsuda K. Resorption of synthetic porous hydroxyapatite and replacement by newly formed bone. J Orthop Sci. 2001, 6, 444–47.

[30]   Tonino AJ, van der Wal BCH, Heyligers IC, Grimm B. Bone remodeling and hydroxyapatite resorption in coated primary hip prostheses. Clin Orthop Relat Res. 2009, 467, 478–84.

[31] Draenert M, Draenert A, Draenert K. Osseointegration of hydroxyapatite and remodeling-resorption of tricalciumphosphate ceramics. Microscopy Res Tech. 2013, 76, 370–80.

[32] Ruigrok TJC. Possible mechanisms involved in the development of the calcium paradox. Gen Physiol Biophys. 1985, 4, 155–65.

[33] McCarron DA, Morris CD, Bukoski R. The calcium paradox of essential hypertension. Am J Med. 1987, 82, 27–33.

[34] Fujita T. Calcium paradox: Consequences of calcium deficiency manifested by a wide variety of diseases. J Bone Miner Metab. 2000, 18, 234–36.

[35] Kheirallah M, Almeshaly H. Bone graft substitutes for bone defect regeneration. A collective review. Int J Dent Oral Sci. 2016; doi: 10.19070/2377-8075-1600051.

[36] Sallent I, Capella-Monsonís H, Procter P, Bozo IY, Deev RV, Zubov D, Vasyliev R, Perale G, Pertici G, Baker J, Gingras P, Bayon Y, Zeugolis DI. The few who made it: Commercially and clinically successful innovative bone grafts. Front Bioeng Biotechnol. 2020, 8; https://doi.org/10.3389/fbioe.2020.00952.

[37] Funda G, Taschieri S, Bruno GA, Grecchi E, Paolo S, Girolamo D, Del Fabbro M. Nanotechnology Scaffolds for alveolar bone regeneration. Materials. 2020, 13, 201; doi: 10.3390/ma13010201.

[38] Wang H, Leeuwenburgh SC, Li Y, Jansen JA. The use of micro- and nanospheres as functional components for bone tissue regeneration. Tissue Eng Part B Rev. 2012, 18, 24–39.

[39] Sakamoto M. Development and evaluation of superporous hydroxyapatite ceramics with triple pore structure as bone tissue scaffold. J Ceram Soc Jpn 2010, 118, 753–57.

[40] Mygind T, Stiehler M, Baatrup A, Li H, Zou X, Flyvbjerg A, Kassem M, Bünger C. Mesenchymal stem cell ingrowth and differentiation on coralline hydroxyapatite scaffolds. Biomaterials. 2007, 28, 1036–47.

[41] Sánchez-Salcedo S, Arcos D, Vallet-Regí M. upgrading calcium phosphate Scaffolds for tissue engineering applications. Key Eng Mater. 2008, 377, 19–42.

[42] Suetsugu Y, Tateishi T. Chapter 6: Implants and biomaterials (hydroxyapatite); www.aqb.jp/englo ish/file/thebasicspart3-6.pdf.

[43] Sheikh Z, Hamdan N, Abdallah M-N, Glogauer M, Grynpas M. Natural and synthetic bone replacement graft materials for dental and maxillofacial applications. Adv Dent Biomater. 2019, 347–76.

[44] Tamimi F, Sheikh Z, Barralet J. Dicalcium phosphate cements: Brushite and monetite. Acta Biomater. 2012, 8, 474–87.

[45] Kao ST, Scott DD. A review of bone substitutes. Oral Maxillofac Surg Clin North Am. 2007, 19, 513–21.

[46] Horowitz RA, Leventis MD, Rohrer MD, Prasad HS. Bone grafting: History, rationale, and selection of materials and techniques. Compend Contin Educ Dent. 2014, 35, 1–6.

[47] Bhatt RA, Rozental TD. Bone graft substitutes. Hand Clin. 2012, 28, 457–68.

[48] Suneelkumar C, Datta K, Srinivasan MR, Kumar ST. Biphasic calcium phosphate in periapical surgery. J Conserv Dent. 2008, 11, 92–96.

[49] Stavropoulos A, Windisch P, Szendröi-Kiss D, Peter R, Gera I, Sculean A. Clinical and histologic evaluation of granular beta-tricalcium phosphate for the treatment of human intrabony periodontal defects: A report on five cases. J Periodontol. 2010, 81, 325–34.

[50] Nakajima Y, Fiorellini JP, Kim DM, Weber HP. Regeneration of standardized mandibular bone defects using expanded polytetrafluoroethylene membrane and various bone fillers. Int J Periodontics Restor Dent. 2007, 27, 151–59.

[51] Fernandez de Grado G, Keller L, Idoux-Gillet Y, Wagner Q, Musset AM, Benkirane-Jessel N, Bornert F, Offner D. Bone substitutes: A review of their characteristics, clinical use, and perspectives for large bone defects management. J Tissue Eng. 2018, 9; doi: 10.1177/2041731418776819.

[52] Hollinger JO, Brekke J. Role of bone substitutes. Clin Orthop. 1996, 324, 55–65.

[53] Liu H, Yazici H, Ergun C, Webster TJ, Bermek H. An in vitro evaluation of the Ca/P ratio for the cytocompatibility of nano-to-micron particulate calcium phosphates for bone regeneration. Acta Biomater. 2008, 4, 1472–79.

[54] Detsch R, Mayr H, Ziegler G. Formation of osteoclast-like cells on HA and TCP ceramics. Acta Biomater. 2008, 4, 139–48.

[55] Kasten P, Luginbühl R, van Griensven M, Barkhausen T, Krettek C, Bohner M, Bosch U. Comparison of human bone marrow stromal cells seeded on calcium-deficient hydroxyapatite, β-tricalcium phosphate and demineralized bone matrix. Biomater. 2003, 24, 2593–603.

[56] Hadi A, Hassan B, Ghizlane A. Bilateral sinus graft with either bovine hydroxyapatite or β tricalcium phosphate, in combination with platelet-rich plasma: A case report. Implant Dent. 2008, 17, 350–59.

[57] Kurkcu M, Mehmet Benlidayi E, Cam B, Sertdemir Y. Anorganic bovine-derived hydroxyapatite versus β-tricalcium phosphate in sinus augmentation. A comparative histomorphometric study. J Oral Implants. 2012, 38, doi: http://dx.doi.org/10.1563/AAID-JOI-D-11-00061.1.

[58] Yang G-L, He FM, En Song E, Hu JA, Wang X-X, Zhao S-F. In vivo comparison of bone formation on titanium implant surfaces coated with biomimetically deposited calcium phosphate or electrochemically deposited hydroxyapatite. Int J Oral Maxillofac Implants. 2010, 25, 669–80.

[59] Xie C, Lu H, Li W, Chen F-M, Zhao Y-M. The use of calcium phosphate-based biomaterials in implant dentistry. J Mater Sci Mater Electron. 2012, 23, 853–62.

[60] Burguera EF, Xu HHK, Sun L. Injectable calcium phosphate cement: Effects of powder-to-liquid ratio and needle size. J Biomed Mater Res Part B Appl Biomater. 2007, 84, 493–502.

[61] Khairoun I, Boltong MG, Driessens FCM, Planell JA. Some factors controlling the injectability of calcium phosphate bone cements. J Mater Sci Mater Electron. 1998, 9, 425–28.

[62] O'Hara RM, Dunne NJ, Orr JF, Buchanan FJ, Wilcox R, Barton DC. Optimisation of the mechanical and handling properties of an injectable calcium phosphate cement. J Mater Sci Mater Electron. 2010, 21, 2299–305.

[63] Xu HH, Wang P, Wang L, Bao C, Chen Q, Weir MD, Chow LC, Zhao L, Zhou X, Reynolds MA. Calcium phosphate cements for bone engineering and their biological properties. Bone Res. 2017, 5, 17056; doi: 10.1038/boneres.2017.56.

[64] Lyu C, Shao Z, Zou D, Lu J. Ridge alterations following socket preservation using a collagen membrane in dogs. BioMed Res Int. 2020, 2020; doi: 10.1155/2020/2567861.

[65] Roberts TT, Rosenbaum AJ. Bone grafts, bone substitutes and orthobiologics: The bridge between basic science and clinical advancements in fracture healing. Organogenesis. 2012, 8, 114–24.

[66] Arinzeh TL, Tran T, McAlary J, Daculsi G. A comparative study of biphasic calcium phosphate ceramics for human mesenchymal stem-cell-induced bone formation. Biomater. 2005, 26, 3631–38.

[67] Jensen SS, Bornstein MM, Dard M, Bosshardt DD, Buser D. Comparative study of biphasic calcium phosphates with different HA/TCP ratios in mandibular bone defects. A long-term histomorphometric study in minipigs. J Biomed Mater Res B. 2009, 90B, 171–81.

[68] Suzuki T, Hukkanen M, Ohashi R, Yokogawa Y, Nishizawa K, Nagata F, Buttery L, Polak J. Growth and adhesion of osteoblast-like cells derived from neonatal rat calvaria on calcium phosphate ceramics. J Biosci Bioeng. 2000, 89, 18–26.

[69] Alam MI, Asahina I, Ohmamiuda K, Takahashi K, Yokota S, Enomoto S. Evaluation of ceramics composed of different hydroxyapatite to tricalcium phosphate ratios as carriers for rhBMP-2. Biomater. 2001, 22, 1643–51.

[70] Jinno T, Davy DT, Goldberg VM. Comparison of hydroxyapatite and hydroxyapatite tricalcium-phosphate coatings. J Arthrop. 2002, 17, 902–09.

[71] De Kok IJ, Peter SJ, Archambault M, van den Bos C, Kadiyala S, Aukhil I, Cooper LF. Hydroxyapatite/ tricalcium phosphate composites; tissue engineering applications. Clin Oral Implant Res. 2003, 14, 481–89.

[72]   Miao X, Tan DM, Li J, Xiao Y, Crawford R. Mechanical and biological properties of hydroxyapatite/
       tricalcium phosphate Scaffolds coated with poly(lactic-co-glycolic acid). Acta Biomater. 2008, 4,
       638–45.
[73]   Yuan H, Yang Z, de Bruijn JD, De Groot K, Zhang X. Material-dependent bone induction by calcium
       phosphate ceramics: A 2.5-year study in dog. Biomater. 2001, 22, 2617–23.
[74]   Dorozhkin SV. Calcium orthophosphate-containing biocomposites and hybrid biomaterials for
       biomedical applications. J Funct Biomater. 2015, 6, 708–832.
[75]   Weiner S, Wagner HD. The material bone: Structure-mechanical function relations. Ann Rev Mater
       Sci. 1998, 28, 271–98.
[76]   Rey C, Combes C, Drouet C, Glimcher MJ. Bone mineral: Update on chemical composition and
       structure. Osteoporos Int. 2009, 20, 1013–21.
[77]   Takagi S, Chow LC, Ishikawa K. Formation of hydroxyapatite in new calcium phosphate cements.
       Biometer. 1998, 19, 1593–99.
[78]   Sugawara A, Fujikawa K, Takagi S, Chow LC. Histopathological and cell enzyme studies of calcium
       phosphate cements. Dent Mat. 2004, 4, 613–20.
[79]   Sugawara A. A new generation of bone regeneration materials. J Oral Implants. 2006, 28, 47–84.
[80]   Slutsky DJ, Osterman AL. Fractures and Injuries of the Distal Radius and Carpus E-Book: The Cutting
       Edge. Saunders, Elsevier. 2009.
[81]   Mohamed KR. Medical Bioceramic Materials. 2017; https://www.academia.edu/40565838/Medical_
       Bioceramic_Materials.
[82]   Skallevold HE, Rokaya D, Khurshid Z, Zafar MS. Bioactive glass applications in dentistry. Int J Mol Sci.
       2019, 20, 5960; doi: 10.3390/ijms20235960.
[83]   Waked W, Grauer J. Silicates and Bone Fusion Orthopedics. 2008, 591–97.
[84]   Esfahanizadeh N, Nourani MR, Bahador A, Akhondi N, Montazeri M. The anti-biofilm activity of
       nanometric zinc doped bioactive glass against putative periodontal pathogens: An in vitro study.
       Biomed Glas. 2018, 4, 95–107.
[85]   Lovelace TB, Mellonig JT, Meffert RM, Jones AA, Nummikoski PV, Cochran DL. Clinical evaluation of
       bioactive glass in the treatment of periodontal Osseous defects in humans. J Periodontol. 1998, 69,
       1027–35.
[86]   Krishnan V, Lakshmi T. Bioglass: A novel biocompatible innovation. J Adv Pharm Technol Res. 2013,
       4, 78; doi: 10.4103/2231-4040.111523.
[87]   Hench L, Hench JW, Greenspan D. Bioglass: A short history and bibliography. J Australas Ceram Soc.
       2004, 40, 1–42.
[88]   El Shazley N, Hamdy A, El-Eneen HA, El Backly RM, Saad MM, Essam W, Moussa H, El Tantawi M, Jain
       H, Marei MK. Bioglass in alveolar bone regeneration in orthodontic patients: Randomized controlled
       clinical trial. JDR Clin Transl Res. 2016, 1, 244–55.
[89]   Ezzat AEM, El-Shenawy HM. Repair of cleft alveolar bone with bioactive glass material using Z-plasty
       flap. Int J Appl Basic Med Res. 2015, 5, 211–13.
[90]   Hing KA. Bone repair in the twenty-first century: Biology, chemistry or engineering? Philos. Transact
       A Math Phys Eng Sci. 2004, 362, 2821–50.
[91]   Nandi SK, Kundu B, Datta S. Development and applications of varieties of bioactive glass
       compositions in dental surgery, third generation tissue engineering, orthopaedic surgery and as
       drug delivery system. In: Pignatello R (Ed.)., Biomaterials Applications for Nanomedicine. InTech.
       2011, 69–116; https://www.intechopen.com/chapters/23619.
[92]   Giannoudis PV, Dinopoulos H, Tsiridis E. Bone substitutes: An update. Injury. 2005, 36, S20–7.
[93]   Jones JR. Review of bioactive glass: From Hench to hybrids. Acta Biomater. 2013, 9, 4457–86.
[94]   Arango-Ospina M, Hupa L, Boccaccini AR. Bioactivity and dissolution behavior of boron-containing
       bioactive glasses under static and dynamic conditions in different media. Biomed Glasses. 2019, 5,
       124–39.

[95]   Valimaki VV, Aro HT. Molecular basis for action of bioactive glasses as bone graft substitute. Scand J Surg. 2006, 95, 95–102.

[96]   Cholewa-Kowalska K, Kokoszka J, Laczka M, Niedźwiedzki L, Madej W, Osyczka AM. Gel-derived bioglass as a compound of hydroxyapatite composites. Biomed Mater. 2009, 4, 055007; doi: 10.1088/1748-6041/4/5/055007.

[97]   Bellucci D, Sola A, Anesi A, Salvatori R, Chiarini L, Cannillo V. Bioactive glass/hydroxyapatite composites: Mechanical properties and biological evaluation. Mater Sci Eng C. 2015, 51, 196–205.

[98]   Khanmohammadi S, Ojaghi-Ilkhchi M, Farrokhi-Rad M. Evaluation of bioglass and hydroxyapatite-based nanocomposite coatings obtained by electrophoretic deposition. Ceramics Intl. 2020, 46, 26069–77.

[99]   Palmer W, Crawford-Sykes A, Rose R. Donor site morbidity following iliac crest bone graft. West Ind Med J. 2008, 57, 490–92.

[100] Garg T, Singh O, Arora SR, Murthy SR. Scaffold: A novel carrier for cell and drug delivery. Crit Rev Ther Drug Carr Syst. 2012, 29, 1–63.

[101] Stamboulis AG, Boccaccini AR, Hench LL. Novel biodegradable polymer/bioactive glass composites for tissue engineering applications. Ad Eng Mater. 2002, 4, 105–09.

[102] Boccaccini AR, Erol M, Stark WJ, Mohn D, Hong Z, Mano JF. Polymer/bioactive glass nanocomposites for biomedical applications: A review. Composites Sci Tech. 2010, 70, 1764–76.

[103] Distler T, Fournier N, Grünewald A, Polley C, Seitz H, Detsch R, Boccaccini AR. Polymer-bioactive glass composite filaments for 3D scaffold manufacturing by fused deposition modeling: Fabrication and characterization. Front Bioeng Biotechnol. 2020, 8; https://doi.org/10.3389/fbioe.2020.00552.

[104] Gerhard LC, Boccaccini AR. Bioactive glass and glass-ceramic Scaffolds for bone tissue engineering. Materials. 2010, 3, 3867–910.

[105] Fu Q, Saiz E, Rahaman MN, Tomsia AP. Bioactive glass Scaffolds for bone tissue engineering: State of the art and future perspectives. Mater Sci Eng C Mater Biol Appl. 2001, 31, 1245–56.

[106] Hench LL, Jones JR. Bioactive glasses: Frontiers and challenges. Front Bioeng Biotechnol. 2005, 3, 194; doi: 10.3389/fbioe.2015.00194.

[107] Barbeck M, Serra T, Booms P, Stojanovic S, Najman S, Engel E, Sader R, Kirkpatrick CJ, Navarro M, Ghanaati S. Analysis of the in vitro degradation and the in vivo tissue response to bi-layered 3D-printed Scaffolds combining PLA and biphasic PLA/bioglass components – Guidance of the inflammatory response as basis for osteochondral regeneration. Bioact Mater. 2017, 2, 208–23.

[108] Russias J, Saiz E, Deville S, Gryn K, Liu G, Nalla RK, Tomsia AP. Fabrication and in vitro characterization of three- dimensional organic/inorganic Scaffolds by robocasting. J Biomed Mat Res Part A. 2007, 83A, 443–45.

[109] Murphy C, Kolan KCR, Long M, Leu M-C, Semon JA, Day DE. 3D printing of a polymer bioactive glass composite for bone repair. In Proceeding, 27th Annual Intl Solid Freeform Fabrication Sym. 2016, 1718–31.

[110] Kolan K, Liu Y, Baldridge J, Murohy C, Semon J, Day D, Lue M. Solvent based 3D printing of biopolymer/Bioactive glass composite and hydrogel for tissue engineering applications. Procedia CIRP. 2017, 65, 38–43.

[111] Kumbar S, Laurencin C, Deng M. Natural and Synthetic Biomedical Polymers. Elsevier, Berlin/ Heidelberg, Germany. 2014.

[112] Kulkarni V, Butte K, Rathod S. Natural polymers – a comprehensive review. Int J Res Pharm Biomed Sci. 2012, 3, 1597–613.

[113] Sergi R, Bellucci D, Cannillo V. A review of bioactive glass/natural polymer composites: State of the art. Materials (Basel). 2020, 13, 5560; https://doi.org/10.3390/ma13235560.

[114] Badylak SF. The extracellular matrix as a biologic scaffold material. Biomaterials. 2007, 28, 3587–93.

[115] D'Ayala GG, Malinconico M, Laurienzo P. Marine derived polysaccharides for biomedical applications: Chemical modification approaches. Molecules. 2008, 13, 2069–106.

[116]   Klouda L, Mikos AG. Thermoresponsive hydrogels in biomedical applications – A review. Eur J Pharm Biopharm. 2011, 68, 34–45.

[117]   Araújo M, Viveiros R, Philippart A, Miola M, Doumett S, Baldi G, Perez J, Boccaccini AR, Aguiar-Ricardo A, Verné E. Bioactivity, mechanical properties and drug delivery ability of bioactive glass-ceramic Scaffolds coated with a natural-derived polymer. Mater Sci Eng C. 2017, 768, 342–51.

[118]   Harini B, Shadamarshan RPK, Rao SH, Selvamurugan N, Balagangadharan K. Natural and synthetic polymers/bioceramics/bioactive compounds-mediated cell signalling in bone tissue engineering. Int J Biol Macromol. 2017, 110, 88–96.

[119]   Bandyopadhyay A, Chowdhury SK, Dey S, Moses JC, Mandel BB. Silk: A promising biomaterial opening new vistas towards affordable healthcare solutions. J Ind Inst Sci. 2019, 99, 445–87; https://link.springer.com/article/10.1007/s41745-019-00114-y.

[120]   Cao Y, Wang B. Biodegradation of silk biomaterials. Int J Mol Sci. 2009, 10, 1514–24.

[121]   Khan MR, Tsukada M, Gotoh Y, Morikawa H, Freddi G, Shiozaki H. Physical properties and dyeability of silk fibers degummed with citric acid. Bioresour Technol. 2010, 101, 8439–45.

[122]   Kwon K-J, Seok H. Silk protein-based membrane for guided bone regeneration. Appl Sci. 2018, 8, 1214; doi: 10.3390/app8081214.

[123]   Mandal BB, Grinberg A, Gil ES, Kaplan DL. High-strength silk protein Scaffolds for bone repair. Appl Biolog Sci. 2012, 109, 7699–704.

[124]   Gil ES, Kluge JA, Rockwood DN, Rajkhowa R, Wang L, Wang X, Kaplan DL. Mechanical improvements to reinforced porous silk scaffolds. J Biomed Mater Res Part A. 2011, 99, 16–28.

[125]   Rockwood DN, Gil ES, Park S-H, Kluge JA, Grayson W, Bhumiratana S, Rajkhowa R, Wang X, Kim SJ, Vunjak-Novakovic G, Kaplan DL. Ingrowth of human mesenchymal stem cells into porous silk particle reinforced silk composite scaffolds: An in vitro study. Acta Biomater. 2011, 7, 144–51.

[126]   Rajkhowa R, Gil ES, Kluge J, Numata K, Wang L, Wang X, Kaplan DL. Reinforced silk Scaffolds with silk particles. Macromol Biosci. 2010, 10, 599–611.

[127]   Meinel L, Fajardo R, Hofmann S, Langer R, Chen J, Snyder B, Vunjak-Novakovic G, Kaplan D. Silk implants for the healing of critical size bone defects. Bone. 2005, 37, 688–98.

[128]   Kundu J, Chung Y-I, Kim YH, Tae G, Kundu SC. Silk fibroin nanoparticles for cellular uptake and control release. Int J Pharm. 2010, 388, 242–50.

[129]   Meinel L, Hofmann S, Karageorgiou V, Zichner L, Langer R, Kaplan D, Vunjak-Novakovic G. Engineering cartilage-like tissue using human mesenchymal stem cells and silk protein scaffolds. Biotechnol Bioeng. 2004, 88, 379–91.

[130]   Saleem M, Rasheed S, Yougen C. Silk fibroin/hydroxyapatite scaffold: A highly compatible material for bone regeneration. Sci Tech Ad Mater. 2020, 21, 242–66.

[131]   Koh KS, Choi JW, Park EJ, Oh TS. Bone regeneration using silk hydroxyapatite hybrid composite in a rat alveolar defect model. Int J Med Sci. 2018, 15, 59–68.

[132]   Cai Y, Guo J, Chen C, Yao C, Chung S-M, Yao J, Lee I-S, Kong X. Silk fibroin membrane used for guided bone tissue regeneration. Mater Sci Eng C. 2017, 70, 148–54.

[133]   Kweon H, Jo Y-Y, Seok H, Kim S-G, Chae W-S, Sapru S, Kundu SC, Park N-R, Che X, Choi J-Y. In vivo bone regeneration ability of different layers of natural silk cocoon processed using an eco-friendly method. Macromol Res. 2017, 25, 806–16.

[134]   Zafar MS, Al-Samadani KH. Potential use of natural silk for bio-dental applications. J Taibah Univ Med Sci. 2014, 9, 171–77.

[135]   Ha Y-Y, Park Y-W, Kweon H, Jo Y-Y, Kim S-G. Comparison of the physical properties and in vivo bioactivities of silkworm-cocoon-derived silk membrane, collagen membrane, and polytetrafluoro-ethylene membrane for guided bone regeneration. Macromol Res. 2014, 22, 1018–23.

[136]   Haugen HJ, Lyngstadaas SP, Rossi F, Perale G. Bone grafts: Which is the ideal biomaterial? J Clin Periodontol. 2019, 46, 92–102.

[137]   Peng C, Shu Z, Zhang C, Chen X, Wang M, Fan L. Surface modification of silk fibroin composite bone scaffold with polydopamine coating to enhance mineralization ability and biological activity for bone tissue engineering. Appl Polym. 2022, https://doi.org/10.1002/app.52900.

[138]   Moses JC, Nandi SK, Mandel BB. Multifunctional cell instructive silk-bioactive glass composite reinforced Scaffolds toward osteoinductive, proangiogenic, and resorbable bone grafts. Ad Healthcare Mater. 2018, 7; https://doi.org/10.1002/adhm.201701418.

[139]   Lee JH, Kweon HY, Oh J-H, Kim S-G. The optimal scaffold for silk sericin-based bone graft: Collagen versus gelatin. Maxillofac Plast Reconstr Surg. 2023, 45; https://doi.org/10.1186/s40902-022-00368-0.

[140]   Wu Z, Meng Z, Wu Q, Zeng D, Guo Z, Yao J, Bian Y, Gu Y, Cheng S, Peng L. Biomimetic and osteogenic 3D silk fibroin composite Scaffolds with nano MgO and mineralized hydroxyapatite for bone regeneration. SAGE Pub J Tissue Eng. 2020, 11; https://doi.org/10.1177/2041731420967791.

[141]   Aranaz I, Alcántara AR, Civera MC, Arias C, Elorza B, Caballero AH, Acosta N. Chitosan: An overview of its properties and applications. Polymers (Basel). 2021, 13; doi: 10.3390/polym13193256.

[142]   Ghormade V, Pathan EK, Deshpande MV. Can fungi compete with marine sources for chitosan production? Int J Biol Macromol. 2017, 104, 1415–21.

[143]   Feng P, Luo Y, Ke C, Qiu H, Wang W, Zhu Y, Hou R, Xu L, Wu S. Chitosan-based functional materials for skin wound repair: Mechanisms and applications. Front Bioeng Biotechnol. 2021, 9; https://doi.org/10.3389/fbioe.2021.650598.

[144]   Ueno H, Mori T, Fujinaga T. Topical formulations and wound healing applications of chitosan. Adv Drug Deliv Rev. 2001, 52, 105–15.

[145]   Alves da Silva ML, Crawford A, Mundy JM, Correlo VM, Sol P, Bhattacharya M, Hatton PV, Reis RL, Neves NM. Chitosan/polyester-based Scaffolds for cartilage tissue engineering: Assessment of extracellular matrix formation. Acta Biomater. 2010, 6, 1149–57.

[146]   Oryan A, Alidadi S, Moshiri A, Maffulli N. Bone regenerative medicine: Classic options, novel strategies, and future directions. J Orthop Surg Res. 2014, 9, 18; doi: 10.1186/1749-799X-9-18.

[147]   Kozusko SD, Riccio C, Goulart M, Bumgardner J, Jing XL, Konofaos P. Chitosan as a bone scaffold biomaterial. J Craniofacial Surg. 2018, 29, 1788–93.

[148]   Menon PR, Mukherjee DP. Development of a composite of hydroxylapatite and chitosan as a bone graft substitute. IEEE. 1995; https://ieeexplore.ieee.org/document/514446.

[149]   van de Graaf GMM, De Zoppa A, Moreira RC, Maestrelli SC, Marques RFC, Campos MGN. Morphological and mechanical characterization of chitosan-calcium phosphate composites for potential application as bone-graft substitutes. Res Biomed Eng. 2015, 31; https://doi.org/10.1590/2446-4740.0786.

[150]   Kjalarsdóttir L, Dýrfjörd A, Dagbjartsson A, Laxdal EH, Örlygsson G, Gíslason J, Einarsson JM, Ng C-H, Jónsson H Jr. Bone remodeling effect of a chitosan and calcium phosphate-based composite. Regener Biomater. 2019, 6, 241–47.

[151]   Huang J, Ratnayake J, Ramesh N, Dias GJ. Development and characterization of a biocomposite material from Chitosan and New Zealand-sourced bovine-derived hydroxyapatite for bone regeneration. ACS Omega. 2020, 5, 16537–46.

[152]   Kowalczyk P, Podgórski R, Wojasiński M, Gut G, Bojar W, Tomasz Ciach T. Chitosan-human bone composite granulates for guided bone regeneration. Int J Mol Sci. 2021, 22, 2324; doi: 10.3390/ijms22052324.

[153]   Xu HH, Simon CG Jr. Fast setting calcium phosphate–chitosan scaffold: Mechanical properties and biocompatibility. Biomaterials. 2005, 26, 1337–48.

[154]   Jebahi S, Oudadesse H, Ben Saleh G, Saoudi M, Mesadhi S, Rebai T, Keskes H, El Feki A, El Feki H. Chitosan-based bioglass composite for bone tissue healing: Oxidative stress status and antiosteoporotic performance in a ovariectomized rat model. Korean J Chem Eng. 2014, 31, 1616–23.

[155]   Shavandi A, Bekhit AE-DA, Sun Z, Ali MA. Injectable gel from squid pen chitosan for bone tissue engineering applications. J Sol Gel Sci Technol. 2015, 77, 675–87.

[156] Nie L, Deng Y, Li P, Hou R, Shavandi A, Yag S. Hydroxyethyl Chitosan-reinforced polyvinyl alcohol/biphasic calcium phosphate hydrogels for bone regeneration. ACS Omega. 2020, 5, 10948–57.

[157] Wattanutchariya W, Changkowchai W. Characterization of porous scaffold from chitosan-gelatin/hydroxyapatite for bone grafting. Proceedings of the International Multiconference of Engineers and Computer Scientists. Hong Kong. 2014; https://www.iaeng.org/publication/IMECS2014/IMECS2014_pp1073-1077.pdf.

[158] Husain S, Al-Samadani KH, Najeeb S, Zafar MS, Khurshid Z, Zohaib S, Qasim SB. Chitosan biomaterials for current and potential dental applications. Materials. 2017, 10, 602; doi: 10.3390/ma10060602.

[159] Aguilar A, Zein N, Harmouch E, Hafdi B, Bornert F, Offner D, Clauss F, Fioretti F, Huck O, Benkirane-Jessel N, Hua G. Application of Chitosan in bone and dental engineering. Molecules. 2019, 24, 3009; doi: 10.3390/molecules24163009.

[160] Jayash SN, Hashim NM, Misran M, Ibrahim N, Al-Namnam NM, Baharuddin NA. Analysis on efficacy of Chitosan-based gel on bone quality and quantity. Front Mater. 2021, 8; https://doi.org/10.3389/fmats.2021.640950.

[161] Fuchs JR, Nasseri BA, Vacanti JP. Tissue engineering: A 21st century solution to surgical reconstruction. Ann Thorac Surg. 2001, 72, 577–91.

[162] Pilipchuk SP, Plonka AB, Monje A, Taut AD, Lanis A, Kang B, Giannobile WV. Tissue engineering for bone regeneration and osseointegration in the oral cavity. Dent Mater. 2015, 31, 317–38.

[163] Fairag R, Rosenzweig DH, Garcialuna JLR, Weber MH, Haglund L. Three-dimensional printed polylactic acid Scaffolds promote bone-like matrix deposition in vitro. ACS Appl Mater Interfaces. 2019, 11, 15306–15.

[164] Danoux CB, Barbieri D, Yuan H, De Bruijn JD, Van Blitterswijk CA, Habibovic P. In vitro and in vivo bioactivity assessment of a polylactic acid/hydroxyapatite composite for bone regeneration. Biomatter. 2014, 4, e27664; doi: 10.4161/biom.27664.

[165] Ai C, Sheng D, Chen J, Cai J, Wang S, Jiang J, Chen S. Surface modification of vascular endothelial growth factor-loaded silk fibroin to improve biological performance of ultra-high-molecular-weight polyethylene via promoting angiogenesis. Int J Nanomed. 2017, 12, 7737–50.

[166] Kashirina A, Yao Y, Liu Y, Leng J. Biopolymers as bone substitutes: A review. Biomater Sci. 2019, 7, 3961–83.

[167] Guduric V, Metz C, Siadous R, Bareille R, Levato R, Engel E, Fricain J-C, Devillard R, Luzanin O, Catros S. Layer-by-layer bioassembly of cellularized polylactic acid porous membranes for bone tissue engineering. J Mater Sci Mater Electron. 2017, 28, 78; doi: 10.1007/s10856-017-5887-6.

[168] Cristache CM, Grose AR, Cristache G, DiDilescu NC, Totu EE. Additive manufacturing and synthetic polymers for bone reconstruction in the maxillofacial region. Materiale Plastice. 2018, 55, 555–62; https://revmaterialeplastice.ro/Articles.asp?ID=5073.

[169] Cheah CW, Al-Namanm M, Lau MN, Lim GS, Raman R, Fairbairn P, Ngeow WC. Synthetic material for bone, periodontal, and dental tissue regeneration: Where are we now, and where are we heading next? Materials (Basel). 2021, 14, 6123; doi: 10.3390/ma14206123.

[170] Palma PJ, Matos S, Ramos J, Guerra F, Figueiredo MH, Kauser J. New formulations for space provision and bone regeneration. Biodental Eng I. 2010, 1, 71–76.

[171] Shi C, Yuan Z, Han F, Zhu C, Li B. Polymeric biomaterials for bone regeneration. Annals of Joint. 2016, 1; doi: 10.21037/aoj.2016.11.02.

[172] Puppi D, Chiellini F, Piras AM, Chiellini E. Polymeric materials for bone and cartilage repair. Prog Polym Sci. 2010, 35, 403–40.

[173] Wang DX, He Y, Bi L, Qu Z-H, Zou J-W, Pan Z, Fab J-J, Chem L, Dong X, Liu X-N, Pei G-X, Ding J-D. Enhancing the bioactivity of Poly(lactic-co-glycolic acid) scaffold with a nano-hydroxyapatite coating for the treatment of segmental bone defect in a rabbit model. Int J Nanomed. 2013, 8, 1855–65.

[174] Kao CT, Lin CC, Chen YW, Yeh C-H, Fang H-Y, Shie M-Y. Poly(dopamine) coating of 3D printed poly (lactic acid) Scaffolds for bone tissue engineering. Mater Sci Eng C Mater Biol Appl. 2015, 56, 165–73.

[175] Noorjahan A, Tan X, Liu Q, Gray MR, Choi P. Study of cyclohexane diffusion in Athabasca Asphaltenes. Energy Fuels. 2014, 28, 1004–11.

[176] Liu X, Ma PX. Polymeric Scaffolds for bone tissue engineering. Ann Biomed Eng. 2004, 32, 477–86.

[177] Warren SM, Fong KD, Nacamuli RP, Song HM, Fang TD, Longaker MT. Biomaterials for skin and bone replacement and repair in plastic surgery. Oper Tech Plast Reconstr Surg. 2002, 9, 10–15.

[178] Corcione CE, Gervaso F, Scalera F, Montagna F, Maiullaro T, Sannino A, Maffezzoli A. 3D printing of hydroxyapatite polymer-based composites for bone tissue engineering. J Polym Eng. 2017, 37, 741–46.

[179] Cao H, Kuboyama N. A biodegradable porous composite scaffold of PGA/beta-TCP for bone tissue engineering. Bone. 2010, 46, 386–95.

[180] Makadia HK, Siegel SJ. Poly lactic-co-glycolic acid (PLGA) as biodegradable controlled drug delivery carrier. Polymers. 2011, 3, 1377–97.

[181] Yun YP, Kim SE, Lee JB, Heo DN, Bae MS, Shin DR, Lim SB, Choi KK, Park SJ, Kwon IK. Comparison of osteogenic differentiation from adipose-derived stem cells, mesenchymal stem cells, and pulp cells on PLGA/hydroxyapatite nanofiber. Tissue Eng Regen Med. 2009, 6, 336–45.

[182] Siracusa V, Maimone G, Antonelli V. State-of-art of standard and innovative materials used in cranioplasty. Polymers. 2021, 13, 1452; doi: 10.3390/polym13091452.

[183] Wang W, Caetano G, Ambler WS, Blaker JJ, Frade MA, Mandal P, Diver C, Bártolo P. Enhancing the hydrophilicity and cell attachment of 3D printed PCL/graphene Scaffolds for bone tissue engineering. Materials. 2016, 9, 992; doi: 10.3390/ma9120992.

[184] Ku JK, Kim YK, Yun PY. Influence of biodegradable polymer membrane on new bone formation and biodegradation of biphasic bone substitutes: An animal mandibular defect model study. Maxillofac Plast Reconstr Surg. 2020, 42, 34; doi: 10.1186/s40902-020-00280-5.

[185] Wuisman PI, Smit TH. Bioresorbable polymers: Heading for a new generation of spinal cages. Eur Spine J. 2006, 15, 133–48.

[186] Al-Namnam N, Nagi S. Recent advances in bone graft substitute for oral and maxillofacial applications: A review. Int J Biosci. 2019, 15, 70–94.

[187] Jeong H-J, Gwak S-J, Seo KD, Lee S, Yun J-H, Cho Y-S, Lee S-J. Fabrication of three-dimensional composite scaffold for simultaneous alveolar bone regeneration in dental implant installation. Int J Mol Sci. 2020, 21, 1863; doi: 10.3390/ijms21051863.

[188] Goh BT, Teh LY, Tan DB, Zhang Z, Teoh SH. Novel 3D polycaprolactone scaffold for ridge preservation: A pilot randomised controlled clinical trial. Clin Oral Implant Res. 2015, 26, 271–77.

[189] Louvrier A, Euvrard E, Nicod LP, Rolin G, Gindraux F, Pazart L, Houdayer C, Risold PY, Meyer F. Odontoblastic differentiation of dental pulp stem cells from healthy and carious teeth on an original PCL-based 3D scaffold. Int Endod J. 2018, 51, e252–63.

[190] Baker MI, Walsh SP, Schwartz ZD. A review of polyvinyl alcohol and its uses in cartilage and orthopedic applications. J Biomed Mater Res B Appl Biomater. 2012, 100, 1451–57.

[191] Benoit DS, Schwartz MP, Durney AR, Anseth KS. Small functional groups for controlled differentiation of hydrogel-encapsulated human mesenchymal stem cells. Nat Mater. 2008, 7, 816–23.

[192] Yang XZ, Sun TM, Dou S, Wu J, Wang Y-C, Wang J. Block copolymer of polyphosphoester and poly(L-lactic acid) modified surface for enhancing osteoblast adhesion, proliferation, and function. Biomacromolecules. 2009, 10, 2213–20.

[193] van Leeuwen AC, Huddleston Slater JJ, Gielkens PFM, de Jong JR, Grijpma DW, Bos RRM. Guided bone regeneration in rat mandibular defects using resorbable poly(trimethylene carbonate) barrier membranes. Acta Biomater. 2012, 8, 1422–29.

[194] Huang MN, Wang YL, Luo YF. Biodegradable and bioactive porous polyurethanes Scaffolds for bone tissue engineering. J Biomed Sci Eng. 2009, 2, 36–40.

[195] Hsieh CY, Tsai SP, Wang DM, Chang Y-N, Hsieh H-J. Preparation of gamma-PGA/chitosan composite tissue engineering matrices. Biomaterials. 2005, 26, 5617–23.

[196] Senra MR, Marques FV. Synthetic polymeric materials for bone replacement. J Composite Sci. 2020, 4, 191; doi: 10.3390/jcs4040191.

[197] Gohil SV, Suhail S, Rose J, Vella T, Nair LS. Polymers and composites for orthopedic applications. In: Bose S, et al. (Ed.)., Materials for Bone Disorders. Academic Press, Cambridge, MA, USA. 2017, 349–403.

[198] Ma Z, He W, Yong T, Ramakrishna S. Grafting of gelatin on electrospun poly(caprolactone) nanofibers to improve endothelial cell spreading and proliferation and to control cell Orientation. Tissue Eng. 2005, 11, 1149–58.

[199] Nonoyama T, Wada S, Kiyama R, Kitamura N, Mredha MTI, Zhang X, Kurokawa T, Nakajima T, Takagi Y, Yasuda K, Gong JP. Double-network hydrogels strongly bondable to bones by Spontaneous Osteogenesis Penetration. Adv Mater. 2016, 28,, 6740–45.

[200] Lee YJ, Lee JH, Cho HJ, Kim HK, Yoon TR, Shin H. Electrospun fibers immobilized with bone forming peptide-1 derived from BMP7 for guided bone regeneration. Biomaterials. 2013, 34, 5059–69.

[201] Bosetti M, Fusaro L, Nicolì E, Borrone A, Aprile S, Cannas M. Poly-L-lactide acid-modified Scaffolds for osteoinduction and osteoconduction. J Biomed Mater Res A. 2014, 102, 3531–39.

[202] Raphel J, Holodniy M, Goodman SB, Heilshorn SC. Multifunctional coatings to simultaneously promote osseointegration and prevent infection of orthopaedic implants. Biomaterials. 2016, 84, 301–04.

[203] Kingshott P, Wei J, Bagge-Ravn D, Gadegaard N, Gram L. Covalent attachment of poly(ethylene glycol) to surfaces, critical for reducing bacterial adhesion. Langmuir. 2003, 19, 6912–21.

[204] Yang F, Williams CG, Wang DA, Lee H, Manson PN, Elisseeff J. The effect of incorporating RGD adhesive peptide in polyethylene glycol diacrylate hydrogel on osteogenesis of bone marrow stromal cells. Biomaterials. 2005, 26, 5991–98.

[205] Schneider OD, Loher S, Brunner TJ, Schmidlin P, Stark WJ. Flexible, silver containing nanocomposites for the repair of bone defects: Antimicrobial effect against E. coli infection and comparison to tetracycline containing scaffolds. J Mater Chem. 2008, 18, 2679–84.

[206] Huang D, Zuo Y, Zou Q, Zhang L, Li J, Cheng L, Shen J, Li Y. Antibacterial chitosan coating on nano-hydroxyapatite/polyamide66 porous bone scaffold for drug delivery. J Biomater Sci Polym Ed. 2011, 22, 931–44.

[207] Tommasi G, Perni S, Prokopovich P. An injectable hydrogel as bone graft material with added antimicrobial properties. Tissue Eng Part A. 2016, 22, 862–72.

[208] Oshida Y, Tominaga T. Nickel-Titanium Materials – Biomedical Applications. De Gruyter Pub. 2020.

[209] Jeng M-D, Chiang C-P. Autogenous bone grafts and titanium mesh-guided alveolar ridge augmentation for dental implantation. J Dent Sci. 2020, 15, 243–48.

[210] Di Stefano DA, Greco G, Gherlone E. A preshaped titanium mesh for guided bone regeneration with an equine-derived bone graft in a posterior mandibular bone defect: A case report. Dent J. 2019, 7, 77; doi: 10.3390/dj7030077.

[211] Xie Y, Li S, Zhang T, Wang C, Cai X. Titanium mesh for bone augmentation in oral implantology: Current application and progress. Int J Oral Sci. 2020, 12, 1–12.

[212] Atef M, Tarek A, Shaheen M, Alarawi RM, Askar N. Horizontal ridge augmentation using native collagen membrane vs titanium mesh in atrophic maxillary ridges: Randomized clinical trial. Clin Implant Dent Res. 2020, 22, 1546–66.

[213] Zhang T, Zhang T, Cai X. The application of a newly designed L-shaped titanium mesh for GBR with simultaneous implant placement in the esthetic zone: A retrospective case series study. Clin Implant Dent Relat Res. 2019, 21, 862–72.

[214] Oshida Y. Magnesium Materials – From Mountain Bikes to Degradable Bone Grafts. De Gruyter Pub. 2021.

[215] Liu C, Wan P, Tan LL, Wang K, Yang K. Preclinical investigation of an innovative magnesium-based bone graft substitute for potential orthopaedic applications. J Orthop Transl. 2014, 2, 139–48.

[216] Nabiyouni M, Brückner T, Zhou H, Gbureck U, Bhaduri SB. Magnesium-based bioceramics in orthopedic applications. Acta Biomater. 2018, 66, 23–43.

[217] Wu Y, Wang Y, Zhao D, Zhang N, Li H, Li J, Zhao Y, Yan J, Zhou Y. In vivo study of microarc oxidation coated Mg alloy as a substitute for bone defect repairing: Degradation behavior, mechanical properties, and bone response. Colloids Surf B Biointerfaces. 2019, 181, 349–59.

6 Scaffold, mesh, and membrane

In this chapter, we will be discussing materials which are used as scaffolds and mesh and/or membranes for both dental and medical implant treatments. In particular, resorbable characteristics in biological environment (aka, bioresorbability) are emphasized.

6.1 Scaffold structure

Scaffolds for tissue engineering are devices or constructions with specific and complex physical and biological functions that interact through biochemical and physical signals with cells and, when implanted, with the body environment [1, 2]. The history of tissue engineering scaffolds development is unique that spans several decades and involves the convergence of various scientific disciplines including biology, materials science, and engineering. Tissue engineering scaffolds are crucial components in the field of regenerative medicine, as they provide a framework for the growth and development of new tissues and organs [3]. Accordingly, numbers of research publications are huge; it was reported that 196,576 results with "Biomaterials and Tissue Engineering Scaffolds" in the full text manner [4].

Any missing piece of bone due to traumas, tumors, avascular necrosis, infections, or extraction in dentistry must be replaced/augmented with a proper functional alternative. Tissue engineering has introduced new hopes as combination of cells, scaffolds, and biofactors for bone regeneration (which is an important phase of healing process). Scaffold device plays an important role in tissue engineering and normally made of polymeric or metallic materials that are engineered to cause desirable cellular interactions to contribute to the formation of new functional tissues for dental and medical purposes. In both dental and medical implant treatments, scaffolds become crucial additional devices to achieve higher success rate and survival rate as well. Scaffold-based bone tissue engineering holds a potential for the future of osseous defects therapies [5]. The main biofunction of bone scaffold allows and stimulates the attachment and proliferation of osteoinductive cells onto its surface and the scaffold is normally 3D structure and should possess desirable characteristics of (i) biocompatibility in terms of cell attachment and proliferation as well as lack of toxicity (i.e., nontoxicity) and inflammatory reactions, (ii) noncarcinogenicity, with excellent osteoconductive and osteoinductive properties as well, (iii) biodegradability (or bioresorbability) for programmed safe substitution of the scaffold material with osteoid deposition, (iv) mechanical properties to bear weight during the amelioration period, (v) proper architecture in terms of porosity and pore sizes for cell penetration, nutrients and waste transfer, and angiogenesis, (vi) sterilibility without loss of bioactivity, and (vii) controlled deliverability of bioactive molecules or drugs [5–9].

https://doi.org/10.1515/9783111136691-006

These requirements assigned to scaffold structure and material (as foreign materials) are directly or indirectly reflected to various properties of receiving host material (aka, bone) that has a hierarchical and complex structure supporting its mechanical, biological, and chemical biofunctions. The heterogeneous and anisotropic structure of bone is composed of optimized irregular arrangement and orientation of macrostructures (such as cancellous and cortical bone), microstructures (like osteons and single trabeculae), sub-microstructures (such as lamellae), nanostructures (like fibrillar collagen), and sub-nanostructures (such as minerals and collagen molecules) [10]. These components are architecturally designed to fulfill the functional needs of each bone. The mechanical properties of bone are made by its component phases and hierarchical structural organization [11]. This transcorporation between host and foreign materials can be explicitly explained by mechanical compatibility and morphological compatibility [12, 13].

There is a material choice for scaffold structure or coating thereon. Normally, osteoconductive materials such as hydroxyapatite (HA) and tricalcium phosphate (TCP) ceramics as well as some biodegradable polymers are suggested and employed. However, recently, much interest has focused on the use of osteoinductive materials like demineralized bone matrix or bone derivatives. However, physiochemical modifications in terms of porosity, mechanical strength, cell adhesion, biocompatibility, cell proliferation, mineralization, and osteogenic differentiation are still required [1].

6.1.1 Polymeric scaffolds

In general, bone tissue engineering requires artificial scaffolds with physicochemical, structural, and biological properties mimicking natural extracellular matrix, which provides a suitable environment for cell recruitment, proliferation, differentiation, and eventually bone regeneration. Facing the complex and sensitive biological systems, ideal scaffolds should not elicit immunological reactions and degrade in a controllable way with nontoxic products that can be excreted through metabolism. Biological agents are also needed to be incorporated to promote the formation of new bone tissues [14]. Furthermore, the macro- and microstructures (e.g., porosity) of the scaffolds should also be carefully designed to provide an optimized microenvironment for cell functions as well as to maintain the diffusion of nutrients and metabolites. To date, diverse materials have been evaluated and utilized as scaffolds for bone regeneration, including metals, bioactive ceramics and glasses, natural and synthetic polymers, and their composites. Among them, polymers and their composites are considered as the most promising candidates for their advantageous biocompatibility and biodegradability over most metals and ceramics. More importantly, polymers possess highly flexible design capacity, and their various properties can be easily tailored to meet specific requirements through manipulating their chemical compositions and structures [14].

Polymeric scaffold materials consist of natural polymers and synthetic polymers. Normally, natural polymeric scaffolds are composed of extracellular biomaterials in three classes: (1) proteins such as collagen, gelatin, fibrinogen, elastin, keratin, and silk, (2) polysaccharides including glycosaminoglycans, cellulose, amylose, dextran, chitin, and chitosan, and (3) polynucleotides such as DNA and RNA [5, 15–17]. They exhibit osteoinductive properties. This group of natural scaffolds could be cell-derived (cells are used to generate new bone tissue or seeded onto a supporting matrix) or tissue-derived (bone tissue is directly used) [18, 19].

Synthetic polymers for scaffold fabrications should include aliphatic polyesters such as poly(lactic acid) (PLA), poly(glycolic acid) (PGA), and poly(caprolactone) (PCL). Their copolymers should be included [20–23]. Other synthetic polymers in bone tissue engineering include poly(methyl methacrylate) (PMMA), PCL, poly hydroxyl butyrate (PHB), polyethylene (PE), polypropylene (PP), polyurethane (PU), poly(ethylene terephthalate) (PET), poly ether ketone, and poly sulfone [24]. Although some synthetic polymers like poly(propylene fumarate) show high compressive strength and a controlled degradation time; however, they lose their strength due to rapid degradation in vivo and created local acidic environment which could make adverse tissue responses [15, 25]. Generally, polymeric materials provide more controllability on physiochemical characteristics of scaffolds such as pore size, porosity, solubility, biocompatibility, enzymatic reactions, and allergic response [26]. They are biocompatible and biodegradable and can be easily fabricated into different shapes [27]. Synthetic polymers were introduced for their excellent mechanical properties so that they also can mechanically support demands for a wide range of applications in orthopedics [28]. Figure 6.1 depicts and compares SEM images of various types of polymeric scaffolds [29], in which A is an SEM image of alginate, B is an alginate-chitosan, C the chitosan, D the chitosan-collagen, E the mesenchymal stem cell cultured on the scaffold collagen (C)-D,L-lactic acid-glycolic acid (PLGA) (P) medium, and F the synthesized porous HA scaffold.

6.1.2 Ceramic scaffolds

As described previously, chemistry of bone tissue (of about 70% of hydroxyapatite and 30% of collagen) promotes to develop bioceramics to mimic bone tissue and provide a higher osteoblasts adherence and proliferation compared to other materials [30, 31]. Calcium phosphate ceramics have been greatly studied for bone tissue repair as tunable bioactive materials [32]. Their physiochemical properties result in osteoconduction and osteoinduction. HA, TCP, and their combination as biphasic and amorphous calcium phosphates (BCPs and ACPs) are common types of calcium phosphate ceramics used in bone tissue engineering [25, 33].

There are several studies to modify the mechanical strength and dissolution rates, and biocompatibility of the scaffold can be done through the addition of calcium phosphate. Drzewiecka et al. [34] evaluated the mechanical and physical properties of com-

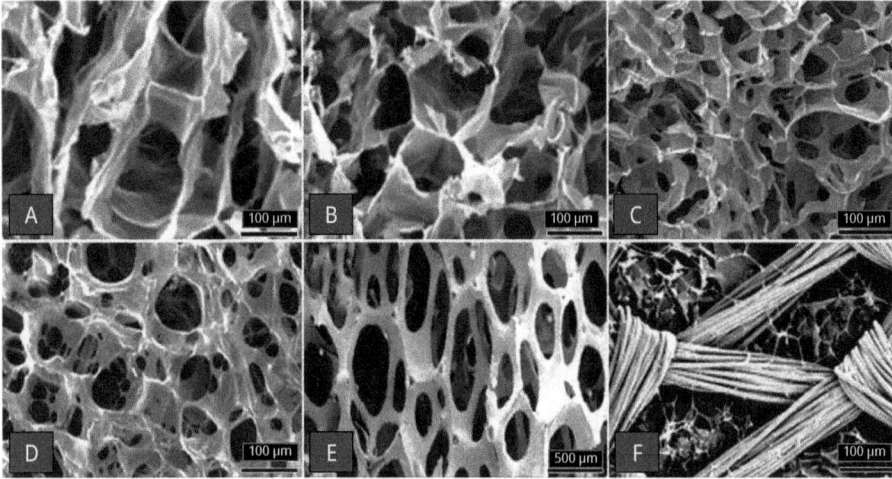

Figure 6.1: SEM images of various types of polymeric scaffolds [29].

posite materials modified with ACP, using two flowable composite materials – an experimental composite material based on di-metacrylic resin filled with colloidal-silica and commercially available composite material. Materials were modified with the addition of ACP as a powder. Diametral tensile strength test and Vickers hardness tests were conducted. It was found that (i) in both composites modified with ACP the decrease of tensile strength was observed, whilst ACP addition to experimental composite increased its hardness and (ii) addition of 5.67 wt% of ACP to SDR resulted in significant decrease of its hardness; suggesting that, in order to improve biological behavior of composite materials, it is possible to modify those materials with the addition of remineralizing agents like ACP [34]. Fielding et al. [35] evaluated the effects of silica (SiO_2) (0.5 wt%) and zinc oxide (ZnO) (0.25 wt%) dopants on the mechanical and biological properties of TCP scaffolds with three-dimensionally interconnected pores. It was obtained that (i) addition of dopants into TCP increased the average density of pure TCP from 90.8% to 94.1% and retarded the $\beta \rightarrow \alpha$ phase transformation at high sintering temperatures, resulting in up to 2.5-fold increase in compressive strength and (ii) in vitro cell-materials interaction studies, carried out using human fetal osteoblastic cells, confirmed that the addition of SiO_2 and ZnO to the scaffolds facilitated faster cell proliferation when compared to pure TCP scaffolds, suggesting that the addition of SiO_2 and ZnO dopants to the TCP scaffolds showed increased mechanical strength as well as increased cellular proliferation [35]. Solubility and surface topography are the most significant factors that influence cell behavior so that, as Samavedi et al. [29] described, designing calcium phosphate cements (CPCs) with suitable physical and chemical properties and osteoinductive potential may improve their in vivo bioactivity.

Wang et al. [36] fabricated porous nanocrystalline hydroxyapatite/chitosan scaffolds via a lyophilization procedure to design a biomimetic and bioactive tissue-engineered bone construct. The nanostructured bone scaffolds were then treated with cold atmosphere plasma (CAP) to create a more favorable surface for cell attachment, proliferation, and differentiation, claiming that the CAP modification method provides a quick one-step process for cell-favorable tissue-engineered scaffold architecture remodeling and surface property alteration. Pasqui et al. [37] developed a new hybrid material consisting of HA in a carboxymethylcellulose (CMC)-based hydrogel: CMC-HA hybrid. The strategy for inserting HA nanocrystals within the hydrogel matrix consists of making the freeze-dried hydrogel to swell in a solution-containing HA microcrystals. When the composite CMC-HA hydrogel was characterized and seeded with osteoblasts MG63 line, it was reported that the scaffold with HA enhanced cell proliferation and metabolic activity and promoted the production of mineralized extracellular matrix more than that observed for the scaffold without HA. Sagar et al. [38] evaluated the complete healing of critical size defect made in the proximal tibia of rabbits, using nanohydroxyapatite with gelatin and carboxymethylated chitin (nHA/gel/CMC) scaffold construct. The architecture indices analyzed by microcomputed tomography showed a significant increase in the percentage of bone volume fraction, with reconciled cortico-trabecular bone formation at the sites treated with nHA/gel/CMC constructs compared to controls. It was also mentioned that, at histology and fluorescence labeling, the uniformly interconnected porous surface of the scaffold construct enhanced osteoblastic activity and mineralization [38].

A combination of collagen and HA has been used in vivo. Recently, Xia et al. [39] developed a biomimetic collagen-apatite scaffold composed of collagen fibers and poorly crystalline bone-like carbonated apatite nanoparticles to improve bone repair and regeneration. Scaffold in vivo enhanced new bone formation in mice. Calvo-Guirado et al. [40] investigated the effect of resorbable collagen membranes (CMs) on critical size defects in rabbit tibiae filled with biphasic calcium phosphate: biphasic calcium phosphate functioned well as a scaffold and allowed mineralized tissue formation. It was found that (i) biphasic calcium phosphate functioned well as a scaffolding material allowing mineralized tissue formation and (ii) the addiction of absorbable CMs enhanced bone gain compared with non-membrane-treated sites [40].

The application of porous HA-collagen as a bone scaffold represents a new trend of mimicking the specific bone extracellular matrix. Application of HA in reconstructive surgery has shown that it is slowly invaded by the host cells [29]. Accordingly, implant compatibility may be augmented by seeding cells before implantation. Vozzi et al. [41] seeded human primary osteoblasts onto innovative collagen-gelatin-genipin (GP)-HA scaffolds and reported that in vitro attachment, proliferation, and colonization of human primary osteoblasts on collagen-GP-HA scaffolds with different percentages (10%, 20%, and 30%) of HA all increased over time in culture, but comparing different percentages of HA, they seem to increase with decreasing of HA component. Jung et al. [42] investigated the role of CMs when used in conjunction with bovine hydroxyapatite particles incorporated with collagen matrix (BHC) for lateral onlay grafts in dogs. It was found that (i) the collagen network of the membranes remained and served as a barrier and (ii) the quantity and quality of bone regeneration were all significantly greater in the membrane group than in the no-membrane group, indicating that the use of barrier membranes in lateral onlay grafts leads to superior new bone formation and bone quality compared with bone-graft alone. There are still several studies on combination of TCP and HA as a hybrid including HA/β-TCP [39, 43], HA/α-TCP [44], and HA/tetracalcium phosphate [45].

6.1.3 Metallic scaffolds

Compared to previous polymeric scaffold and ceramic scaffolds, metallic scaffolds exhibit a wider range of mechanical properties such as high strength, ductility, fracture toughness, hardness, formability, as well as corrosion resistance, high biotribology resistance, and biocompatibility, indicating that metallic scaffolds can be feasible for the most load-bearing applications in joint arthroplasty, bone replacement, and dental implant treatments [46]. However, there are many disadvantages of metallic biomaterials as bone scaffold [47]. They should include: (1) The main disadvantage of metallic biomaterials is their lack of biological recognition on the material surface. To overcome this restraint, surface coating or surface modification presents a way to preserve the mechanical properties of established biocompatible metals improving the surface biocompatibility. Moreover, in order to enhance communication between cells, facilitating their organization within the porous scaffold, it is desired to integrate cell-recognizable ligands and signaling growth factors on the surface of the scaffolds. Indeed, biofactors that influence cell proliferation, differentiation, migration, morphologies, and gene expression can be incorporated in the scaffold design and fabrication to enhance cell growth rate and direct cell functions; (2) Another limitation of the current metallic biomaterials is the possible release of toxic metallic ions and/or particles through corrosion or wear possible that lead to inflammatory cascades and allergic reactions, which reduce the biocompatibility and cause tissue loss. A proper treatment of the material surface may help to avoid this problem and create a direct bonding with the tissue; (3) The magnitude of elastic modulus for bulk metallic implant materials surpasses that of cortical bone by far and results in a failed stress transmission from biomaterial to bone, the so-called stress-shielding effect. The stress shielding may lead to bone resorption or even fretting due to micro-motions occurring at the bone/implant interface. The ideal porous metal coating would have an open-cell structure, high porosity, and microstructure resembling that of cancellous bone. Additionally, it would possess a similar modulus of elasticity and high frictional characteristics. Thus, it would be more biologically compatible and result in earlier and increased levels of bone ingrowth into the implant [46].

Several metallic scaffolds are used to provide support for bone defect regeneration such as titanium and its alloys [12, 48], iron and its alloy like stainless steel [49], or magnesium and its alloy [50, 51].

Fe and Fe-based alloy

Pure iron (Fe) possesses the modulus of elasticity of 211 GPa, which is higher than that of Mg (41 GPa) and 316 L stainless steel (190 GPa) [52]. However, inflammatory response and systemic toxicity have been observed with in vivo implantation of Fe stents in descending aorta of rabbits [49]. Peuster et al. [53] determined whether corrodible materials may be safely used as biodegradable cardiovascular implants. Corrodible iron stents (>99.8% iron) were produced from pure iron and laser cut with a stent design similar to a commercially available permanent stent (PUVA-AS16). A total of 16 NOR-I stents were implanted into the native descending aorta of 16 New Zealand white rabbits. It was obtained that (i) no thromboembolic complications and no adverse events occurred during the follow-up of 6–18 months and (ii) all stents were patent at repeat angiography after 6, 12, and 18 months with no significant neointimal proliferation, no pronounced inflammatory response, and no systemic toxicity; suggesting that degradable iron stents can be safely implanted without significant obstruction of the stented vessel caused by inflammation, neointimal proliferation, or thrombotic events [53]. Farack et al. [54] coated biocorrodible iron foams with different calcium phosphate phases to obtain a bioactive surface and controlled degradation. Further adhesion, proliferation, and differentiation of SaOs-2 (sarcoma osteogenic) and human mesenchymal stem cells were investigated under both static and dynamic culture conditions. It was found that (i) HA-coated foams released 500 µg/g iron per day for Dulbecco's modified eagle medium (DMEM) and 250 µg/g iron per day for McCoys, the unmodified reference 1,000 µg/g iron per day for DMEM and 500 µg/g iron per day for McCoys, while no corrosion could be detected on brushite-coated foams, and (ii) using a perfusion culture system with conditions closer to the in vivo situation, cells proliferated and differentiated on iron foams coated with either brushite or HA while in static cell culture cells could proliferate only on Fe-brushite. It was then concluded that the degradation behavior of biocorrodible pure iron foams can be varied by different calcium phosphate coatings, offering opportunities for design of novel bone implants [54].

Hermawan et al. [55] prepared Fe-35Mn (wt%) alloy which was intended to be used as a metallic degradable biomaterial for stent applications through a powder metallurgy route. The effects of processing conditions on the microstructure, mechanical properties, magnetic susceptibility, and corrosion behavior were investigated and the results were compared to those of the 316 L stainless steel, a gold standard for stent applications. Fe-35Mn alloy was ductile with a strength approaching that of wrought 316 L SS. It was reported that its corrosion rate was evaluated in a modified Hank's solution (containing no antibacterial/antifungal preservative, calcium chloride, magnesium chloride, phenol red, or sodium bicarbonate) and found superior to that of pure iron (slow in vivo degradation rate); concluding that the mechanical, magnetic, and corrosion characteristics of the Fe-35Mn alloy are considered suitable for further development of a new class of degradable metallic biomaterials. It has been recognized that selective laser melting (SLM) can produce complex hierarchical architectures paving the way for highly customizable biodegradable load-bearing bone scaffolds [55]. Carluccio et al. [56] prepared SLM-manufactured Fe-35Mn bone scaffolds suitable for load-bearing applications. It was reported that (i) the microstructure of the scaffold consisted primarily of γ-austenite, leading to high ductility, (ii) the mechanical properties of the scaffold were sufficient for load-bearing applications even after 28 days immersion in SBF, (iii) corrosion tests showed that the corrosion rate was much higher than bulk pure iron, attributed to a combination of the manufacturing method, the addition of Mn to the alloy, and the design of the scaffold, (iv) in vitro cell testing showed that the scaffold had good biocompatibility and viability toward mammalian cells, and (v) the presence of filopodia showed good osteoblast adhesion. The in vivo analysis showed successful bone integration with the scaffold, with new bone formation observed after 4 weeks of implantation; concluding that (vi) the SLM manufactured porous Fe-35Mn implants showed promise for biodegradable load-bearing bone scaffold applications.

Today, stainless (typically, 316 L: Fe-18Cr-10Ni-2Mo with low level of carbon content) is one of the most frequently used biomaterials for internal fixation devices because of a favorable combination of mechanical properties, corrosion resistance, and cost effectiveness when compared to other metallic implant materials. The biocompatibility of implant quality stainless steel has been proven by successful human for decades [49]. New nickel-free stainless steels have been recently developed primarily to address the issue of nickel sensitivity. These stainless steels also have superior mechanical properties and better corrosion resistance. The Ni-free compositions appear to possess an extraordinary combination of attributes for potential implant applications in the future. The area of biomaterials is a continuous research area. Among others, stainless steels are also used in medical applications such as fracture plates, wires, sutures, and implants. Surgical grade 316 L stainless steel is specifically used for medical application. It primarily contains nickel, chromium, and molybdenum wherein the purpose of nickel is to retain the FCC structure of stainless steel almost at any temperature unlike conventional steel which has ferrite (BCC structure) in ambient temperature. Although nickel imparts a polished and glossy finish to surgical grade stainless steel which is an important factor to keep it hygienic, the harmful effect of nickel on human body has been investigated and reported for quite some time now. The adverse effect of nickel (Ni) ions that release from stainless steel upon crevice and pitting corrosion has led to the evolution of Ni-free stainless steel with high nitrogen content [57]. There are newly developed Ni-free stainless steels, including Fe-18Cr-12Mn austenitic stainless steel [58], Fe-17Cr-3Mo-11Mn stainless steel [59], or surface modified by low-temperature nitriding treated Fe-18Cr-12Mn stainless steel [60]. All claimed excellent corrosion resistance and the newly developed high-nitrogen nickel-free stainless steel is a reliable substitute for the conventional medical stainless steels.

Tantalum (Ta)

Tantalum (Ta) is widely used in bone tissue engineering and knee replacement surgeries [60]. The similar elasticity of Ta to bone can decrease the imposed stress levels (in other words, reduced stress shielding). Bobyn et al. [61] investigated the characteristics of bone ingrowth of a new porous (75–80% by volume) tantalum biomaterial in a simple transcortical canine model using cylindrical implants. Histological studies were conducted on two types of material, one with a smaller pore size averaging 430 μm at 4, 16, and 52 weeks and the other with a larger pore size averaging 650 μm at 2, 3, 4, 16, and 52 weeks. It was reported that (i) the extent of filling of the pores of the tantalum material with new bone increased from 13% at 2 weeks to between 42% and 53% at 4 weeks, (ii) by 16 and 52 weeks the average extent of bone ingrowth ranged from 63% to 80%, (iii) the tissue response to the small and large pore sizes was similar, with regions of contact between bone and implant increasing with time and with evidence of Haversian remodelling (which is the coordinated activity of osteoblasts and osteoclasts to resorb and replace existing cortical bone) within the pores at later periods, and (iv) mechanical tests at 4 weeks indicated a minimum shear fixation strength of 18.5 MPa, substantially higher than has been obtained with other porous materials with less volumetric porosity, suggesting that (v) the porous tantalum biomaterial has desirable characteristics for bone ingrowth; further studies are warranted to ascertain its potential for clinical reconstructive orthopedics [61].

Although tantalum metal is recognized as a candidate for use as an implant material in high load-bearing bony defects, due to its attractive features such as high fracture toughness and high workability, Ta material does not have bone-bonding ability, that is, bioactivity, and therefore the development of bioactive tantalum metal is highly desirable [62]. Accordingly, Miyazaki et al. [62] evaluated the bonding strength of the apatite layer to the substrate in comparison with that to the untreated tantalum metal. It was found that (i) the apatite layer formed on the NaOH- and heat-treated tantalum metal shows higher adhesive strength than that formed on the untreated metal, (ii) the amorphous sodium tantalate layer formed on the tantalum metal by NaOH followed by heat treatments has a smooth graded structure where its concentration gradually changes from the surface into the interior metal, and (iii) smooth graded structure with complex of apatite is constructed after soaking in SBF, showing the higher bonding strength of the apatite layer formed on the treated metal [62]. Hacking et al. [63] determined the soft tissue attachment strength and extent of ingrowth to a porous tantalum biomaterial. Eight dorsal subcutaneous implants (in two dogs) were evaluated at 4, 8, and 16 weeks. Upon retrieval, all implants were surrounded completely by adherent soft tissue. Implants were harvested with a tissue flap on the cutaneous aspect and peel tested in a servo-hydraulic tensile test machine at a rate of 5 mm/min. Following testing, implants were dehydrated in a solution of basic fuschin, defatted, embedded in methylmethacrylate, and processed for thin-section histology. At 4, 8, and 16 weeks, the attachment strength to porous tantalum was 61, 71, and 89 g/mm respectively. It was also mentioned that (i) histologic analysis showed complete tissue ingrowth throughout the porous tantalum implant, (ii) blood vessels were visible at the interface of and within the porous tantalum material, and (iii) the tissue attachment strength to porous tantalum was three- to sixfold greater than was reported in a similar study with porous beads, demonstrating that porous tantalum permits rapid ingrowth of vascularized soft tissue and attains soft tissue attachment strengths greater than with porous beads. Porous tantalum trabecular metal has recently been incorporated in titanium dental implants as a new form of implant surface enhancement. Bencharit et al. [64] reviewed the concept on the porous tantalum metal in implant dentistry and described that (i) porous tantalum metal is used to improve the contact between osseous structure and dental implants and therefore presumably facilitate osseointegration, (ii) success of porous tantalum metal in orthopedic implants led to the incorporation of porous tantalum metal in the design of root-form endosseous titanium implants, (iii) the porous tantalum three-dimensional enhancement of titanium dental implant surface allows for combining bone ongrowth together with bone ingrowth or osseoincorporation, and (iv) while little is known about the biological aspect of the porous tantalum in the oral cavity, there seems to be several possible advantages of this implant design.

Cobalt-chromium alloys

The CoCr beads produced from spherical cobalt chrome metal are sintered to the implant substrate to create a macro- and microporous surface for bone ingrowth. The CoCr-beaded porous coatings are applied in acetabular cups, femoral stems, and total knee arthroplasty components [47]. Recently, SLM has been extensively employed to fabricate metallic scaffold devices. SLM is one of the metal additive manufacturing (AM) technologies that use a bed of powder with a source of heat to create metal parts.

Han et al. [65] evaluated the effects of unit cell topology on the compression properties of porous Co-Cr scaffolds fabricated by SLM technique; with four different topologies, that is, cubic close packed (CCP), face-centered cubic (FCC), body-centered cubic (BCC), and spherical hollow cubic (SHC), designed, and fabricated via SLM process. It was found that (i) the Mises stress predicted by finite element simulations showed that different unit cell topologies resulted in distinct stress distributions on the bearing struts of scaffolds, whereas the unit cell size directly determined the stress value, (ii) comparisons on the stress results for four topologies showed that the FCC unit cell has the minimum stress concentration due to its inclined bearing struts and horizontal arms, and (iii) simulations and experiments both indicated that the compression modulus and strengths of FCC, BCC, SHC, and CCP scaffolds with the same cell size presented in a descending order [65]. Over the last decade, advances in AM have allowed to obtain complex 3D porous lattice in materials suitable for orthopedic applications [62]. Caravaggi et al. [66] evaluated the feasibility of bone-replicating CoCr porous scaffolds manufactured via SLM and studied the effect of topography on bone cells viability and proliferation. Hence, small cylindrical porous lattices were modeled from micro-CT images of human trabecular bone and from the repetition of spherical-hollow and BCC unit cells and manufactured via SLM from CoCr powder. Macro- and micro-characterization of the porous samples were assessed using optical microscope, micro-CT, and SEM. It was mentioned that (i) osteoblast-like cells proliferation and viability were assessed in vitro, and compared to those cultured on a standard nonporous implant-to-bone interface, showing steady increase on all geometries over time, (ii) SEM analysis confirmed the quality of cells morphology, spread, and organization on all lattices, and (iii) the SLM process appeared not to alter the biocompatibility of CoCr; however, 15–100 μm irregularities and macroalterations were observed in the porous scaffolds with respect to the 3D nominal models [66].

Lu et al. [67] fabricated Co-^{29}Cr-9W-^{3}Cu porous scaffolds with different unit cell types including octahedron (OCT), FCC, and hexahedron (HCP) by SLM, aiming to reduce the stress shielding effect and promote the early bone ingrowth. It was reported that (i) the effect of pore size and unit cell types on the compressive stress and elastic modulus were investigated, and regulating pore size led to the compressive stress and elastic modulus ranging from 3.7 to 467 MPa and from 2.6 to 36.3 GPa, respectively, (ii) in vitro results demonstrated that the ability to support the osteogenesis and angiogenesis was for the most part similar between the FCC-65, OCT-65, and HCP-65 scaffolds (porosity of 65%), and (iii) after implantation in goats for 12 weeks, the FCC-65, OCT-65, and HCP-65 scaffolds showed sufficient results for guiding bone ingrowth, indicating good biocompatibility and osteointegration; interestingly, the scaffold with OCT unit cell structure offered the strongest ability to accelerate bone mineralization compared with the FCC-65, HCP-65 scaffold, possibly due to mechanical-adapted properties. It was suggested that the possibility of the SLM-produced Co^{29}Cr^{9}W^{3}Cu scaffold achieved osseointegration by guiding early bone ingrowth into the bone contact porous surface of the ankle joint [67].

Titanium and its alloys
Due to their unique properties (including high specific strength, low weight, excellent corrosion resistance, and biocompatibility), CpTi (commercially pure titanium) has been employed as dental implants. However, for bone replacement components, the strength of CpTi is not sufficient so that Ti alloys (such as Ti-6Al-4 V and Ti-6Al-7Nb) are preferred due to their superior mechanical properties [12, 68]. Titanium may be used for sintered beads and fiber metal mesh coatings. Plasma spray titanium coatings for total hip and knee arthroplasty have proven to be a safe, predictable material in long-term follow-up studies. However coating possesses a low porosity ranging from 30% to 50% which limits the maximum interfacial strength that can form via bone ingrowth. Autopsy retrieval studies have demonstrated an average ingrowth of 15–30%. These traditional materials do not have the characteristics that would permit their use as bulk structural materials for bone implants or as bone-graft substitutes. Similarly, these traditional implants require a significant proportion of host bone-

implant contact for a successful outcome [47]. The metallic foam has an overall porosity ranging from 60% to 80% with pore sizes from 100 to 600 μm and may be used in bulk form as well as a coating. These new porous metals show characteristics resembling those of cancellous bone with high surface friction properties, improved porosity levels, and relatively low modulus of elasticity; thus, potentially, providing a surface that will result in a long-lasting bond and substantial levels of bone ingrowth. Titanium-based foams are the extensively materials used to fabricate porous metal implant in primary and revision total hip, knee, and shoulder arthroplasty [47].

Ti foam scaffolds for dental implant applications are designed based on many interrelated factors including geometry, alloys, surface properties, and various pore characteristics. The stability of the implant-bone interface, strength and rate of osseointegration, and the long-term success of dental implants are achieved by appropriate design and proper material selection. Although there is no specific implant design that meets all the requirements in dental applications, implants can be engineered to maximize strength, interfacial stability, and load transfer [69]. Success rates of endosseous implants depend on the site of the implant, bruxism (grinding of the teeth) or other large occlusal forces, hygiene in the mouth, preoperative or postoperative infection, placement in the bone of inadequate quality or quantity, the clinical status of the patient, the skill and experience of the surgeon, biomechanical elements, and the type and size of implant material [69]. In fact, all of these factors interact and determine success or failure [70, 71]. Though efforts to improve the osseointegration properties of implants have inspired much of the literature around dental implants for the last 50 years, it becomes apparent that there are numerous known and possibly several unknown factors that need to be considered in the fabrication of dental implants in order to adequately address these needs. For instance, while osseointegration of implants in regions of poor bone quality (i.e., low bone density, in the case of highly cancellous bone, or low vascularity, in the case of primarily cortical bone) or insufficient quantity of bone (in terms of the width or height of the alveolar ridge) remains problematic, new design and/or material innovations need to be explored to further promote osseointegration of dental implants in these areas [72, 73]. The question is posed as to whether the Ti foam scaffold would potentially be a suitable candidate to address these limitations.

Balla et al. [74] attested feasibility of porous Ta structure as a potential bone implant and compared it with porous Ti control implants. Since Ti element is not biodegradable and does not integrate with biomolecules, surface modifications have been suggested to improve Ti bioactivity [75]. It can include surface oxidation to form TiO_2 (rutile type titanium dioxide) or titania (titanium dioxide) coating [76]. Moreover, a combination of Cr-Co alloys and stainless steel with titanium alloys can improve its biocompatibility [1]. Ti-6Al-4V or Ti-6Al-7Nb (V-free Ti-based alloy) possess better mechanical properties compared to CpTi and can be used in joint implants [12, 13]. Nontoxic alloys of beta titanium like Nb, Ta, and Zr are also offered [77]. Biocompatibility has been increased in the second generation of titanium alloys like Ti-15 Mo-^{5}Zr-^{3}Al,

Ti-^{15}Zr-^{4}Nb-^{2}Ta-$^{0.2}$Pd, Ti-^{12}Mo-^{6}Zr-^{2}Fe, and Ti-^{29}Nb-^{13}Ta-$^{4.6}$Zr [5, 78]. Porous titanium and its alloys can be used in permanent implants due to their good mechanical properties [79, 80]. Faria et al. [80] compared the potential of porous Ti sponge rods with synthetic HA for the healing of bone defects in a canine model and reported that (i) the Ti foam exhibited good biocompatibility, and its application resulted in improved maintenance of bone height compared with control sites and (ii) the Ti foam in a rod design exhibited bone ingrowth properties suitable for further exploration in other experimental situations. Although the use of Ni-Ti alloys has been banned in America and Europe due to allergic response and toxicity problems of Ni ions [81], nitinol (-^{50}Ni-^{50}Ti equiatomic weight) has shown high biocompatibility and significant plasticity for bone scaffolding [68, 82, 83].

Bystedt et al. [84] investigated titanium granules as bone substitute in patients planned for augmentation of the sinus floor prior to or in conjunction with placement of dental implants. Sixteen patients with uni- or bilateral edentulism and need for augmentation of the sinus floor were included. Grafting and installation of the dental implants (18 fixtures) were carried out in the same session if primary stability of the implants could be achieved (12 patients). It was reported that (i) the patients have been followed 12–36 months after prosthetic loading. Three implants were found mobile and were removed (13.0%), (ii) the implants were found mobile at abutment connection and were removed, and (iii) one patient in the single-stage group had a postoperative sinus infection, which was successfully treated with antibiotics; however, one out of two implants in this patient was found mobile and was removed after 1 year in function. It was, hence, concluded that (i) titanium granules seem to function as augmentation material in the sinus floor and (ii) it is, however, not clear if the material can be safely used for two-stage procedures. Although there is a paucity of literature regarding the clinical outcomes and result of porous titanium scaffolds, longer follow-up periods and a larger sample of patients are still required to obtain reliable clinical success rates [85]. Combining two or more materials such as ceramics and polymers [86–88], the structure and biochemical properties can be modified to achieve more favorable characteristics like biodegradability [8, 89].

Magnesium and its alloys

High-purity biodegradable magnesium (Mg) is believed to exhibit excellent biocompatibility and coating of Mg element is expected to show appropriate for implant applications and to improve the interaction between the implant and the biological environment [51, 90]. In recent years, research on magnesium alloys had increased significantly for hard tissue replacement and stent application due to their outstanding advantages because (i) Mg alloys have mechanical properties similar to bone which avoid the stress shielding phenomenon, (ii) they are biocompatible essential to the human metabolism as a factor for many enzymes, (iii) main degradation product Mg is an essential trace element for human enzymes, and (iv) the most important reason is they are perfectly biodegradable in the body fluid [90–92]. Mg-Y-Nd evaluated its feasibility as bone implant applications are mainly focused on biocompatibility and corrosion resistance [93]. It was concluded that the biodegradable Mg-Y-Nd implant is superior to the control Ti-6Al-4V with respect to both bone/implant interface strength and osseointegration. Mg-Zr-Sr alloys have recently been developed as biodegradable implant materials [94]. It was reported that (i) both Zr and Sr are evaluated excellent alloying elements in manufacturing biodegradable Mg alloy implants, (ii) Zr addition refined the grain size, improved the ductility, smoothed the grain boundaries, and enhanced the corrosion resistance of Mg alloys, (iii) Sr addition led to an increase in compressive strength, better in vitro biocompatibility, and significantly higher bone formation in vivo, and (iv) Mg–xZr–ySr alloys with x and $y \leq 5$ wt% would make excellent biodegradable implant materials for load-bearing applications [94]. Biodegradable Mg-Zn-Zr was also developed as an orthopedic implant material due to its biodegradable feature and suitable mechanical properties [94]. Amorphous Si film (which was prepared by plasma enhanced chemical vapor deposition of SiH_4) was coated on Mg-Y-Nd alloy for biomedical application [95]. The hemolysis test and blood platelets adhesion test were conducted, and it was reported that the hemolysis of Mg-Y-Nd alloy decreased after being coated by Si and the platelets attached on the Si film were at the inactivated stage with a round shape [96].

Mg and its alloys have a wide range of elongation (from 3% to 21.8%) and tensile strength (from 86.8 to 280 MPa). Its elastic modulus (41–45 GPa) is closer to that of the bone compared to other metals [51, 97], which leads to reducing a risk of the stress shielding to achieve the biomechanical compatibility [12].

Porous Mg has better degradation behavior (slower hydrogen evolution) and slower decrement of compressive yield strength in simulated body fluid (SBF) immersion tests [98]. Both porosity and pore size modifications can also adjust its stiffness and strength range to that of bone; however, higher porosity decreases the corrosion resistance of Mg. Cerium, neodymium, calcium, and praseodymium are used in orthopedic applications with Mg alloys [99, 100]. High corrosion and toxins of Mg have limited the application of this metal in medicine [101]. Early stages of in vivo biocompatibility studies of Mg scaffolds have recently been started [102]. Qin et al. [103] additively manufactured biodegradable Zn-xMg ((x = 1, 2, and 5 wt%) alloy porous scaffolds by laser powder bed fusion (L-PBF) and investigated the influence of Mg content on microstructure, mechanical properties, in vitro corrosion, cytocompatibility, in vivo degradation, biocompatibility, and osteogenic effect. It was reported that (i) the compressive strength and elastic modulus of Zn-1 Mg porous scaffolds reached the highest as 40.9 ± 0.4 MPa and 1.17 ± 0.11 GPa, respectively, equivalent to those of cancellous bone, (ii) the corrosion rate and cell viability slightly rose with increasing the Mg content, and (iii) histological analysis after 6- and 12-week implantation in rabbit femurs showed enhanced bone formation around the Zn-1 Mg porous scaffolds compared with pure Zn counterparts, indicating that (iv) Zn-1 Mg porous scaffolds produced by L-PBF presented promising results to fulfill customized requirements of biodegradable bone implants [103].

An ideal bone substituting material should be bone-mimicking in terms of mechanical properties, present a precisely controlled and fully interconnected porous structure, and degrade in the human body to allow for full regeneration of large bony defects. However, simultaneously satisfying all these three requirements has so far been highly challenging. In order to overcome these issues, Li et al. [104] introduced topologically ordered porous magnesium (WE43: Mg-4Y-3RE, where Y stans for yttrium and RE for rare earth elements) scaffolds based on the diamond unit cell that were fabricated by SLM and satisfy all the requirements which were subjected to test the in vitro biodegradation behavior (up to 4 weeks), mechanical properties, and biocompatibility. It was mentioned that AM of porous Mg may provide distinct possibilities to adjust biodegradation profile through topological design and open up unprecedented opportunities to develop multifunctional bone substituting materials that mimic bone properties and enable full regeneration of critical-size load-bearing bony defects [104].

Magnesium and its alloys are other metals that are used in bone tissue engineering. Bioresorbabililty, high biodegradability, suitable mechanical properties, noninflammatory responses, and bone cells activation support have been counted as its characteristics [50, 51, 102–108]. It is essential to control surface corrosion rate to control biodegradation of medical magnesium materials. For this end, there are several ways proposed including alloying or surface coating. Li et al. [99] fabricated binary Mg-xCa (x = 1–3 wt%) alloys with various Ca contents. It was shown that (i) the in vitro corrosion test in SBF indicated that the microstructure and working history of Mg-xCa alloys strongly affected their corrosion behaviors; an increasing content of Mg$_2$Ca phase led to a higher corrosion rate whereas hot rolling and hot extrusion could reduce it and (ii) the cytotoxicity evaluation using L-929 cells revealed that Mg-1Ca alloy did not induce toxicity to cells, and the viability of cells for Mg-1Ca alloy extraction medium was better than that of control, suggesting that (iii) Mg-1Ca alloy had the acceptable biocompatibility as a new kind of biodegradable implant material [99]. Shadanbaz and Dias [109] recognized the effect of coating calcium phosphate on magnesium materials to control rapid degradation. Wang et al. [110] fabricated calcium phosphate (Ca–P) coating on the surface of an AZ31 (Mg-3Al-1Zn) alloy by a chemical deposition process and conducted the in vitro and in vivo studies on a Ca–P-coated and -uncoated AZ31 alloy to determine the effect of Ca–P coating on the corrosion behavior and biocompatibility of the AZ31 alloy. It was mentioned that (i) the Ca–P coating reduced the in vitro and in vivo corrosion rates of the AZ31 alloy, (ii) cell experiments showed significantly good adherence and high proliferation on the Ca–P-coated AZ31 alloy than those on the uncoated AZ31 alloy, (ii) the blood cell aggregation tests showed that the Ca–P-coated AZ31 alloy had decreased the blood cell aggregation compared to the uncoated AZ31 alloy, and (iii) the animal experiments showed that the uncoated AZ31 alloy degraded more rapidly than the Ca–P-coated AZ31 alloy and the Ca–P coating provided significantly good biocompatibility, thus suggesting that the Ca–P coating not only slowed down the corrosion rate of the AZ31 alloy but also improved its biocompatibility. It was, therefore, concluded that the Ca–P-coated AZ31 alloy can be considered as a promising biomaterial for orthopedic applications [110].

6.1.4 Composite scaffolds

Recently, bioactive composite materials have been suggested to combine the advantages of two or more different materials (metallic, ceramic, and polymeric materials) [24]. Composite materials improve the scaffold properties and allow controlled degradation for tissue engineering applications [5, 111, 112]. Composites of main natural bone bioceramics including CP, HA, and TCP with poly(L-lactic acid), collagen, gelatin, and chitosan have been greatly used as scaffolding materials for bone repair studies [113–115]. Excellent mechanical properties and osteoconductivity have made polymer/ceramic composite as promising materials for bone tissue engineering [25, 116–118].

There are furthermore some composites proposed and tested [85]. Combining two or more materials such as ceramics and polymers was fabricated by solid free-form fabrication technology [86], controlled deposition method [87], or mechanical mixing polymer and ceramics [88], the structure and biochemical properties can be modified to achieve more favorable characteristics like biodegradability [8, 89]. For example, HA/PLGA composites possess the osteoconductive properties of hydroxyapatite and biodegradability of PLGA. Biodegradable polymer/bioceramic composites scaffold can overcome the limitation of conventional ceramic bone substitutes such as brittleness and difficulty in shaping. To better mimic the mineral component and the microstructure of natural bone, novel nano-hydroxyapatite (NHA)/polymer composite scaffolds with high porosity and well-controlled pore architectures as well as high exposure of the bioactive ceramics to the scaffold surface is developed for efficient bone tissue engineering [119]. Hence, Huang et al. [119] fabricated regular and highly interconnected porous poly(lactide-*co*-glycolide) (PLGA)/NHA scaffolds by thermally induced phase separation technique. It was reported that (i) pore size of the PLGA/NHA scaffolds decreases with the increase of PLGA concentration and NHA content, (ii) the introduction of NHA greatly increases the mechanical properties and water absorption ability which greatly increase with the increase of NHA content, and (iii) mesenchymal stem cells are seeded and cultured in three-dimensional (3D) PLGA/NHA scaffolds to fabricate in vitro tissue engineering bone, which is investigated by adhesion rate, cell morphology, cell numbers, and alkaline phosphatase assay, indicating that the PLGA/NHA scaffolds exhibit significantly higher cell growth, alkaline phosphatase activity than PLGA scaffolds, especially the PLGA/NHA scaffolds with 10 wt% NHA. Hence, it was concluded that the newly developed PLGA/NHA composite scaffolds may serve as an excellent 3D substrate for cell attachment and migration in bone tissue engineering [119]. Reichert et al. [120] demonstrated that polycaprolactone (PCL/TCP) scaffolds were combined with recombinant human bone morphogenetic protein (BMP-7)) to completely bridge a critical size of tibial defect in a sheep model. Mizuno [121] also showed that adult stem cells have been used to generate new tissue in combination with scaffold matrices. Currently tissue engineering has developed smart delivery system which act in a sequential manner (one agent appears while another disappears) to achieve sequential delivery of BMP-2 and BMP-7, where nanocapsule of

PLGA releases one of the growth factors and then coentrapped in chitosan fiber or PCL 3D-plotted scaffolds [122]. Coskun et al. [122] prepared poly(3-hydroxybutyrate), its copolymers with 3-hydroxyvalerate (HV) (PHBV8 and PHBV22), and their HA-containing composites (5 and 15%) by injection molding. It was mentioned that (i) smart systems can be used for controlled drug delivery using a group of polymers that physically or chemically respond to environmental stimuli such as light, temperature, or pH value and (ii) the mechanism of delivery suggests that upon decrease of temperature of the target site, swelling of nanospheres leads to the release of their content, maximizing delivery at the target site.

6.1.5 Design and fabrication methods

There are several unique and characteristic concepts for scaffold design and fabrication thereof. They should principally include customized (or personalized) bone scaffold design and structure with various degrees of porosity in surface zone. Oshida [12, 13] described that there are three major requirements for successful implant treatment; namely, they are (1) biochemical compatibility, (2) biomechanical compatibility, and (3) morphological compatibility. Later, Milovanovic et al. [123] mentioned that the personalized bone scaffold should be easily manufactured (manufacturability) and of course acceptable implantability, which should be confirmed and compatible with respect to biochemically reaction (simply, excellent biocorrosion resistance, although in some clinical cases, it should be time-controlled biodegradable to some extent), biomechanical compatible so that unwanted stress shielding can be controlled and minimized, and geometrically congruent to the patient's specific anatomy (aka, morphological compatibility). Figure 6.2 illustrates an interrelation among these parameters (which is similar to the figure presented by Milovanovic et al. [123]).

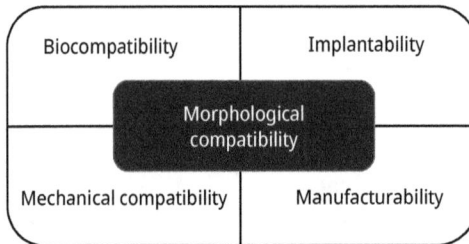

Figure 6.2: Interrelationship among five important parameters for acceptable scaffold design.

A variety of porous metallic scaffolds have been employed to support biological fixation of implants and to improve longevity of orthopedic implants. For the long-term success of cementless implants bone ingrowth is crucial around and with the porous surfaces [47]. Matassi et al. [47] listed several characteristics for ideal porous metallic structure:

(1) Biocompatibility: it should support normal cellular activity without any local and systemic toxic effects to the host tissue; it must be osteoconductive and osteoinductive and be able to induce blood vessels formation within or around the implant. Furthermore it should be nonimmunogenic [25].

(2) Mechanical properties: it should offer mechanical properties similar to the host bone with a sufficient mechanical strength. Bone responds to the absence and presence of physical load. In response to these loads, the body either resorbs or forms bone [124]. Given this principle, it is important to design a matrix that possesses mechanical properties that are similar to the tissue in the immediate surrounding area of the defect [125]. An over engineered matrix may result in bone resorption around the implant site, while an under engineered matrix may fail as a mechanical support to the skeleton.

(3) Pore size: scaffolds should have macro- (pore size > 100 mm) and microporosity (pore size < 20 mm) and pores must be interconnected. Multiscale porous scaffolds involving both micro and macroporosities can perform better than only one-dimensional porosity scaffold [126]. Unfortunately, porosity reduces mechanical properties such as compressive strength and resistance to corrosion [127].

(4) Physical properties: initial strength for safe handling during sterilizing, packaging, transportation to surgery, as well as survival through physical forces in vivo and sterile environment for cell seeding.

(5) The material should be reproducibly processable into three-dimensional structure and it must tolerate sterilization according to the required international standards for clinical use. In addition, the ideal porous metal would be able to stand alone as an independent structure rather than solely as a porous coating.

A major classification of porous metals, or metal foams, is between open-cell and closed-cell. In closed-cell foams each cell is completely enclosed by a thin wall or membrane of metal, whilst in open-cell foams, the individual cells are interconnected, allowing tissue to infiltrate the foam and anchor it into position [128]. Ryan et al. [128] classified porous implants into three distinct types: (i) partly or fully porous-coated solid substrates, (ii) fully porous materials, and (iii) porous metal segment joined to a solid metallic part. Porosity and pore size of biomaterial scaffolds play a critical role in bone formation in vitro and in vivo. Karageorgiou and Kaplan [129] reviewed the relationship between porosity and pore size of biomaterials used for bone regeneration. It was mentioned that (i) in vitro, lower porosity stimulates osteogenesis by suppressing cell proliferation and forcing cell aggregation; in contrast, in vivo, higher porosity and pore size result in greater bone ingrowth, and (ii) based on studies, the minimum requirement for pore size is considered to be approximately 100 μm due to cell size, migration requirements, and transport. However, pore sizes > 300 μm are recommended due to enhanced new bone formation and the formation of capillaries. Because of vascularization, pore size has been shown to affect the progression of osteogenesis. Small pores favored hypoxic conditions and induced osteochondral formation before osteogenesis, while large pores, that are well-vascularized, lead to direct osteogenesis (without preceding cartilage formation). It was also described that gradations in pore sizes are recommended on the formation of multiple tissues and tissue interfaces [129]. Porous implant structures involve porosity, pore size, etc., as major morphological design parameters [130]. The level of these factors has a significant impact on the mechanical and biological properties of the implants. Bone implant porosity and stiffness/strength require optimized trade-off to improve long-term load sharing while simultaneously promoting osseointegration [131].

After designing and optimizing the microstructure (in particular, pore size and porosity) of the scaffolds, the next step is to manufacture the designed structures. Due to the complication of internal structures and purpose of customized (or personalized) design of scaffolds, the AM technology has been widely employed to produce such scaffolds. Figure 6.3 depicts an artistic digital rendering 3D image of complex organic scaffolding of healthy bone tissue [132]. When fabricating synthetic bone-grafting material, special efforts are made to mimic the conditions of the native bone through the manipulation of variables such as porosity and surface topography.

Figure 6.3: Digital rendering 3D image of complex organic scaffold structure [132].

X-ray micro-computer tomography (µCT) is an excellent technique for obtaining 3D images of scaffold pore networks [133]. Figure 6.4 shows a µCT image of a bioactive glass (BAG) scaffold made by the sol-gel foaming process. The pore structure is very similar to trabecular bone [134].

A large variety of techniques have been used in the fabrication of three-dimensional scaffolds, sometimes in combination. In general, it is difficult to create complex scaffold micro-architectures with precise control using conventional techniques. As to conventional techniques, they should include solvent casting/particulate leaching, gas foaming, emulsification freeze-drying, phase separation, or electrospinning [135–137].

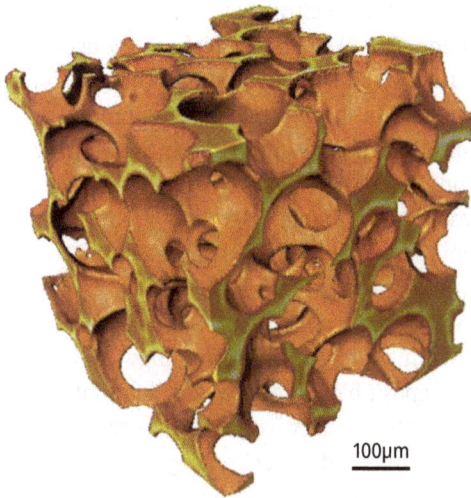

100μm

Figure 6.4: Reconstructed X-ray μCT image of a bioactive glass foam scaffold [134].

Additively manufacturing technology is an emerging technique, which enables the production of nonhomogeneous and irregular structures [123, 138–142]. Additively manufactured absorbable porous metals provide unparalleled opportunities to realize the challenging requirements for bone-mimetic implants. First, multiscale geometries of such implants can be customized to mimic the micro-architecture and mechanical properties of human bone. The interconnected porous structure additionally increases the surface area to facilitate adhesion and proliferation of bone cells. Finally, their absorption properties are tunable to maintain the structural integrity of the implant throughout the bone healing process, ensuring sufficient loadbearing when needed and full disintegration after their job is done. Such a combination of properties paves the way for complete bone regeneration and remodeling [143]. Among the various AM techniques (through mainly the 3D-printing method), the selective laser sintering (SLS), the SLM, the direct metal laser sintering (DMLS), the electron beam melting (EBM), the binder jetting (BJ), the fused deposition modeling (FDM), and 3D-bioprinting method have been successfully used to produce the porous bone implants [138–144]. The SLS is to use the laser beam to scan the powders in the powder bed according to the path specified by the computer and then bond and solidify the raw powder material on the working table. SLM is another widely used AM technology, in which the metal powder is heated to complete melting and formed in layer-by-layer fashion until the entire product is formed. Unlike the SLS, the SLM has higher laser energy and the powders can be completely melted during the processing without the binder. The EBM is an AM technology using an electron gun to generate an electron beam to melt the metal powders. Compared with the SLS and SLM techniques, the primary advantage of EBM is its high beam-material coupling efficiency, which makes it easy when processing metals with an extreme high melting point so that it has been utilized to produce the porous metal scaffolds. The BJ is the AM technology using the powder and liquid-binding agent. In the first step of this technique, the powder is bound through adhesion and chemical reaction. In the second step, the binder is removed, followed by the post-treatment process such as sintering, infiltration with a second material, and HIP for densification. The DMLS is an AM technique for metal 3D printing. In this process, the metal powder (20 µm diameter), free of binder or fluxing agent, is completely melted by the scanning of a high-power laser beam. The resulting part has properties like the original material. FDM is a material extrusion method of AM where materials are extruded through a nozzle and joined together to create 3D objects. Probably, the most promising current AM technology is FDM for two reasons: (1) it may be easily adapted to deposit biocompatible, biodegradable materials, and even the bone-graft mixture and (2) it is relatively cheap regarding any other AM technologies. The low accuracy of this kind of technology does not have great importance since the geometric and dimensional tolerances of the PBS and other orthopedic implants are not so tight as for the parts related to conventional mechanical engineering [145, 146]. Finally, as an emerging technology, 3D bioprinting offers a potential solution to help ease the burden of arthritis and other cause of bone defects within orthopedics. Bioprinting can be used to deposit living cells, extracellular matrices, and other biomaterials in user-defined patterns to build complex tissue constructs "from the bottom up." The potential to create inherent vascular structures is also improved by bioprinting, as internal channels containing vascular cells can be printed into constructs, fostering the ingrowth of blood vessels in vivo. By contrast, the conventional tissue engineering method of seeding cells onto a prefabricated scaffold does not allow for precise 3D placement of cells or biological content, limiting capacity to create complex hierarchical tissue constructs [135, 147]. Commonly used bioprinting techniques should include inkjet, laser-assisted, microvalve, and extrusion bioprinting [124].

6.2 Mesh – membrane

6.2.1 Membrane for GBR treatments

As bone defects heal, there is a competition between soft-tissue and bone-forming cells to invade the surgical site. In general, soft-tissue cells migrate at a much faster rate than bone-forming cells. Therefore, the primary goal of barrier membranes is to allow for selective cell repopulation and guide the proliferation of various tissues during the healing process [148]. Underneath the membrane, the regeneration process occurs, which involves angiogenesis and migration of osteogenic cells. The initial blood clot is replaced by woven bone after vascular ingrowth, which later is transformed into load-bearing lamellar bone. This ultimately supports hard- and soft-tissue regeneration [149].

If a barrier membrane is not utilized, lack of space maintenance will result in soft-tissue integration and compromised bone growth. Guided bone regeneration (GBR) is used to enhance bone growth of the alveolus for implant placement and around peri-implant defects [150–152]. The GBR characteristics are required to provide the maximum membrane function and mechanical support to the tissue during bone formation. In general, membranes are used in GBR procedures to act as biological and mechanical barriers against the invasion of cells not involved in bone formation, such as epithelial cells, and allow for the migration of the slower-migrating bone-forming cells into the defect sites [153–155]. Therefore, the primary goal of barrier membranes is to allow for selective cell repopulation and guide the proliferation of various tissues during the healing process. The membrane used for GBR is an essential component of the treatment and is a crucial and widely used component of implant dentistry. Membranes act as a biological and mechanical barrier against the invasion of cells not involved in bone formation, allowing for the migration of slower-migrating bone-forming cells into the defect sites [13]. The desirable characteristics of the membrane utilized for GBR therapy can include biocompatibility, cell-occlusion properties, integration by the host tissues, clinical manageability, space-making ability, and adequate mechanical and physical properties [148].

A barrier membrane should satisfy the following conditions: tissue adhesion without mobility, block soft tissue in-growth, easy to use, maintaining an appropriate space, and biocompatibility. Currently, barrier membranes are of two types: nonresorbable and resorbable [152]. The type of membrane used for GBR is based on several factors, including the morphology of the osseous defect, rigidity for space maintenance, barrier function, and handling characteristics. Another important consideration is the type of bone-grafting material used under the membrane [150, 155].

6.2.2 Required properties

There are several properties required as an ideal membrane material and structure [148, 156]. These should include:
(1) The membrane should have high mechanical properties (stiffness and plasticity) to protect the blood clot and resist passage of unwanted cells and bacteria. The amount of regenerated bone in the bone defect would be reduced if the membranes collapse into the defect space. Therefore, the ideal GBR membrane should be sufficiently rigid to withstand the compression of the overlying soft tissue. It should also possess a degree of plasticity in order to be easily contoured and mold to the shape of the defect. A balance between these mechanical properties is required to achieve an adequate space-making capacity.
(2) Ideally, the membrane should be biocompatible, resulting in no inflammation or interaction between the membrane and the host tissue to avoid wound dehiscence or infection.

(3) The membrane should have the ability to maintain space to allow for the bone regeneration process. Adequate membrane stiffness is paramount to maintain space and prevent defect collapse.

(4) Porosity is an important property of the GBR membrane. Studies have addressed the role of this property in the biological response in vivo using nonresorbable and resorbable membranes. The pore size of the membrane influences the degree of bone regeneration in the underlying secluded space.

(5) The membrane should provide stabilization of the blood clot, which allows the regeneration process to progress and reduce connective-tissue integration into the defect.

(6) The surface of the membrane should prevent fibrous tissue from invading the graft site, which directly correlates to the membrane porosity. A larger pore size may inhibit bone formation by allowing the overpopulation of faster-growing cells. When the pore size is too small, cell migration is limited for collagen deposition and the formation of avascular tissue results.

(7) The resorption time of the membrane should coincide with the regeneration rate of bone tissue, which is dependent on the location of the graft, vascularity, and the quantity of graft material.

(8) The membrane should be capable of size and shape alteration while maintaining adequate stiffness to prevent collapse of the graft site, leading to personalized shape forming [150, 156].

6.2.3 Nonresorbable membranes

There are two types of membranes: nonresorbable type and resorbable type. Barrier membranes composed of nonresorbable material is commonly used for relatively large-scale tissue regeneration owing to the ease of control of the shielding period [157, 158]. Since nonresorbable membranes need a second surgical intervention for membrane removal, subsequently, a second generation of membranes made of resorbable materials was developed and became widely used in different clinical situations. Furthermore, they have the advantage that degradation by-products of the base materials do not need to be considered [159]. However, these membranes require surgical removal after tissue regeneration. It has been reported that nonbiodegradable membranes showed higher risk of complications related to membrane exposure during implantation than biodegradable membranes [160, 161]. Nonresorbable membranes exhibit excellent biocompatibility, superior mechanical strength, and increased rigidity and generally achieve more favorable space maintenance than resorbable membranes. However, wound dehiscence is more common with nonresorbable membranes, and these membranes have the disadvantage of the need for a second surgery, resulting in increased morbidity, costs, and patient discomfort. The most common types of nonresorbable membranes include polytetrafluoroethylene (PTFE) and titanium (Ti) mesh [13]. They can also be used in combination with metal pins and mini screws to avoid the collapse of their morphology [162, 163].

It has been demonstrated, in various clinical and experimental studies, that excellent outcomes were reported using nonresorbable membranes in GTR and GBR treatments. However, several drawbacks and complications are recognized. Early exposure of barrier membranes to the oral environment and subsequent bacterial colonization can necessitate premature retrieval of the membranes [164]. Wound infection following the exposure of e-PTFE membranes can compromise the results of grafting [165]. Another major disadvantage of nonresorbable membranes is the need for a second surgery to remove the bio-inert membrane, causing discomfort and increased costs for the patients, as well as the risk of losing some of the regenerated bone, because flap elevation results in a certain amount of crestal bone resorption [166–168]. Last, due to the rigidity of the nonresorbable membranes, extra stabilization of the membrane with miniscrews and tacks is often required [158].

Polytetrafluoroethylene (PTFE)

As nonabsorbable membrane materials, there are polymers and metal. PTFE is an example of a material used in nonbiodegradable membranes. PTFE is a stable polymer in vivo and it is categorized as a bioinert material [169]. This chemical stability, which counts in favor of biocompatibility, allows PTFE to endure biodegradation and prevents host immune responses. PTFE has high barrier function between tissues; therefore it tends to reduce the blood supply resulting in dehiscence of the gingiva [170, 171].

PTFE membranes can be differentiated by expanded (e-PTFE), high-density (d-PTFE), and titanium-reinforced forms. The first reported synthetic polymer used for GBR was e-PTFE (expanded form of PTFE) and it is considered to be one of the most inert, stable polymers in the biological system. It resists breakdown by host tissues and does not elicit immunological reactions [172]. The first reported barrier membrane was made from expanded PTFE (e-PTFE) [173–175]. The expanded PTFE membrane (e-PTFE) was the first type of membrane used in implant dentistry and the e-PTFE membrane is advantageous as it prevents fibroblasts and connective-tissue cells from invading the bone defect and yet allows the osteogenic cells to repopulate the graft area [176]. The e-PTFE membranes have a high incidence of exposure, thereby resulting in an increased infection rate because of the ingrowth of bacteria into the highly porous structure. The e-PTFE membrane is sintered with pores between 5 and 20 μm in the structure of the material, allowing for soft-tissue ingrowth to lead to increased difficulty in removal [152]. The e-PTFE membrane acts as a mechanical hindrance. Fibroblasts and other connective-tissue cells are prevented from entering the bone defect so that the presumably slower-migrating cells with osteogenic potential are allowed to repopulate the defect [152]. The biologic principle of osteopromotion by exclusion has proved to be predictable for ridge enlargement or defect regeneration [177]. The e-PTFE membrane has been shown to produce bone predictably in localized bony defects around implants with or without bone-grafts [178]. Additionally the efficacy of e-PTFE barrier membranes to preserve and regenerate bone around implants placed in fresh extraction sockets were also validated in several other studies [179, 180].

It was notified that the e-PFTE exposed to the oral cavity resulted in migration of microorganisms through the highly porous membrane. The average pore size of 5–20 μm and the diameter of pathogenic bacteria generally less than 10 μm, migration of micro-organisms through the highly porous e-PTFE membrane at exposure is a common complication [152]. To address this problem, d-PTFE membrane with a nominal pore size of less than 0.3 μm was developed. The increased efficacy of d-PTFE membranes in guided tissue regeneration has been proven with animal and human studies [181]. Even when the membrane is exposed to the oral cavity, bacteria is excluded by the membrane while oxygen diffusion and transfusion of small molecules across the membrane is still possible. Thus, the d-PTFE membranes can result in good bone regeneration even after exposure [182, 183]. Because the larger pore size of e-PTFE membranes allows tight soft tissue attachment, it usually requires sharp dissection at membrane removal. On the contrary, removal of d-PTFE is simplified due to lack of tissue ingrowth into the surface structure [184].

PTFE only membranes can be used for treatment; however titanium-reinforced membranes are common owing to their effective space-making [159]. To increase the rigidity of e-PTFE and d-PTFE membranes, titanium was added to the PTFE membranes. Because of the increased structural rigidity, these membranes are easier to modify to conform to the defect. These types of membranes are especially useful in the treatment of large osseous defects. The e-PTFE membrane and d-PTFE membrane are also available as titanium-reinforced e-PTFE or d-PTFE. The embedded titanium framework allows the membrane to be shaped to fit a variety of defects without rebounding and provides additional stability in large, nonspace maintaining osseous defects [159].

Titanium

Both the pure metal (CpTi: commercially pure titanium) and alloys (such as Ti-6Al-4V or Ti-6Al-7Nb) possess good biocompatibility, mechanical strength, durability, low density, and corrosion resistance. In addition, titanium is a bioinert material that can be used as a stable metal owing to the rapid formation of a passive layer [12].

Guided bone regenerative membranes can help in treating moderate-to-severe osseous defects, but the inherent physical property of the membrane to collapse toward the defect due to the pressure of the overlying soft tissues (thus reducing the space required for regeneration) makes the overall amount of regenerated bone questionable [152]. The use of titanium mesh which can maintain the space can be a predictable and reliable treatment modality for regenerating and reconstructing a severely deficient alveolar ridge [185, 186]. The main advantages of the titanium mesh are that it maintains and preserves the space to be regenerated without collapsing and it is flexible and can be bent. It can be shaped and adapted so it can assist bone regeneration in nonspace-maintaining defects. Due to the presence of holes within the mesh, it does not interfere with the blood supply directly from the periosteum to the underlying tissues and bone-grafting material. It is also completely biocompatible to oral tissues [186, 187]. Titanium mesh can be used before placing dental implants (staged approach) to gain bone volume or in conjunction with dental implant placement (non-staged approach) [152].

The use of titanium mesh which can maintain the space can be a predictable and reliable treatment modality for regenerating and reconstructing a severely deficient alveolar ridge [186, 188, 189]. The main advantages of the titanium mesh are that it maintains and preserves the space to be regenerated without collapsing and it is flexible and can be bent. It can be shaped and adapted so it can assist bone regeneration in non-space-maintaining defects. Due to the presence of holes within the mesh, it does not interfere with the blood supply directly from the periosteum to the underlying tissues and bone-grafting material. It is also completely biocompatible to oral tissues [186, 189]. Titanium mesh can be used before placing dental implants (staged approach) to gain bone volume or in conjunction with dental implant placement (non-staged approach) [189]. Jovanovic et al. [190] developed titanium-reinforced e-PTFE membrane and reported that the reinforcement of e-PTFE membrane with titanium was able to maintain a large, protected space for blood clot stabilization without the addition of bone-grafts and provided superior preservation of the original form of the regenerated ridge during the healing period. Jaquiéry et al. [191] have used titanium meshes and autogenous bone-graft into 26 patients with small and midsize orbital defects and described that titanium meshes provided stability and can support the orbital content. For improving healing and bone regeneration, titanium mesh can be coupled with several membranes including CMs, platelet-rich fibrin (PRF), and polytetrafluoroethylene (PTFE). Azar et al. [153] reviewed to ascertain the effects of adding vertical bone to titanium mesh following GBR with a CM, PRF, and PTFE for dental implant treatment. It was found that 13 publications were found and chosen which discuss the usage of titanium mesh in conjunction with CMs, PRF, and PTFE as barrier membranes for GBR and concluded that the GBR technique, which combines titanium mesh with PTFE and CMs, can improve alveolar bone vertical addition, and the combination of titanium mesh with PRF can help the healing process move faster. GBR technique with titanium mesh was initially used to fix bone defects for maxillofacial bone reconstruction. Gradually, it is applied to bone reconstruction on alveolar bone defects [192]. Since sufficient bone volume is still considered the most important prerequisite for the osseointegration of dental implants, von Arx et al. [193] evaluated the TIME technique (autogenous bone-grafting combined with stabilization using a titanium mesh) for localized alveolar ridge augmentation in 20 patients who had insufficient bone volumes for the primary placement of dental implants and mentioned that (i) the stabilizing titanium mesh was best suited for vertical ridge augmentations and (ii) another feature of the mesh was the excellent tissue compatibility, with little clinical or histologic inflammation, even when the mesh had become exposed. Due to its superior mechanical strength and biocompatibility, titanium mesh can achieve a desirable effect in horizontal, vertical, and three-dimensional bone defects in bone augmentation surgery. Currently, there are various clinical procedures for bone augmentation with titanium mesh, which can be roughly divided into titanium mesh bone augmentation in simultaneous implantation, titanium mesh bone augmentation with delayed implantation, and GBR with titanium mesh in combination with other bone augmentation methods [192]. Customized biodegradable mesh membranes were developed for dental bone-grafts from magnetic resonance imaging or computed tomography using three-dimensional (3D) printing [194]. A biodegradable PCL polymer was used. Various designs of mesh membrane and fabrication procedure were studied. First, a mesh membrane was designed and customized by using a 3D model. Then, an inverse compression mold was made of an acrylic-based resin using LCD-based stereolithography-printing technique. Next, the PCL sheet was placed and compressed in the designed mold to create the customized membrane. The results show that the customized membrane prototype exhibited good biocompatibility, biodegradability, and mechanical properties; suggesting that the developed membrane customization and fabrication procedure could be practically useful for dental bone-graft application.

In recent years, the surface modification of titanium mesh has become a research trend to obtain better biological activity. Bioactive coatings for accelerating bone regeneration by improving the differentiation and proliferation of osteoblasts have been widely developed in tissue engineering [192]. Nguyen et al. [195] compared the effects of titanium mesh covered with calcium–phosphorus coating and untreated titanium mesh in GBR on the rat model and mentioned that (i) when compared to the untreated titanium mesh group, there is no soft tissue intervention under the titanium mesh of experimental group and (ii) the bone density in experimental group is significantly higher, which improves the structural durability of bone regeneration. GBR using a perforated titanium membrane is actively used in oral and orthopedic surgeries to provide space for the subsequent filling of a new bone in the case of bone defects and to achieve proper bone augmentation and reconstruction. The surface modification of a titanium membrane using a strontium-substituted calcium phosphate coating has become a popular trend to provide better bioactivity and biocompatibility on the membrane for improving the bone regeneration because strontium can not only stimulate the differentiation of osteoblasts but also inhibit the differentiation of osteoclasts. Based on this background, Nguyen et al. [196] formed the strontium-doped calcium phosphate coating on the titanium mesh by the cyclic precalcification method and evaluated its effects on bone regeneration by in vitro analysis of osteogenesis-related gene expression and in vivo evaluation of osteogenesis of the titanium mesh using the rat calvarial defect model. It was reported that the strontium-doped calcium phosphate-treated mesh showed a higher expression of all genes related to osteogenesis in the osteoblast cells and resulted in new bone formation with better osseointegration with the mesh in the rat calvarial defect in comparison with the results of untreated and calcium phosphate-treated meshes; indicating that proper coating treatment on the titanium mesh can form stable new bone without specific adverse reactions and shorten the time taken for GBR.

Studies have shown that the titanium mesh surface is covered with hydrocarbon molecules, which will gradually accumulate over time, causing the loss of titanium's hydrophilicity and damaging the bone conduction and osseointegration ability of titanium [192]. This process is called "titanium biological aging" [197]. However, ultraviolet (UV) treatment can change the surface properties of aged titanium: during the UV treatment, the hydrocarbon pollution on the surface of titanium can be removed [51, 192]. Meanwhile, the titanium mesh has obtained super-hydrophilic, and its surface charge is changed from negative to positive [198]. This UV treatment is defined as the titanium's photofunctionalization, which can enhance the biologic capabilities and stability of titanium mesh and increase the adhesion and retention of osteoblasts on the titanium mesh, and it is considered as a method to solve the biological aging problem. Due to the simple procedure of photofunctionalization technology to treat titanium surface and the consistently positive results in clinical studies of photofunctionalized titanium implants, clinical trials of photofunctionalized titanium mesh for bone augmentation are expected to be carried out soon [199]. Recently, Miyazaki et al. [200] investigated the effectiveness of hydrogen peroxide treatment and UV photofunctioning on early osseointegration of implant placement site which was also prepared by the piezosurgery. It was concluded that UV surface alteration and enough blood supply by piezosurgery preparation exhibited synergistic effects on improvement of implant stability quotient scales, indicating that these dual techniques appear applicable to implant treatments.

Titanium membranes provide effective shape and can be applied for vertical and severe horizontal bone loss in combination with bone substitutes [201, 202]. Recently, Hasegawa et al. [202] designed a regular hexagonal honeycomb structure with 1 mm of inner circle in the titanium membrane and placed autologous bone to the bony defects created in the Beagle dogs and covered with the prototype membrane. It was shown that (i) mature osseous tissue was formed after 26 weeks of implantation; however, microperforations might be a difficulty when retrieving the membrane at the second surgery and (ii) epithelial and connective tissue that invades the perforations will inevitably be removed with the membrane, potentially causing discomfort to the patient and prolonging the healing period. Otawa et al. [203] investigated to verify the modeling accuracy of various products and to produce custom-made devices for bone augmentation in individual patients requiring implantation. Two-(2D) and three-dimensional (3D) specimens and custom-made devices that were designed as membranes for GBR were produced using a computer-aided design (CAD) and rapid prototyping (RP) method. It was reported (i) the accuracy of the 2D and 3D specimens indicated precise results in various parameters, (ii) the error of overlapped images between the CAD and scanned data indicated that accuracy was sufficient for GBR, and (iii) in integrating area of all devices, the maximum and average error were 292 and 139 µm, respectively. It was, therefore, concluded that high modeling accuracy can be achieved in various products using the CAD/RP-SLM method, suggesting the possibility of clinical applications.

Co-Cr alloy
The use of cobalt and cobalt-chromium alloy in barrier membranes has been reported [204, 205]. Decco et al. [204] applied a cobalt-chromium membrane to a noncritical bone defect in rabbit tibia and reported that the tested membrane showed solid space-making and favorable bone augmentation. Cobalt-chromium alloy is a bioinert metal like titanium, but biotribological processes can lead to the release of toxic chromium and cobalt ions into the body [206]. Cobalt-chromium alloy has the advantages of mechanical strength and low cost compared with titanium; however it shows poorer biocompatibility than titanium. Therefore there have been no clinical trials using cobalt-chromium membrane implants in humans.

Furthermore, Co-Cr-Mo alloy has also been suggested for GBR [204]. Although this alloy is known to be less biocompatible than titanium and titanium alloy, it has superior mechanical properties (e.g., stiffness and toughness). The potential use of Co-Cr-Mo alloy for GBR has been evaluated in a recent animal study but it has not yet been documented in any clinical report. It was mentioned that placement of Co-Cr-Mo membrane on a rabbit tibial defect provides sufficient space and promotes bone regeneration [204]. Ceramic materials such as calcium sulfate [207, 208], hydroxyapatite [209–211], and beta-TCP [212, 213] have been incorporated in resorbable membranes.

6.2.4 Resorbable membranes

Currently there are two kinds of resorbable membranes: polymeric and collagen derived from different animal sources. The advantages of bioresorbable membranes should include the elimination of the need for membrane removal, greater cost-effectiveness, and decreased patient morbidity [142, 152]. Resorbable membranes exhibit the advantage of no second-stage surgery for removal, thus decreasing discomfort and morbidity to the patient.

Polymeric membranes
Polymeric membranes are valuable in preserving alveolar bone in extracting sockets and preventing alveolar ridge defects as well as ridge augmentation around exposed implants. Polymeric membranes are made up of synthetic polyesters, PGAs, polylactides (PLAs), or copolymers [214]. A clinical advantage of PGA, PLA, and their copolymers is their ability to be completely biodegraded to carbon dioxide and water via the Krebs cycle; thus they do not need to be removed at a second surgery [214]. Lekovic et al. [215] evaluated the clinical effectiveness of a bioabsorbable membrane made of glycolide and lactide polymers in preserving alveolar ridges following tooth extrac-

tion using a surgical technique based on the principles of GBR. Sixteen patients requiring extractions of two anterior teeth or bicuspids participated. Following the elevation of buccal and lingual full-thickness flaps and extraction of teeth, experimental
sites were covered with bioabsorbable membranes; control sites did not receive any
membrane. Titanium pins served as fixed reference points for measurements. Flaps
were advanced in order to achieve primary closure of the surgical wound. No membrane became exposed in the course of healing. Reentry surgeries were performed at
6 months. It was shown that experimental sites were presented with significantly less
loss of alveolar bone height, more internal socket bone fill, and less horizontal resorption of the alveolar bone ridge, suggesting that the treatment of extraction sockets
with membranes made of glycolide and lactide polymers is valuable in preserving alveolar bone in extraction sockets and preventing alveolar ridge defects. Simon et al.
[216] examined whether the amount of osseous structure 4 months postoperatively
after GBR was significantly less than the amount surgically created and if this change
was uniform over the area treated using polyglactide membrane over demineralized
freeze-dried bone allograft (DFDBA) for ridge preservation in 19 extraction sites of 10
patients. Results indicated that (i) loss in width of supplemented bone after 4 months
of healing ranged from 52.1% to 58.0% 3 mm from the crest, 47.6% to 67.4% 5 mm
from the crest, and 39.1% to 46.7% 10 mm from the crest and (ii) loss of augmented
height averaged 14. 7% in the center of the edentulous area but ranged from 60.5% to
76.3% 3 mm mesial and distal to the midpoint; indicating significant nonuniform loss
of augmented alveolar height and width during GBR healing, and the implications of
these findings impact preoperative augmentation planning for endosseous implantology. Evaluating the early reactions of the tissues to the insertion of polylactic membranes used in connection with titanium implants, Piattelli et al. [217] described that
the usage of biodegradable polymeric membranes has been associated with inflammatory reactions in the body. Simion et al. [218] evaluated the characteristics of resorption and the pattern of bacterial collonization of polyglycolic and polylactic
resorbable membranes under controlled experimental conditions. It was found that
(i) once exposed, PLA/PGA membranes started to resorb almost instantly and (ii) the
resorption process last for 3–4 weeks; resulting in spontaneous healing and closure of
the wound; whilst a degradation process that is too fast could reduce the barrier function time and the space-making ability of the membrane, which could negatively affect the outcome of bone regeneration.

Biodegradable membranes, which are almost exclusively polymer-based (natural
and synthetic polymers), have the advantages of few complications and low cost as
well as secondary surgeries not being necessary [219, 220]. Therefore, biodegradable
membranes are regarded as a first-choice material when the treatment outcome is expected to be the same as that using nonbiodegradable materials [158]. However, biodegradable membranes are liable to show tissue regeneration failure due to the
volume loss of the membrane and its degradation by-products [221, 222]. For animal-
derived CMs, residual virus and cross-linker also present concerns [221, 223]. In gen-

eral, biodegradable membranes show lower mechanical strength and are therefore less efficient at space-making than nonbiodegradable PTFE and titanium mesh. In GBR process, bone substitutes are used in combination with biodegradable membranes to maintain the membrane shape and lead to space-making [158]. As to biodegradable synthetic polymers, biodegradable aliphatic polyesters – such as polylactic acid (PLA), polyglycolic acid (PGA), PCL, and their copolymers (e.g. PLGA and poly(lactide-*co*-caprolactone)) – are also used as barrier membranes [224–230].

Collagen membranes
Most of the commercially available CMs are developed from type I collagen or a combination of type I and type III collagen. The source of collagen comes from tendon, dermis, skin, or pericardium of bovine, porcine, or human origin [231]. Liu et al. [152] listed several advantages of collagen materials for use: a barrier membrane, including hemostasis [232], chemotaxis for periodontal ligament fibroblasts [233] and gingival fibroblasts [234], weak immunogenicity [235], easy manipulation and adaption, a direct effect on bone formation [236], and ability to augment tissue thickness [237]. Hence, collagen material appears to be an ideal choice for a bioresorbable GTR or GBR barrier [152].

CMs have been widely utilized in bone regeneration procedures. Chung et al. [238] investigated a bioresorbable type I CM as a barrier for GBR. Ten human subjects with at least one pair of contralateral periodontal lesions with probing pocket depths of greater than or equal to 5 mm and radiographic evidence of greater than or equal to 40% bone loss were included. Each patient underwent contralateral surgical flap procedures. A collagen barrier was adapted to the tooth in the experimental defect and the flap replaced and sutured. The controls consisted of the same procedure without the placement of the barrier. It was found that (i) standardized measurements of change in probing attachment levels and fill of intrabony defects were obtained at the time of surgery, and 1 year later at the time of surgical re-entry, (ii) the mean probing attachment gain in the test sites was 0.56 ± 0.57 mm, and there was a mean probing attachment loss of 0.71 ± 0.91 mm in the control sites, and (iii) the gain of bone in test lesions was 1.16 ± 0.95 mm, while no gain was observed in the control lesions (*P* less than 0.01), demonstrating that (iv) sites treated with a collagen barrier comprised cross-linked bovine Type I collagen exhibited significantly better healing as compared to control sites over the 1-year period of the study. Blumenthal and Steinberg [239] evaluated the clinical efficacy of a combined graft of autolyzed antigen-extracted allogeneic (AAA) bone and microfibrillar collagen covered with a resorbable CM in human infrabony defects to compare at 1 year with debrided controls, AAA bone-grafts alone, combined AAA bone-collagen grafts (without membrane), and debrided defects covered only with CMs. It was reported that (i) all treatment modalities showed improvement over the debrided controls, (ii) similar advantages to using bone-collagen grafts with and without membranes were found in reducing probing depths and gaining new attachment, (iii) significant differences were found when comparing the multifaceted bone-graft collagen-membrane technique to all others in achieving superior defect fill, and (iv) 93% of all defects treated resulted in 50% or greater fill. CMs can also be used for regeneration in periodontal furcation defects [240–242]. Chitosan [243–246] and alginate [247–249] have also been introduced for use as barrier membranes in the study phase.

Compared to (reinforced) nonresorbable barrier membranes, both collagen and synthetic polyester membranes lack space-making ability. These membranes are often used with tenting or supporting materials (different bone-grafts or bone fillers) to prevent space collapse [153]. When grafting materials are used with bioresorbable membranes, the results of GBR procedures are generally favorable and even comparable to the results achieved with nonresorbable barriers [250–252]. Grafting material alone seems to be less effective than the combination of a supporting material and a barrier [250].

GBR is widely employed to overcome insufficient bone quantity and anatomical problems. During the procedure, the need of a membrane has been a question to many clinicians. The use of a membrane in GBR is thought to be advantageous, achieving mechanical stabilization and preventing micromovement of the bone-graft material. Although resorbable membranes are widely used as they do not need to be removed after placement, it is known that removal is nearly impossible when the membrane is only partially resorbed and that there can be giant cell reactions during the process of resorption. Moreover, the early exposure of the membrane necessitates frequent postoperative observation followed by potential additional fees. In contrast, there are reports showing no significant difference in bone regeneration when groups of with and without membranes were compared with intact periosteum. The use of membrane in preventing soft tissue ingrowth during GBR procedure for better clinical results is controversial. Based on the background, Lee and Kim [253] compared and analyzed the clinical results of GBR using the autogenous tooth bone-graft (AutoBT; Korea Tooth Bank Co., Seoul, Korea) material with and without the resorbable membrane (Bio-Arm, ACE Surgical Supply Company, Inc., USA). It was obtained that there was no statistically significant difference in pre- and postoperative reduction of bone defect height, bone level change, and bone regeneration in percentage between the two groups and concluded that (i) both groups showed clinically acceptable bone regeneration without any eventful complications and (ii) the use of resorbable membrane is not a critical factor in GBR when using AutoBT.

6.3 Bioresorption and biodegradation

6.3.1 Terminology and characteristics

There are several terms related to similar phenomenon; namely, they are biocorrosion, biodegradation, and bioresorption [13, 51], although no clear distinction among these can be found.

As Liu and Zheng [254] pointed out that the term "biodegradable" has been not only used in biomaterials but also in ecology waste management and even natural environment. Meanwhile, the term "resorbable" has long been used in biological reaction (osteoclast-driven bone resorption), but is inappropriate for implants that do not carry the potential to grow back into their original form. The term "absorbable" focuses more on the host metabolism to the foreign biodegradation products of the implanted material/device compared with the term "degradable/biodegradable." Meanwhile the coherence and normalization of the term "absorbable" carried by its own in laws and standards contribute as well. Hence, there is still confusion and misuse. Even with this situation, there are somewhat widely accepted definition and terminology. Biodegradation can be explained in the way that it is degraded into macromolecules, but they can stay in the body and can migrate; for example, ultra-high molecular weight polyethylene is generated from joint orthopedic prosthesis due to an action of biotribology [254]. Merits of biodegradable bone plates should include (1) they do not require a second surgery for removal, (2) they can avoid unwanted stress shielding, and (iii) they can achieve controlled drug delivery function. Bioresorption should be an entire bulk degraded in vivo, and it is eliminated from the body into low-molecular weight molecules such as that PLAs is eliminated faster. Metals are classified as bioresorbable as biocorrosion and/or biotribology leads to the total elimination of the material, but the time required is much longer than other resorbable polymers. Resorption of a material causes (i) water sorption, (ii) reduction of mechanical properties, (iii) reduction of molar mass, and (iv) complete loss of weight. Bioresorption can be initiated by (1) solubilization (e.g., dextran, polyvinyl alcohol, and polyethylene oxide), (2) inonization followed by solubilization (e.g., polyacrylic acid and polyvinyl acetate), (3) enzymatically catalyzed hydrolysis (e.g., polysaccharides and polyamides), and (4) simple hydrolysis (e.g., aliphatic polyesters) [255].

Without clear differentiation between biodegradation and bioresorption, both terms are utilized in scientific report [256, 257]. For example, it was mentioned that biodegradation or bioresorption of calcium phosphate materials implies cell-mediated degradation in vitro or in vivo environment. Cellular activity during biodegradation or bioresorption occurs in acid media; thus the factors affecting the solubility or the extent of dissolution (which, in turn, depends on the physicochemical properties) of the Ca-P materials are important. Enrichment of the microenvironment due to the release of calcium and phosphate ions from the dissolving Ca-P materials affects the proliferation and activities of the cells. The increase in the concentrations of calcium and phosphate ions promotes the formation of carbonate apatite which is similar to the bone apatite [257].

The human body is a composite structure, completely constructed of biodegradable materials. This allows the cells of the body to remove and replace old or defective tissue with new material.

Consequently, artificial resorbable biomaterials have been developed for application in regenerative medicine [258]. Moreover, there is a clear statement on these terms [259, 260]. For example, Forrestal et al. [259] clearly defined the term "bioresorbability" by stating that metallic drug-eluting stents have led to significant improvements in clinical outcomes but are inherently limited by their caging of the vessel wall. Fully bioresorbable scaffolds (BRS) have emerged in an effort to overcome these limitations, allowing a "leave nothing behind" approach. Although theoretically appealing, the initial experience with BRS technology was limited by increased rates of scaffold thrombosis compared with contemporary stents.

6.3.2 Rate and its control

The material for biodegradable device (such as biostent) is requested to have at least the following characteristics: (1) it must be biocompatible, (2) degradation rate must be controlled, (3) degradation products of the material must also be biocompatible, (4) the material must stay in the place for several months before its complete bioabsorption, and (5) the radial force of the resultant stent must be enough for scaffolding effect during the requested period [261, 262]. Based on these requirements, two metallic elements including iron and magnesium have been utilized along with plastics. Accordingly, implantable medical and dental devices (mesh, membrane, scaffold, and others) should experience two competing processes, that is, degradation process and deterioration of mechanical properties and these form an ideal compromise [51]. Referring to Figure 6.5, part A illustrates the ideal compromise between mechanics and corrosion of absorbable metals for coronary stent application. Corrosion rate stays low during the first 6–8 months, while mechanical integrity stays high to allow vessel remodelling [257]. Similar illustration is valid for absorbable bone implant, but the mechanical integrity should remain high for the first 3–6 months to allow bone repair process that takes place [263, 264]. Part B shows relationship between corrosion rate-dependent mechanical support (of an absorbable implant [265] and projected mechanical support by bone regeneration). Dependent on the defect site and size, mechanical implant integrity should remain high enough for about 3–6 months to ensure initial mechanical support prior to implant failure by progressive corrosion.

Times for complete in vivo degradation of current biodegradable stents are mapped at the bottom. Taking an example case when the stent is used for arterial vessel, it begins with a very slow degradation rate to keep the optimal mechanical integrity during arterial vessel remodelling. This can be achieved, for example, by coating the stent with a degradable polymer. Thereafter, the degradation progresses while the mechanical integrity decreases. The degradation ideally occurs at a sufficient rate that will not cause an intolerable accumulation of degradation product around the implantation site. Mg alloy stent degrades completely before the remodelling period; conversely, polymeric stents continue their presence thereafter. The degradation of Fe stents, which completes after the remodelling period but is not as long as that of polymeric stents, could be considered as approaching the ideal degradation time [266]. Hence, depending on the anticipated healing period, the stent material type should be accordingly selected.

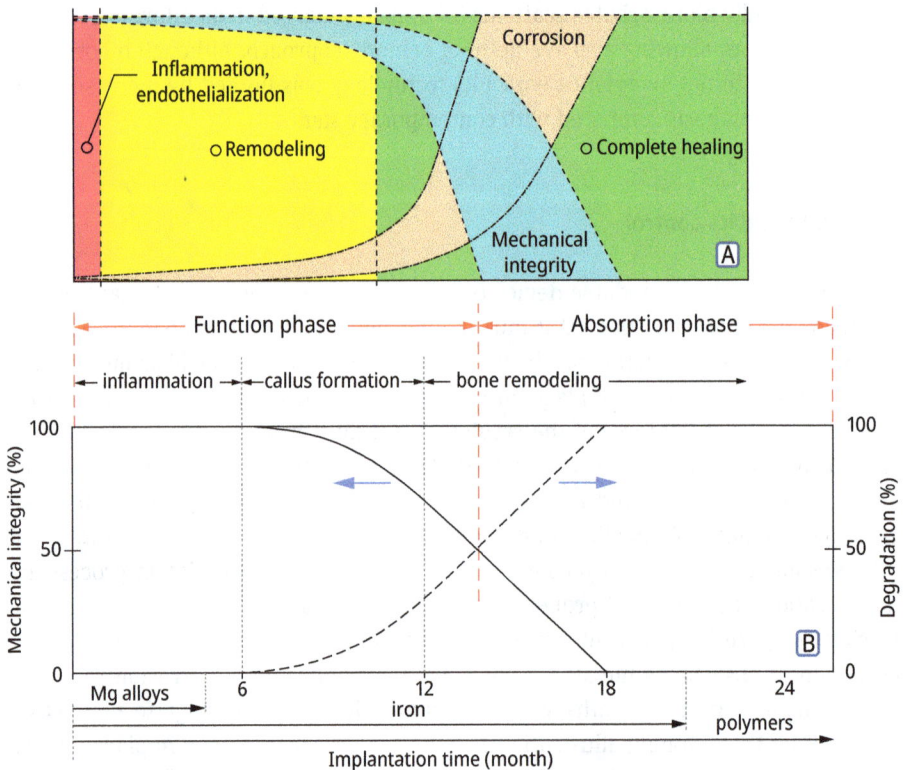

Figure 6.5: The ideal compromise between degradation and mechanical integrity during in vivo implantation of a biodegradable stent [51, 263–265].

Uncontrolled degradation of biomaterials could result in loss of their mechanical integrity, metal contamination in the body, and intolerable hydrogen evolution by tissue. As promising applicability of biodegradable Mg materials, it is crucial to understand degradation mechanism in biological environment and to manage its rate when Mg materials are chosen for orthopedic stent or implants or scaffold frameworks for tissue engineering.

A majority of orthopedic implants for repairing fractured bone or joint is normally made of metallic materials including austenitic stainless steels, Co-Cr-Mo alloy, commercially pure Ti and Ti-6Al-4V or Ti-6Al-7Nb alloys, due to their biocompatibility and adequate mechanical behavior. These are called (semi-)permanent implants. Usually, once the patient has recovered from a traumatic injury was completely healed, a revision surgery is necessary to remove the implant from the body with careful operation not to cause unwanted problems associated with osteopenia, inflammation of adjacent tissues, or sarcoma. Alternatively, to avoid such postextraction of the implant, intensive efforts are being made in recent years to develop new classes of so-called biodegradable implants, which are composed of nontoxic materials that become reabsorbed by the human body after a reasonable period of time. Conventionally these implants are polymeric materials. However, there are drawbacks associated with polymeric implants; (1) they are often rather costly, (2) exhibit relatively low mechanical strength, so biomechanical compatibility will not be able to establish, and (3) polymers can also react with human tissues, leading to osteolysis, particularly unpolymerized monomer would be toxic to surrounding tissues [267, 268]. For these reasons, it is highly desirable to develop cost-effective biodegradable metallic alloys, with better mechanical performance than polymers. Although Fe-based and Mg-based alloys are materials' choice since they exhibit relatively fast biodegradation and strong enough of biomechanical performance, Mg alloys are preferred because their stiffness (i.e., Young's modulus) is closer to that of human bone, leading to established biomechanical compatibility [269].

However semipermanent metallic implants (such as Ti-based alloys, Co-Cr alloys, and stainless steels) are removed after the completion of the healing process to avoid diverse side effects. Long-term disadvantages of this practice include the failure to adapt to rapid growth in young children, bone degradation by stress shielding, microbial implant infections, excessive fibrosis, or persistent inflammation [270]. Novel bioresorbable metal implants (such as iron, Zn-based alloys, and Mg-based alloys) could provide support during the healing process and then disappear to avoid long-term side effects without requiring surgical removal [271, 272]. Degradation of Mg materials in vivo environment is compared with other biodegradable materials as follows; it is 0.16 mm/year for pure annealed Fe, 0.2 mm/year for as cast pure Zn, and 407 mm/year for pure as cast pure Mg [273]. As Sanchez et al. [274] pointed out, there are several factors influencing degradation rate, including alloy factors (type of material, grain size, alloying element and purity, other metallurgical parameters), in vivo factors (tissue pH value, vascularization of peri-implant zone, chlorine ion concentration, and others), and in vitro factors (solution pH and temperature, test methods, and others). Since iron and zinc material exhibit slow degradation rate, it would result in (1) complete degradation may require several years or (2) lifetime by far exceeds expected healing periods [275–277]. On the other hand, due to very rapid degradation rate (in other words, very high corrosion rate) of Mg-based alloys, there is a risk of mechanical failure before the healing process is completed [103, 278, 279]. Being metallic materials, Mg and Mg alloys made for scaffolds provide the necessary mechanical support for tissue healing and cell growth in the early stage, while natural degradation and reabsorption by surrounding tissues in the later stage make an unnecessarily follow-up removal surgery. However, uncontrolled degradation may collapse the scaffolds resulting in premature implant failure, and there has been much research in controlling the degradation rates of Mg alloys [280]. To this end, there are mainly two ways: (1) alloying and (2) surface modification.

Degradation rate, controlled by alloying effects

Degradation behavior of Mg-based alloys is affected by various factors including aqueous corrosive environment and temperature, composition, micro- and macrostructure, surface structure, alloying elements, impurities, heat treatment, and secondary phase evolution and manufacturing method and its resultant altered microstructures [281, 282]. Since biodegradable implants become reabsorbed by the human body after a certain period of time, they should be composed of biocompatible alloying elements so that some alloying element (if alloyed in Mg-based alloys) could exhibit a potential cytotoxicity. For example, elements such as Ni, Cr, and V are not suitable to be in contact with human tissues since these three are well known as heavy toxic metallic elements. Their substitution by nontoxic elements such as Zn and Ca has permitted the fabrication of biocompatible Mg-based alloys with potential use as biomaterials. Alloying element concentrations, the corrosion behavior of Mg phase can be tuned as a function of the concentration of elements in solid solution. Depending on the nature and distribution of these elements within the matrix phase, the occurrence of microgalvanic cells can be either mitigated or favored. For example, an Al-containing Mg matrix phase becomes more passive as the Al content increases and consequently the corrosion rate decreases. On the contrary, in Zr-containing Al-free Mg alloys, the central areas of the grains (which are enriched in Zr) do not corrode while the grain boundaries severely corrode [267, 268]. Agarwal et al. [283] found that inclusion of alloying elements such as Al, Mn, Ca, Zn, and rare earth elements provides improved corrosion resistance (in other words slowing the degradation rate) of Mg-based alloys. Figure 6.6 illustrates various alloying effects of metallic elements in terms of decreasing (beneficial) corrosion current density or increasing (adverse) corrosion current density and increasing (more noble) corrosion potential or decreasing (more less noble) corrosion potential, proving a guideline for appropriate selection of alloying element(s) for Mg-based material [280].

Degradation rate, controlled by surface modification

The crucial problem associated with Mg-based alloys is high corrosion (or degradation) rates in physiological conditions, which makes their biodegradability to be faster than the time required to heal the bone. Hence, it is important to control the degradation rate and to keep their mechanical integrity until the bone heals. Another drawback of magnesium and its alloys is that corrosion is accompanied by intense hydrogen evolution. This gas can be accumulated in pockets next to the implants or can form subcuta-

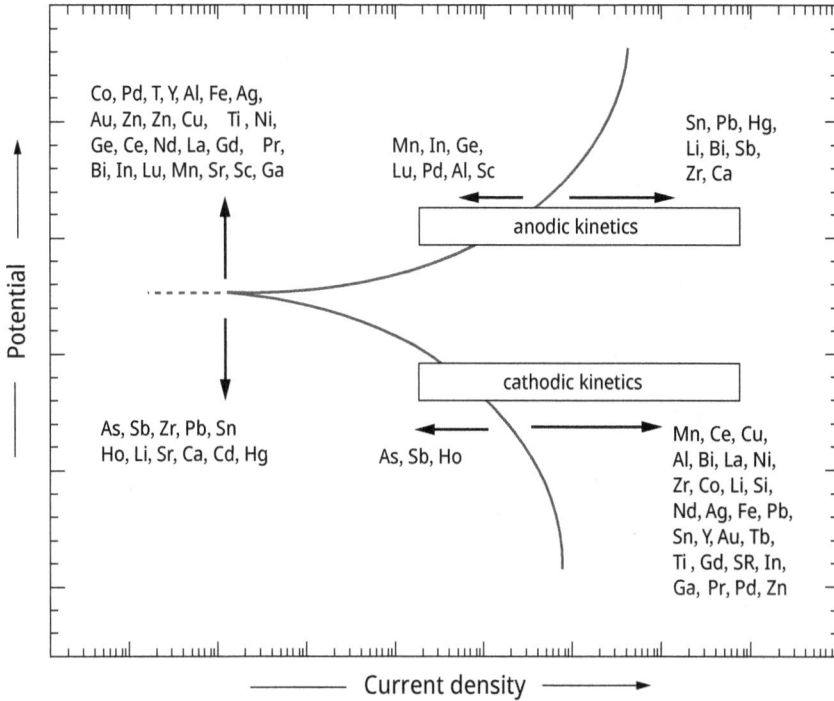

Figure 6.6: Effects of alloying elements to control corrosion behaviors [51, 280].

neous gas bubbles [283]. Accordingly, a well-balanced property between controlled rate of biodegradation and maintained mechanical integrity during biodegradation becomes extremely important for biodegradable-graded Mg-based alloys. There are two methods principally employed for controlling surface degradation rate: apatite coating and/or surface oxidation treatment. It is believed that preparing stabilized apatite on biodegradable Mg alloy may improve biocompatibility and promote osteointegration and such apatite includes Ca-P coatings, brushite ($CaHPO_4 \cdot 2H_2O$), hydroxyapatite ($Ca_{10}(PO_4)_6$ $(OH)_2$), and fluoridated hydroxyapatite ($Ca_5(PO_4)_3(OH)_{1-x}F_x$) [284]. If surface layer is subjected to oxidation to form dense and protective oxide film, the degradation rate (in other words, corrosion rate) can be suppressed and such treatment can include plasma electrolytic oxidation [285] and microarc oxidation [286].

6.3.3 Biodegradable materials

Biodegradable materials belong to the second generation of biomaterials, which have been closely related to bone defect repair for nearly half a century [287]. Biodegradable materials are widely used in bone tissue engineering because of their biodegradability. As the graft degrades, bone tissue grows into the graft's interior, and the small biomolecules produced by the degradation can regulate the regenerative microenvironment to adapt to the growth of bone tissue [288]. At the same time, the mechanical properties of the graft gradually decrease, and the biological stress of the body moves from the graft to the new bone tissue, which avoids the stress-shielding effect while stimulating tissue regeneration [289]. Therefore, the degradable biomaterial avoids the injury and related economic burden caused by a second operation. According to the current research status, biodegradable materials are mainly composed of biodegradable polymers, biodegradable ceramics, and biodegradable magnesium-based materials.

Biodegradable polymers
Depending on their source, polymers can be classified as natural or synthetic. Natural biodegradable polymers, such as chitosan, silk fibroin, fibrinogen, collagen, and hyaluronic acid, have been extensively studied as bone defect repair materials due to their biodegradability, bioactivity, and biocompatibility [288]. There are several drawbacks associated with these polymeric materials, including high water solubility, poor mechanical properties, and possible denaturation during processing and possible immunogenicity [290]. When some synthetic polymers are degraded in vivo, their degradation products are acidic and thus change the local pH value, which in turn accelerates the implant degradation rate and induces inflammatory reactions [291].

Natural biodegradable polymers which are extensively used can include collagen (playing an important role in regulating the extracellular matrix of the cellular microenvironment [292]), chitosan (enhances cell adhesion, proliferation, osteoblast differentiation and mineralization, due to its biological activity, biodegradability, antibacterial and biocompatibility, and hydrophilic surface [293]), fibrin (forming in the last step of the coagulation cascade by thrombin acting on fibrinogen [294]), and silk fibroin (produced by silkworms, spiders, and some insects to form silk fiber [295]).

Synthetic biodegradable polymers should include PCL (despite its biodegradability and biocompatibility, after a large number of long-term experiments, researchers found that the degradation rate of PCL was slow and the mechanical properties were poor, so it proved not to be an ideal bone defect repair material [296]); PGA: due to its excellent tensile modulus and controlled solubility, PGA has been used as the first biodegradable suture in clinical practice for many years [297]); PLA: PLA is a biodegradable polymer made from starch sourced from renewable plant resources (such as sugar cane and corn [271], and at present, L-PLA and D,L-PLA (mixture of L-and D-lactic acid) are the most widely used PLA in clinical [298]); PLGA: PLGA is a widely used biodegradable polymer that has the advantages of safety, biocompatibility, noncytotoxicity, ideal mechanical properties, and controllable degradation [271]); and poly (*para*-dioxanone) (PDS): It is a biodegradable polyester obtained by the ring-opening polymerization of a hydroquinone monomer and with excellent biodegradability and biocompatibility, PDS is very popular in tissue engineering and fracture repair [299]); and HYAFF-11 (hyaluronan-based biodegradable polymer; it can be commonly found in the extracellular matrix, and due to its good biocompatibility and a degradation rate that can be controlled by the esterification degree; HYAFF-11 is a very promising material for tissue repair [300]).

Biodegradable ceramics

They should include HA (hydroxyapatite: $Ca_{10}(PO_4)_6(OH)_2$, is a widely used bioactive and biodegradable calcium phosphate that accounts for almost 65% of the total bone mass and constitutes most of the inorganic components of bone tissue [271]); TCP, Ca3 $(PO_4)_2$, is a common absorbable bioactive ceramic material with a calcium/phosphorus ratio of 1.5. TCP has three crystalline forms: α-TCP, β-TCP, and α'-TCP [301]); and DCP is a kind of acid calcium phosphate with a basic calcium source and acidic phosphorus source [271]. As the main component of CPC, DCP has two forms, namely, monetite ($CaHPO_4$, dicalcium phosphate anhydrous (DCPA)) and brushite ($CaHPO_4 \cdot 2H_2O$, dicalcium phosphate dihydrate) [302]), and calcium sulfate and silicate-based bioceramics as well as composite materials based on bioactive ceramics can be included.

Bioactive glasses

In the early 1970s, a silicate-based 45S5 glass based on the system of SiO_2 (45%)-Na_2O (24.5%)-CaO (24.5%)-P_2O_5 (6%) was developed [303]. Since then, BAG began to enter people's field of vision and played an important role in the repair of bone defects [304]. Bioactive glasses are considered highly reactive surfaces formed by melt or sol-gel techniques. Bioactive glass forms a hydroxy-carbonated apatite layer when immersed in biological fluid, which enhances protein adsorption to the surface of the implant and integration with surrounding bone. The rate of ion release from the bioglass surface is determined by the Ca:P ratio, composition, and microstructure. The initial reaction of some BAGs with biological fluids causes local pH to increase; some studies propose that this is beneficial to cell activity and HA production [305, 306].

Biodegradable metal materials

Metal implants (e.g., 304 L or 316 L stainless, Ti-6Al-4V or Ti-6Al-7Nb, or Co-Cr-Mo alloys) have a long history of application in orthopedic surgery, especially in the field of bone repair. However, these materials have many shortcomings, such as nonbiodegradability and stress-shielding effects due to a large difference in modulus of elasticity between these metallic materials and bone, which limit their application in bone defect repair. In recent years, biodegradable metals have attracted extensive attention from researchers due to their excellent biocompatibility and degradability [307]. Specifically, the most widely studied biodegradable metals include magnesium, iron, zinc, and their-based alloys. At the same time, these three metals are essential elements for maintaining the normal function of the human body, which has been confirmed by many studies to have good biocompatibility to human cells and tissues [308–310].

Biodegradable Mg-based alloys: With good biocompatibility, suitable mechanical strength and biodegradability, Mg and Mg-based alloys are widely favored by researchers in the field of bone regeneration [106]. Among the cations in the human body, magnesium is ranked fourth and is mainly stored in bone tissues, participating in many metabolic processes in the body [271]. The biomechanical properties of magnesium are suitable for bone tissue. The density of Mg-based alloys (1.7–1.9 g/cm^3) is very similar to that of human cortical bone (1.75 g/cm^3) [311]. The elastic modulus of magnesium-based metals is about 45 GPa, which is relatively close to that of natural bone (3–20 GPa), while the density of titanium alloy and stainless steel is 4.47 and 7.8 g/cm^3, respectively, and the elastic modulus is 110 and 200 GPa [51, 312]. Therefore, compared with commonly used titanium alloys and stainless steels, Mg-based alloys only have a negligible stress-shielding effect. A variety of biodegradable grade Mg-based alloys have been developed and clinically used. Coronary stents improve immediate and late results of balloon angioplasty by tacking up dissections and preventing wall recoil. These goals are achieved within weeks after angioplasty, but with current technology stents permanently remain in the artery, with many limitations including the need for long-term antiplatelet treatment to avoid thrombosis. Based on this background, Erbel et al. [308] reported a prospective multicenter clinical trial of coronary implantations of absorbable magnesium stents. In total 63 patients (44 men and 19 female; mean age 61.3) in eight centers with single de novo lesions in a native coronary artery in a multicenter, non-randomized prospective were subjected to the study. Follow-up included coronary angiography and intravascular ultrasound at 4 months and clinical assessment at 6 months and 12 months. The primary endpoint was cardiac death, nonfatal myocardial infarction, or clinically driven target lesion revascularization at 4 months. It was found that (i) 71 stents, 10–15 mm in length and 3.0–3.5 mm in diameter, were successfully implanted after predilatation in 63 patients, (ii) the overall target lesion revascularization rate was 45% after 1 year, (iii) after serial intravascular ultrasound examinations, only small remnants of the original struts were visible, well embedded into the intima, and (iv) neointimal growth and negative remodelling were the main operating mechanisms of restenosis, indicating that biodegradable magnesium stents can achieve an immediate angiographic result similar to the result of other metal stents and can be safely degraded after 4 months [313].

Biodegradable Fe-based alloys: Iron is an essential trace element in the human body. Iron material, with excellent mechanical properties close to 316 L stainless steels, plays an important role in the field of tissue engineering [254]. Compared with pure magnesium, pure iron has stronger mechanical properties and does not release hydrogen during the biodegradation process after implantation [314]. However, research shows that the main disadvantage of pure Fe and Fe-based alloys is the slow degradation rate [315]. Peuster et al. [316] evaluated the safety of a corrodible stent produced from pure iron in a peripheral stent design (6–12 mm diameter) in a slotted tube design similar to a commercially available 316 L stent which served as control. Both stents were implanted into the descending aorta of 29 minipigs with an over-stretch injury without technical problems. It was reported that (i) histomorphometry and quantitative angiography showed no difference with regard to the amount of neo-intimal proliferation between 316 L and iron stents, (ii) histopathological examination of heart, lung, spleen, liver, kidney, and para-aortic lymphatic nodes demonstrated no signs of iron overload or iron-related organ toxicity, and (iii) adjacent to the iron stent struts, there was no evidence for local toxicity due to corrosion products, concluding that (iv) iron is a suitable metal for the production of a large-size degradable stent with no local or systemic toxicity, and (v) a faster degradation rate, however, is desir-able and further studies have to focus on the modification of the composition and de-sign of the stent to expedite the degradation process. Therefore, improving iron degradation rate is an urgent task to promote the use of iron-based stents in clinical practice. To this end, researchers have made a variety of attempts such as surface modification [317], alloying [318], and adding a second phase [319]. Additively manu-factured (AMed) porous biomaterials can increase the surface area of the material [320]. Generally speaking, a larger surface area usually leads to a higher biodegrada-tion rate. Therefore, for iron and its alloys, increasing the surface area may be a promising way to accelerate its biodegradation rate [321]. Li et al. [322] reported on AM functionally graded biodegradable porous metallic biomaterials, using SLM. It was found that (i) the topological design with functional gradients controlled the fluid flow, mass transport properties and biodegradation behavior of the AMed porous iron specimens, as up to fourfold variations in permeability and up to threefold varia-tions in biodegradation rate were observed for the different experimental groups, and (ii) after 4 weeks of biodegradation, the mechanical properties of the specimens (i.e., $E = 0.5$–2.1 GPa, $\sigma y = 8$–48 MPa) remained within the range of the values reported for trabecular bone, indicating that (iii) topological design in general, and functional gradients in particular can be used as an important tool for adjusting the biodegrada-tion behavior of AMed porous metallic biomaterials, and (iv) the biodegradation rate and mass transport properties of AMed porous iron can be increased while maintain-ing the bone-mimicking mechanical properties of these biomaterials.

Biodegradable Zn-based alloys: Similar to iron, zinc is also an important trace ele-ment required by the human body and plays an important role in many physiological activities (such as growth, immunity, and wound healing) [323]. It is reported that about 85% of zinc is present in muscles and bones, so zinc is essential for bone devel-opment and growth [324]. A series of in vitro studies have shown that zinc ions can promote stem cell osteogenesis and increase mineral deposits as well as promote oste-oblast adhesion, proliferation and differentiation [325, 326]. Bowen et al. [327] pro-posed zinc as an exciting new biomaterial for use in bioabsorbable cardiac stents. It was mentioned that (i) Zn is not only a physiologically relevant metal with behavior that promotes healthy vessels, but also it combines the best behaviors of both current bioabsorbable stent materials: iron and magnesium, and (ii) the biodegradation rate of pure zinc stent was faster than that of Fe and Mg alloy. Although Zn as an essential element with osteogenic potential of human body is soft, brittle, and has low mechanical strength in practice, it needs further improvement in order to meet the clinical requirements. Li et al. [328] prepared Zn-based alloys (with Mg, Ca, and Sr) such as cast, rolled, and extruded Zn-1Mg, Zn-1Ca, and Zn-1Sr. It was reported that (i) Zn-1X (Mg, Ca, and Sr) alloys have profoundly modified the mechanical properties and biocompatibility of pure Zn. Zn-1X (Mg, Ca, and Sr) alloys showed great potential for use in a new generation of biodegradable implants, opening up a new avenue in the area of biodegradable metals. In addition, in vivo experiments show that Zn-Sr alloy has a good role in promoting new bone formation. Recently, Tiffany et al. [329] added zinc to the mineralized collagen suspension and then lyophilized to form a porous zinc-containing mineralized collagen bone scaffold. Zhang et al. [330] developed a new type of scaffold made of an Mg-Zn-Ca alloy with a shape that mimics cortical bone and can be filled with morselized bone and evaluated its durability and efficacy in a rabbit ulna-defect model. Different types of scaffold-surface coating were evaluated: group A, no coating; group B, a 10 μm microarc oxidation coating; group C, a hydrothermal duplex composite coating; and group D, an empty-defect control. It was reported that calcein fluorescence and histology revealed that greater mineral densities and better bone responses were achieved for groups B and C than for group A, with group C providing the best response, concluding that (i) newly developed Mg-Zn-Ca-alloy scaffold effectively aided bone repair, (ii) the group C scaffold exhibited the best corrosion resistance and osteogenesis properties, making it a candidate scaffold for repair of bone defects.

Sheikh et al. [271] reviewed the recent developments and advances in the use of biodegradable materials for bone repair purposes. It was mentioned that (i) the choice between using degradable and nondegradable devices for orthopedic and maxillofacial applications must be carefully weighed, (ii) although traditional biodegradable devices for osteosynthesis have been successful in low or mild load-bearing applications, continuing research and recent developments in the field of material science have resulted in development of biomaterials with improved strength and mechanical properties; hence, biodegradable materials, including polymers, ceramics, and magnesium alloys have attracted much attention for osteologic repair and applications, and (iii) the next generations of biodegradable materials for bone repair and regeneration applications require better control of interfacing between the material and the surrounding bone tissue and the mechanical properties and degradation/resorption profiles of these materials require further improvement to broaden their use and achieve better clinical results.

References

[1] Mozafari M, Sefat F, Atala A (Eds.). Handbook of Tissue Engineering Scaffolds: Volume Two. Woodhead Pub. 2019.

[2] Scaffold for Tissue Engineering; https://www.sciencedirect.com/topics/materials-science/scaffold-for-tissue-engineering.

[3] Tissue Engineering; https://en.wikipedia.org/wiki/Tissue_engineering.

[4] https://www.academia.edu/search?q=Biomaterials%20and%20tissue%20engineering%20scaffolds&tab=1.

[5] Ghassemi T, Shahroodi A, Ebrahimzadeh MH, Mousavian A, Movaffagh J, Moradi A. Current concepts in scaffolding for bone tissue engineering. Arch Bone Jt Surg. 2018, 6, 90–99.

[6] Shimba K, Chang C-H, Asahina T, Moriya F, Kotani K, Jimbo Y, Gladkov A, Antipova O, Pigareva Y, Kolpakov V, Mukhina I, Kazantsev V, Pimashkin A. Functional scaffolding for brain implants: Engineered neuronal network by microfabrication and iPSC technology. Front Neurosci. 2019, 13, 890; doi: 10.3389/fnins.2019.00890.

[7] Brown BN, Valentin JE, Stewart-Akers AM, McCabe GP, Badylak SF. Macrophage phenotype and remodeling outcomes in response to biologic scaffolds with and without a cellular component. Biomaterials. 2009, 3, 1482–91.

[8] Hutmacher DW. Scaffolds in tissue engineering bone and cartilage. Biomaterials. 2000, 21, 2529–43.

[9] Porter JR, Ruckh TT, Popat KC. Bone tissue engineering: A review in bone biomimetics and drug delivery strategies. Biotechnol Prog. 2009, 25, 1539–60.

[10] Rho JY, Kuhn-Spearing L, Zioupos P. Mechanical properties and the hierarchical structure of bone. Med Eng Phys. 1998, 20, 92–102.

[11] Weiner S, Traub W. Bone structure: From angstroms to microns. FASEB J. 1992, 6, 879–85.

[12] Oshida Y. Bioscience and Bioengineering of Titanium Materials. Elsevier. 2007.

[13] Oshida Y, Miyazaki T. Biomaterials and Engineering for Implantology. De Gruyter. 2022.

[14] Shi C, Yuan Z, Han F, Zhu C, Li B. Polymeric biomaterials for bone regeneration. Ann Joint. 2016, 1; doi: 10.21037/aoj.2016.11.02.

[15] Yan J, Li J, Runge MB, Dadsetan M, Chen Q, Lu L, Yaszemski MJ. Cross-linking characteristics and mechanical properties of an injectable biomaterial composed of polypropylene fumarate and polycaprolactone co-polymer. J Biomater Sci Polym. 2011, 22, 489–504.

[16] Short AR, Koralla D, Deshmukh A, Wissel B, Stocker B, Calhoun M, Dean D, Winter JO. Hydrogels that allow and facilitate bone repair, remodeling, and regeneration. J Mater Chem B. 2015, 3, 7818–30.

[17] Lee SH, Shin H. Matrices and scaffolds for delivery of bioactive molecules in bone and cartilage tissue engineering. Adv Drug Deliv Rev. 2007, 59, 339–59.

[18] Pei M, Li J, Shoukry M, Zhang Y. A review of decellularized stem cell matrix: A novel cell expansion system for cartilage tissue engineering. Eur Cell Mater. 2011, 22, 333–43.

[19] Koob S, Torio-Padron N, Stark GB, Hannig C, Stankovic Z, Finkenzeller G. Bone formation and neovascularization mediated by mesenchymal stem cells and endothelial cells in critical-sized calvarial defects. Tissue Eng Part A. 2010, 17, 311–21.

[20] Kretlow JD, Mikos AG. Mineralization of synthetic polymer scaffolds for bone tissue engineering. Tissue Eng. 2007, 13, 927–38.

[21] Moradi A, Dalilottojari A, Pingguan-Murphy B, Djordjevic I. Fabrication and characterization of elastomeric scaffolds comprised of a citric acid-based polyester/hydroxyapatite microcomposite. Mater Des. 2013, 50, 446–50.

[22] Ali Akbari Ghavimi S, Ebrahimzadeh MH, Solati-Hashjin M, Osman A, Azuan N. Polycaprolactone/starch composite: Fabrication, structure, properties, and applications. J Biomed Mater Res A. 2015, 103, 2482–98.

[23] Ghavimi SA, Ebrahimzadeh MH, Shokrgozar MA, Solati-Hashjin M, Osman NA. Effect of starch content on the biodegradation of polycaprolactone/starch composite for fabricating in situ pore-forming scaffolds. Polymer Test. 2015.

[24] Wang M. Developing bioactive composite materials for tissue replacement. Biomaterials. 2003, 24, 2133–51.

[25] Bose S, Roy M, Bandyopadhyay A. Recent advances in bone tissue engineering scaffolds. Trends Biotechnol. 2012, 30, 546–54.

[26] Fuchs JR, Nasseri BA, Vacanti JP. Tissue engineering: A 21st century solution to surgical reconstruction. Ann Thorac Surg. 2001, 72, 577–91.

[27] Ishaug SL, Yaszemski MJ, Bizios R, Mikos AG. Osteoblast function on synthetic biodegradable polymers. J Biomed Mater Res. 1994, 28, 1445–53.

[28] Athanasiou KA, Agrawal CM, Barber FA, Burkhart SS. Orthopaedic applications for PLA-PGA biodegradable polymers. Arthroscopy. 1998, 14, 726–37.

[29] Oryan A, Alidadi S, Moshiri A, Maffulli N. Bone regenerative medicine: Classic options, novel strategies, and future directions. J Orthop Surg Res. 2014, 9; doi: 10.1186/1749-799X-9-18.

[30] Stevens B, Yang Y, Mohandas A, Stucker B, Nguyen KT. A review of materials, fabrication methods, and strategies used to enhance bone regeneration in engineered bone tissues. J Biomed Mater Res B Appl Biomater. 2008, 85, 573–78.

[31] Ducheyne P, Qiu Q. Bioactive ceramics: The effect of surface reactivity on bone formation and bone cell function. Biomaterials. 1999, 20, 2287–303.

[32] Dorozhkin SV, Epple M. Biological and medical significance of calcium phosphates. Angew Chem Int Ed Engl. 2002, 41, 3130–46.

[33] Samavedi S, Whittington AR, Goldstein AS. Calcium phosphate ceramics in bone tissue engineering: A review of properties and their influence on cell behavior. Acta Biomater. 2013, 9, 8037–45.

[34] Drzewiecka K, Krasowski J, Krasowski M, Łapińska B. Mechanical properties of composite material modified with amorphous calcium phosphate. J Achiev Mater Manufact Eng. 2016, 74, 22–28.

[35] Fielding GA, Bandyopadhyay A, Bose S. Effects of silica and zinc oxide doping on mechanical and biological properties of 3D printed tricalcium phosphate tissue engineering scaffolds. Dent Mater. 2012, 28, 113–22.

[36] Wang M, Cheng X, Zhu W, Holmes B, Keidar M, Zhang LG. Design of biomimetic and bioactive cold plasma modified nanostructured scaffolds for enhanced osteogenic differentiation of bone marrow derived mesenchymal stem cells. Tissue Eng Part A. 2014, 20, 1060–71.

[37] Pasqui D, Torricelli P, De Cagna M, Fini M, Barbucci R. Carboxymethyl cellulose-hydroxyapatite hybrid hydrogel as a composite material for bone tissue engineering applications. J Biomed Mater Res A. 2014, 102, 1568–79.

[38] Sagar N, Pandey AK, Gurbani D, Khan K, Singh D, Chaudhari BP, Soni VP, Chattopadhyay N, Dhawan A, Bellare JR. In-vivo efficacy of compliant 3D nano-composite in critical-size bone defect repair: A six month preclinical study in rabbit. PLoS One. 2013, 8, e77578; doi 10.1371/journal.pone.0077578.

[39] Xia Y, Zhou P, Cheng X, Xie Y, Liang C, Li C, Xu S. Selective laser sintering fabrication of nano-hydroxyapatite/poly-ε-caprolactone scaffolds for bone tissue engineering applications. Int J Nanomed. 2013, 8, 4197–213.

[40] Calvo-Guirado JL, Ramírez-Fernández MP, Delgado-Ruíz RA, Maté-Sánchez JE, Velasquez P, de Aza PN. Influence of biphasic β-TCP with and without the use of collagen membranes on bone healing of surgically critical size defects. A radiological, histological, and histomorphometric study. Clin Oral Implants Res. 2014, 11, 1228–38.

[41] Vozzi G, Corallo C, Carta S, Fortina M, Gattazzo F, Galletti M, Giordano N. Collagen-gelatin-genipin-hydroxyapatite composite scaffolds colonized by human primary osteoblasts are suitable for bone tissue engineering applications: In vitro evidences. J Biomed Mater Res A. 2014, 102, 1415–21.

[42] Jung UW, Lee JS, Lee G, Lee IK, Hwang JW, Kim MS, Choi SH, Chai JK. Role of collagen membrane in lateral onlay grafting with bovine hydroxyapatite incorporated with collagen matrix in dogs. J Periodontal Implant Sci. 2013, 43, 64–71.

[43] Eleftheriadis E, Leventis MD, Tosios KI, Faratzis G, Titsinidis S, Eleftheriadi I, Dontas I. Osteogenic activity of β-tricalcium phosphate in a hydroxyl sulphate matrix and demineralized bone matrix: A histological study in rabbit mandible. J Oral Sci. 2010, 52, 377–84.

[44] Velasquez P, Luklinska ZB, Meseguer-Olmo L, de Val Mate-Sanchez JE, Delgado-Ruiz RA, Calvo-Guirado JL, Ramirez-Fernandez MP, de Aza PN. αTCP ceramic doped with dicalcium silicate for bone regeneration applications prepared by powder metallurgy method: In vitro and in vivo studies. J Biomed Mater Res A. 2013, 101, 1943–54.

[45] Piccinini M, Rebaudi A, Sglavo VM, Bucciotti F, Pierfrancesco R. A new HA/TTCP material for bone augmentation: An in vivo histological pilot study in primates sinus grafting. Implant Dent. 2013, 22, 83–90.

[46] Nouri A, Hodgson PD, Wen C. Biomimetic porous titanium scaffolds for orthopedic and dental applications. In: Mukherjee A (Ed.)., Biomimetics Learning from Nature. 2010, Vol. 21, 415–50.

[47] Matassi F, Botti A, Sirleo L, Carulli C, Innocenti M. Porous metal for orthopedics implants. Clin Cases Miner Bone Metab. 2013, 10, 111–15.

[48] Cachinho SCP, Correia RN. Titanium scaffolds for osteointegration: Mechanical, in vitro and corrosion behaviour. J Mater Sci Mater Med. 2008, 19, 451–57.

[49] Disegi JA, Eschbach L. Stainless steel in bone surgery. Injury. 2000, 31, 2–6.

[50] Di Mario C, Griffiths H, Goktekin O, Peeters N, Verbist J, Bosiers M, Deloose K, Heublein B, Rohde R, Kasese V, Ilsley C, Erbel R. Drug-eluting bioabsorbable magnesium stent. J Inter Cardiol. 2004, 17, 391–95.

[51] Oshida Y. Magnesium Materials. De Gruyter. 2021.

[52] Song G. Control of biodegradation of biocompatable magnesium alloys. Corros Sci. 2007, 49, 1696–701.

[53] Peuster M, Wohlsein P, Brügmann M, Ehlerding M, Seidler K, Fink C, Brauer H, Fischer A, Hausdorf G. A novel approach to temporary stenting: Degradable cardiovascular stents produced from corrodible metal-results 6–18 months after implantation into New Zealand white rabbits. Heart. 2001, 86, 563–69.

[54] Farack J, Wolf-Brandstetter C, Glorius S, Nies B, Standke G, Quadbeck P, Worch H, Scharnweber D. The effect of perfusion culture on proliferation and differentiation of human mesenchymal stem cells on biocorrodible bone replacement material. Mater Sci Eng. 2011, 176, 1767–72.

[55] Hermawan H, Alamdari H, Mantovani D, Dube D. Iron-manganese: New class of metallic degradable biomaterials prepared by powder metallurgy. Powder Metall. 2008, 51, 38–45.

[56] Carluccio D, Xu C, Venezuela J, Cao Y, Kent D, Bermingham M, Demir AG, Previtali B, Ye Q, Dargusch M. Additively manufactured iron-manganese for biodegradable porous load-bearing bone scaffold applications. Acta Biomater. 2020, 103, 346–60.

[57] Patnaik L, Maity SR, Kumar S. Status of nickel free stainless steel in biomedical field: A review of last 10 years and what else can be done. Mater Today. 2020, 26, 638–43.

[58] Romanczuk E, Perkowski K, Oksiuta Z. Microstructure, mechanical, and corrosion properties of Ni-Free Austenitic stainless steel prepared by Mechanical Alloying and HIPing. Materials. 2019, 12, 3416; https://doi.org/10.3390/ma12203416.

[59] Yang K, Ren Y. Nickel-free austenitic stainless steels for medical applications. Sci Technol Adv Mater. 2010, 11; https://doi.org/10.1088/1468-6996/11/1/014105.

[60] Borgioli F, Galvanetto E, Bacci T. Surface modification of a Nickel-Free Austenitic stainless steel by low-temperature nitriding. Metals. 2021, 11, 1845; https://doi.org/10.3390/met11111845.

[61] Bobyn J, Stackpool GJ, Hacking SA, Tanzer M, Krygier JJ. Characteristics of bone ingrowth and interface mechanics of a new porous tantalum biomaterial. J Bone Joint Surg Br. 1999, 81, 907–14.

[62] Miyazaki T, Kim HM, Kokubo T, Ohtsuki C, Kato H, Makamura T. Enhancement of bonding strength by graded structure at interface between apatite layer and bioactive tantalum metal. J Mater Sci Mater Med. 2002, 13, 651–55.

[63] Hacking SA, Bobyn JD, Toh K, Tanzer M, Krygier JJ. Fibrous tissue ingrowth and attachment to porous tantalum. J Biomed Mater Res. 2000, 52, 631–38.

[64] Bencharit S, Byrd WC, Altarawneh S, Hosseini B, Leong A, Reside G, Morelli T, Offenbacher S. Development and applications of porous tantalum trabecular metal-enhanced titanium dental implants. Clin Implant Dent Relat Res. 2014, 16, 817–26.

[65] Han C, Yan S, Wen S, Xu T. Effects of the unit cell topology on the compression properties of porous Co-Cr scaffolds fabricated via selective laser melting. Rapid Prototyp J. 2017, 23, 16–27.

[66] Caravaggi P, Liverani E, Leardini A, Fortunato A, Belvedere C, Baruffaldi F, Fini M, Parrilli A, Mattioli-Belmonte M, Tomesani L, Pagani S. CoCr porous scaffolds manufactured via selective laser melting in orthopedics: Topographical, mechanical, and biological characterization. Biomed Mater Res Part B. 2019, 107B, 2343–53.

[67] Lu Y, Zhou Y, Liang X, Zhang X, Zhang C, Zhu M, Tang K, Lin J. Early bone ingrowth of Cu-bearing CoCr scaffolds produced by selective laser melting. An in vitro and in vivo study. Mater Des. 2013, 228, 111822; https://doi.org/10.1016/j.matdes.2023.111822.

[68] Oshida Y, Tominaga T. Nickel-Titanium Materials. De Gruyter, 2020.

[69] Nouri A. Titanium foam scaffolds for dental applications. In: Wen C (Ed.)., Metallic Foam Bone Processing, Modification and Characterization and Properties. 2017; https://www.academia.edu/30241861/Titanium_foam_scaffolds_for_dental_applications.

[70] Porter JA, von Fraunhofer JA. Success or failure of dental implants? A literature review with treatment considerations. Gen Dent. 2005, 53, 423–32.

[71] Park JM, Kim HJ, Park EJ, Kim MR, Kim SJ. Three dimensional finite element analysis of the stress distribution around the mandibular posterior implant during non-working movement according to the amount of cantilever. J Adv Prosthodont. 2014, 6, 361–71.

[72] Simon Z, Watson PA. Biomimetic dental implants –New ways to enhance osseointegration. J Can Dent Assoc. 2002, 68, 286–88.

[73] Adell R, Eriksson B, Lekholm U, Brånemark PI, Jemt T. Long-term follow-up study of osseointegrated implants in the treatment of totally edentulous jaws. Int J Oral Maxillofac Implants. 1990, 5, 347–59.

[74] Balla VK, Bodhak S, Bose S, Bandyopadhyay A. Porous tantalum structures for bone implants: Fabrication, mechanical and in vitro biological properties. Acta Biomater. 2010, 6, 3349–59.

[75] Das K, Balla VK, Bandyopadhyay A, Bose S. Surface modification of laser-processed porous titanium for load-bearing implants. Scripta Mater. 2008, 59, 82–85.

[76] Rao X, Li J, Feng X, Chu CJ. Bone-like apatite growth on controllable macroporous titanium scaffolds coated with microporous titania. Mech Behav Biomed Mater. 2018, 77, 225–33.

[77] Okazaki Y. A new Ti-15Zr-4Nb-4Ta alloy for medical applications. Curr Opin Solid State Mater Sci. 2001, 5, 45–53.

[78] Zdeblick TA, Phillips FM. Interbody cage devices. Spine. 2003, 28, S2–S7.

[79] Crowninshield RD. Mechanical properties of porous metal total hip prostheses. Instr Course Lect. 1985, 35, 144–48.

[80] Faria PE, Carvalho AL, Felipucci DN, Wen C, Sennerby L, Salata LA. Bone formation following implantation of titanium sponge rods into humeral osteotomies in dogs: A histological and histometrical study. Clin Implant Dent Relat Res. 2010, 12, 72–79.

[81] Assad M, Chernyshov A, Leroux MA, Rivard CH. A new porous titanium-nickel alloy: Part 1. Cytotoxicity and genotoxicity evaluation. Biomed Mater Engin. 2002, 12, 225–37.

[82] Prymak O, Bogdanski D, Köller M, Esenwein SA, Muhr G, Beckmann F, Donath T, Assad M, Epple M. Morphological characterization and in vitro biocompatibility of a porous nickel-titanium alloy. Biomaterials. 2005, 26, 580–87.

[83] Greiner C, Oppenheimer SM, Dunand DC. High strength, low stiffness, porous NiTi with superelastic properties. Acta Biomater. 2005, 1, 705–16.

[84] Bystedt H, Rasmusson L. Porous titanium granules used as osteoconductive material for sinus floor augmentation: A clinical pilot study. Clin Implant Dent Relat Res. 2009, 11, 101–05.

[85] Kheirallah M, Almeshaly H. Bone graft substitutes for bone defect regeneration. A collective review. Int J Dent Oral Sci. 2016, 3, 247–57.

[86] Hutmacher DW, Cool S. Concepts of scaffold-based tissue engineering – the rationale to use solid freeform fabrication techniques. J Cell Mol Med. 2007, 11, 654–69.

[87] Tanner KE. Bioactive composites for bone tissue engineering. Proc Inst Mech Eng H. 2010, 224, 1359–72.

[88] Miyazaki T. Design of bone integrating organic-inorganic composite suitable for bone repair. Front Biosci. 2013, 1, 333–40.

[89] Hutmacher DW, Schantz JT, Lam CX, Tan KC, Lim TC. State of the art and future directions of scaffold-based bone engineering from a biomaterials perspective. J Tissue Eng Regen Med. 2007, 1, 245–60.

[90] Salunke P, Shanov V, Witte F. High purity biodegradable magnesium coating for implant application. Mater Sci Eng B. 2011, 176, 1711–17.

[91] Yang J, Cui F, Lee IS. Surface modifications of magnesium alloys for biomedical applications. Annals of Biomed Eng. 2011, 39, 1857–71.

[92] Kirkland NT, Birbilis N, Staiger MP. Assessing the corrosion of biodegradable magnesium implants: A critical review of current methodologies and their limitations. Acta Biomater. 2012, 8, 925–36.

[93] Yuen CK, Ip WY. Theoretical risk assessment of magnesium alloys as degradable biomedical implants. Acta Biomater. 2010, 6, 1808–12.

[94] Li Y, Wen C, Mushahary D, Sravanthi R, Harishankar N, Pande G. Hodgson P.Mg–Zr–Sr alloys as biodegradable implant materials. Acta Biomater. 2012, 8, 3177–88.

[95] Yang X, Li M, Lin X, Tan L, Lan G, Li L, Yin Q, Xia H, Zhang Y, Yang K. Enhanced in vitro biocompatibility/bioactivity of biodegradable Mg–Zn–Zr alloy by micro-arc oxidation coating contained Mg2SiO4. Surf Coat Technol. 2013, 233, 65–73.

[96] Li M, Cheng Y, Zheng YF, Zhang X, Xi TF, Wei SC. Plasma enhanced chemical vapor deposited silicon coatings on Mg alloy for biomedical application. Surf Coat Technol. 2012, 228, S262–5.

[97] Gu XN, Zheng YF. A review on magnesium alloys as biodegradable materials. Front Mater Sci China. 2010, 4, 111–15.

[98] Gu XN, Zhou WR, Zheng YF, Liu Y, Li YX. Degradation and cytotoxicity of lotus-type porous pure magnesium as potential tissue engineering scaffold material. Mater Lett. 2010, 64, 1871–74.

[99] Li Z, Gu X, Lou S, Zheng Y. The development of binary Mg-Ca alloys for use as biodegradable materials within bone. Biomaterials. 2008, 29, 1329–44.

[100] Witte F, Fischer J, Nellesen J, Crostack HA, Kaese V, Pisch A, Beckmann F, Windhagen H. vitro and in vivo corrosion measurements of magnesium alloys. Biomaterials. 2006, 27, 1013–118.

[101] Saris NE, Mervaala E, Karppanen H, Khawaja JA, Lewenstam A. Magnesium: An update on physiological, clinical and analytical aspects. Clin Chim Acta. 2000, 294, 1–26.

[102] Witte F, Ulrich H, Palm C, Willbold E. Biodegradable magnesium scaffolds: Part II: peri-implant bone remodeling. J Biomed Mater Res. 2007, 81, 757–65.

[103] Qin Y, Liu A, Guo H, Shen Y, Wen P, Lin H, Xia D, Voshage M, Tian Y, Zheng Y. Additive manufacturing of Zn-Mg alloy porous scaffolds with enhanced osseointegration: In vitro and in vivo studies. Acta Biomater. 2022, 145, 403–15.

[104] Li Y, Zhou J, Pavanram P, Leeflang MA, Fockaert LI, Pouran B, Tümer N, Schröder KU, Mol JMC, Weinans H, Jahr H, Zadpoor AA. Additively manufactured biodegradable porous magnesium. Acta Biomater. 2018, 67, 378–92.

[105] Witte F, Kaese V, Haferkamp H, Switzer E, Meyer-Lindenberg A, Wirth CJ, et al. In vivo corrosion of four magnesium alloys and the associated bone response. Biomaterials. 2005, 26, 3557–63.

[106] Staiger MP, Pietak AM, Huadmai J, Dias G. Magnesium and its alloys as orthopedic biomaterials: A review. Biomaterials. 2006, 27, 1728–34.

[107] Heublein B, Rohde R, Kaese V, Niemeyer M, Hartung W, Haverich A. Biocorrosion of magnesium alloys: A new principle in cardiovascular implant technology? Heart. 2003, 89, 651–56.

[108] Walker J, Shadanbaz S, Kirkland NT, Stace E, Woodfield T, Staiger MP, Dias GJ. Magnesium alloys: Predicting in vivo corrosion with in vitro immersion testing. J Biomed Mater Res. 2012, 100B, 1134–41.

[109] Shadanbaz S, Dias GJ. Calcium phosphate coatings on magnesium alloys for biomedical applications: A review. Acta Biomaterialia. 2012, 8, 20–30.

[110] Wang Y, Zhu Z, Xu X, He Y, Zhang B. Improved corrosion resistance and biocompatibility of a calcium phosphate coating on a magnesium alloy for orthopedic applications. SAGE J. 2016; https://journals.sagepub.com/doi/10.1177/1721727X16677763.

[111] Cascone M, Barbani N, Cristallini C, Giusti P, Ciardelli G, Lazzeri L. Bioartificial polymeric materials based on polysaccharides. J Biomater Sci Polym Ed. 2001, 12, 267–81.

[112] Ciardelli G, Chiono V, Vozzi G, Pracella M, Ahluwalia A, Barbani N, Cristallini C, Giusti P. Blends of poly-(ε-caprolactone) and polysaccharides in tissue engineering applications. Biomacromolecules. 2005, 6, 1961–76.

[113] Dhandayuthapani B, Yoshida Y, Meakawa T, Kumar DS. Polymeric scaffolds in tissue engineering application: A review. Int J Polymer Sci. 2011, 2011; https://doi.org/10.1155/2011/290602.

[114] Bigi A, Boanini E, Panzavolta S, Roveri N, Rubini K. Bonelike apatite growth on hydroxyapatite–gelatin sponges from simulated body fluid. J Biomed Mater Res. 2002, 59, 709–15.

[115] Zhang Y, Zhang M. Synthesis and characterization of macroporous chitosan/calcium phosphate composite scaffolds for tissue engineering. J Biomed Mater Res. 2001, 55, 304–12.

[116] Kang HG, Kim SY, Lee YM. Novel porous gelatin scaffolds by overrun/particle leaching process for tissue engineering applications. J Biomed Mater Res B Appl Biomater. 2006, 79, 388–97.

[117] Roether J, Boccaccini AR, Hench L, Maquet V, Gautier S, Jérôme R. Development and in vitro characterisation of novel bioresorbable and bioactive composite materials based on polylactide foams and Bioglass®for tissue engineering applications. Biomaterials. 2002, 23, 3871–78.

[118] Banerjee SS, Tarafder S, Davies NM, Bandyopadhyay A, Bose S. Understanding the influence of MgO and SrO binary doping on the mechanical and biological properties of β-TCP ceramics. Acta Biomater. 2010, 6, 4167–74.

[119] Huang YX, Ren J, Chen C, Ren TB, Zhou XY. Preparation and properties of ploy(lactide-co-glycolide) (PLGA)/Nano-Hydroxyapatite (NHA) Scaffolds by thermally induced phase separation and rabbit MSCs culture on scaffolds. J Biomater Appl. 2008, 22, 409–32.

[120] Reichert JC, Cipitria A, Epari DR, Saifzadeh S, Krishnakanth P, Berner A, Woodruff MA, Schell H, Mehra M, Schuetz MA, Duda GN, Hutmacher DW. A tissue engineering solution for segmental defect regeneration in loadbearing long bones. Sci Transl Med. 2012, 4, 141ra93; doi: 10.1126/scitranslmed.3003720.

[121] Mizuno H. Adipose derived stem cells for tissue repair and regeneration: Ten years of research and a literature review. J Nippon Med Sch. 2009, 76, 56–66.

[122] Coskun S, Korkusuz F, Hasirci V. Hydroxyapatite reinforced poly(3 hydroxybutyrate) and poly(3-hydroxybutyrate- co-3-hydroxyvalerate) based degradable composite bone plate. J Biomater Sci Polym E. 2005, 16, 1485–502.

[123] Milovanovic JR, Stojkovic MS, Husain KN, Korunovic ND, Arandjelovic J. Holistic approach in designing the personalized bone scaffold: The case of reconstruction of large missing piece of mandible caused by congenital anatomic anomaly. J Healthcare Eng. 2020; https://doi.org/10.1155/2020/6689961.

[124] Matassi F, Nistri L, Chicon Paez D, Innocenti M. New biomaterials for bone regeneration. Clin Cases Miner Bone Metab. 2011, 8, 21–24.

[125] Khan Y, Yaszemski MJ, Mikos AG, Laurencin CT. Tissue engineering of bone: Material and matrix considerations. J Bone Joint Surg Am. 2008, 90, 36–42.

[126] Woodard JR, Hilldore AJ, Lan SK, Park CJ, Morgan AW, Eurell JA, Clark SG, Wheeler MB, Jamison RD, Wagoner Johnson AJ. The mechanical properties and osteoconductivity of hydroxyapatite bone scaffolds with multi-scale porosity. Biomaterials. 2007, 28, 45–54.

[127] Dabrowski B, Swieszkowski W, Godlinski D, Kurzydlowski KJ. Highly porous titanium scaffolds for orthopaedic applications. J Biomed Mater Res B Appl Biomater. 2010, 95, 53–61.

[128] Ryan G, Pandit A, Apatsidis DP. Fabrication methods of porous metals for use in orthopaedic applications. Biomaterials. 2006, 27, 2651–70.

[129] Karageorgiou V, Kaplan D. Porosity of 3D biomaterial scaffolds and osteogenesis. Biomaterials. 2005, 26, 5474–91.

[130] Melancon D, Bagheri Z, Johnston R, Liu L, Tanzer M, Pasini D. Mechanical characterization of structurally porous biomaterials built via additive manufacturing: Experiments, predictive models, and design maps for load-bearing bone replacement implants. Acta Biomater. 2017, 63, 350–68.

[131] Kelly CN, Wang T, Crowley J, Wills D, Pelletier MH, Westrick ER, Adams SB, Gall K, Walsh WR. High-strength, porous additively manufactured implants with optimized mechanical osseointegration. Biomaterials. 2021, 279, 121206; https://doi.org/10.1016/j.biomaterials.2021.121206.

[132] Synthetic Bone Grafting: An Overview of the Process, Terms, and Materials. 2022; https://www.himed.com/blog/synthetic-bone-grafting-process-definition-materials#:~:text=Synthetic%20bone%20graft%20materials%20can,bioglasses%2C%20metals%2C%20and%20composites; nicolasprimola/stock.adobe.com.

[133] Toda H, Ohgaki T, Uesugi K, Kobayashi M, Kuroda N, Kobayashi T, Niinomi M, Akahori T, Makii K, Aruga Y. Quantitative assessment of microstructure and its effects on compression behavior of aluminum foams via high-resolution synchrotron x-ray tomography. Metall Mater Trans A. 2006, 37, 1211–19.

[134] Jones JR. Observing cell response to biomaterials. Mater Today. 2006, 9, 34–43.

[135] Turnbull G, Clarke J, Picard F, Riches P, Jia L, Han F, Li B, Shu W. 3D bioactive composite scaffolds for bone tissue engineering. Bioactive Mater. 2018, 3, 278–314.

[136] Johnson T, Bahrampourian R, Patel A, Mequanint K. Fabrication of highly porous tissue-engineering scaffolds using selective spherical porogens. Biomed Mater Eng. 2010, 20, 107–18.

[137] Liao C-J, Chen C-F, Chen J-H, Chiang S-F, Lin Y-J. Fabrication of porous biodegradable polymer scaffolds using a solvent merging/particulate leaching method. J Biomed Mater Res. 2002, 59, 676–81.

[138] Huo Y, Lu Y, Meng L, Wu J, Gong T, Zou J, Bosiakov S, Cheng L. A critical review on the design, Manufacturing and assessment of the bone scaffold for large bone defects. Front Bioeng Biotechnol. 2021, 9; https://doi.org/10.3389/fbioe.2021.753715.

[139] Koju N, Niraula S, Fotovvati B. Additively manufactured porous Ti6Al4V for bone implants: A review. Metals. 2022, 12, 687; https://doi.org/10.3390/met12040687.

[140] Yuan L, Ding S, Wen C. Additive manufacturing technology for porous metal implant applications and triple minimal surface structures: A review. Bioact Mater. 2019, 4, 56–70.

[141] Gao C, Wang C, Jin H, Wang Z, Li Z, Shi C, Leng Y, Yang F, Liu H, Wang J. Additive manufacturing technique-designed metallic porous implants for clinical application in orthopedics. RSC Adv. 2018, 8, 25210–27.

[142] Rana M, Karmakar SK, Pal B, Datta P, Roychowdhury A, Bandyopadhyay A. Design and manufacturing of biomimetic porous metal implants. J Mater Res. 2021, 36, 3952–62.

[143] Jahr H, Li Y, Zhou J, Zadpoor AA, Schröder K-U. Additively manufactured absorbable porous metal implants – processing, alloying and corrosion behavior. Front Mater. 2021, 8; https://doi.org/10.3389/fmats.2021.628633.

[144] Ackland DC, Robinson D, Redhead M, Lee PVS, Moskaljuk A, Dimitroulis G. A personalized 3D-printed prosthetic joint replacement for the human temporomandibular joint: From implant design to implantation. J Mech Behav Biomed Mater. 2017, 69, 404–11.

[145] Stojkovic M, Trajanovic M, Vitkovic N. Personalized orthopedic surgery design challenge: Human bone redesign method. Proc CIRP. 2019, 84, 701–06.

[146] Vitković N, Stojkovic M, Majstorovic V, Trajanovic M, Milovanovic J. Novel design approach for the creation of 3D geometrical model of personalized bone scaffold. CIRP Ann. 2018, 67, 177–80.

[147] Fennema E, Rivron N, Rouwkema J, van Blitterswijk C, de Boer J. Spheroid culture as a tool for creating 3D complex tissues. Trends Biotechnol. 2013, 31, 108–15.

[148] Nishibori M, Betts NJ, Salama H, Listgarten MA. Short-term healing of autogenous and allogeneic bone grafts after sinus augmentation: A report of 2 cases. J Periodontol. 1994, 65, 958–66.

[149] Pitaru S, Tal H, Soldinger M, Grosskopf A, Noff M. Partial regeneration of periodontal tissues using collagen barriers. Initial observations in the canine. J Periodontol. 1988, 59, 380–86.

[150] Bunyaratavej P, Wang HL. Collagen membranes: A review. J Periodontol. 2001, 72, 215–29.

[151] Hammerle CH, Jung RE. Bone augmentation by means of barrier membranes. Periodontology 2000. 2003, 33, 36–53.

[152] Liu J, Kerns DG. Mechanisms of Guided. Bone Regeneration. Open Dent J. 2014, 8, 56–65.

[153] Azhar IS, Ayulita D, Laksono H, Margaretha TA. The efficiency of PRF, PTFE, and titanium mesh with collagen membranes for vertical alveolar bone addition in dental implant therapy: A narrative review. J Intl Oral Health. 2022, 14, 543–50.

[154] Lee S-W, Kim S-G. Membranes for the Guided Bone Regeneration. Maxillofac Plast Reconstr Surg. 2014, 36, 239–46.

[155] Misch CM, Misch-Haring MA. Guided Bone Regeneration: Membrane Selection. Dent Today. 2021; https://www.dentistrytoday.com/guided-bone-regeneration-membrane-selection/.

[156] Aprile P, Letourneur D, Simon-Yarza T. Membranes for Guided Bone Regeneration: A Road from Bench to Bedside. Adv Healthcare Mater. 2020, 9, 2000707; https://doi.org/10.1002/adhm.202000707.

[157] Soldatos NK, Stylianou P, Koidou VP, Angeloomanos GE. Limitations and options using resorbable versus nonresorbable membranes for successful guided bone regeneration. Quintessence Int. 2017, 48, 131–47.

[158] Sasaki J-I, Abe GL, Thongthsi P, Tsuboi R, Kohno T, Imazato S. Barrier membranes for tissue regeneration in dentistry. Biomater Invest Dent. 2021, 8, 43–63.

[159] Evans GH, Yukna RA, Cambre KM, Gardiner DL. Clinical regeneration with guided tissue barriers. Curr Opin Periodontol. 1997, 4, 75–81.

[160] Murphy KG. Postoperative healing complications associated with Gore-Tex periodontal material. Part II. Effect of complications on regeneration. Int J Period Restor Dent. 1995, 15, 548–61.

[161] Machtei EE. The effect of membrane exposure on the outcome of regenerative procedures in humans: A meta-analysis. J Periodontol. 2001, 72, 512–16.

[162] Leinonen S, Suokas E, Veiranto M, Törmälä P, Ashammakhi N. Holding power of bioabsorbable ciprofloxacin-containing self-reinforced poly-L/DL-lactide 70/30 bioactive glass 13 miniscrews in human cadaver bone. J Craniofac Surg. 2002, 13, 212–18.

[163] Chasioti E, Chiang TF, Drew HJ. Maintaining space in localized ridge augmentation using guided bone regeneration with tenting screw technology. Quintessence Int. 2013, 44, 763–71.

[164] Tempro PJ, Nalbandian J. Colonization of retrieved polytetra-fluoroethylene membranes: Morphological and microbiological observations. J Periodontol. 1993, 64, 162–68.

[165] Misch CM, Misch CE. The repair of localized severe ridge defects for implant placement using mandibular bone grafts. Implant Dent. 1995, 4, 261–67.

[166] Nevins M, Mellonig JT. Enhancement of the damaged edentulous ridge to receive dental implants: A combination of allograft and the GORE-TEX membrane. Int J Periodont Restorat Dent. 1992, 12, 96–111.

[167] Pihlstrom BL, McHugh RB, Oliphant TH, Ortiz-Campos C. Comparison of surgical and nonsurgical treatment of periodontal disease: A review of current studies and additional results after 61/2 years. J Clin Periodontol. 1983, 10, 524–41.

[168] Rasmusson L, Sennerby L, Lundgren D, Nyman S. Morphological and dimensional changes after barrier removal in bone formed beyond the skeletal borders at titanium implants: A kinetic study in the rabbit tibia. Clin Oral Implants Res. 1997, 8, 103–16.

[169] Korzinskas T, Jung O, Smeets R, Stojanovic S, Najman S, Glenske K, Hahn M, Wenisch S, Scnettler R, Barbeck M. In vivo analysis of the biocompatibility and macrophage response of a non-resorbable PTFE membrane for guided bone regeneration. Int J Mol Sci. 2018, 19, 2952; doi: 10.3390/ijms19102952.

[170] Chiapasco M, Zaniboni M. Clinical outcomes of GBR procedures to correct peri-implant dehiscences and fenestrations: A systematic review. Clin Oral Implants Res. 2009, 20, 113–23.

[171] Garcia J, Dodge A, Luepke P, Wang H-L, Kapila Y, Lin G-H. Effect of membrane exposure on guided bone regeneration: A systematic review and meta-analysis. Clin Oral Implants Res. 2018, 29, 328–38.

[172] Sheikh Z, Abdallah M, Hamdan N, Javaid M, Khurshid Z, Matilinna K. Barrier membranes for tissue regeneration and bone augmentation techniques in dentistry. In: Matinlinna J (Ed.), Handbook of Oral Biomaterials. Pan Stanford Publishing, Singapore. 2014, 605–36.

[173] Nyman S, Lindhe J, Karring T, Rylander H. New attachment following surgical treatment of human periodontal disease. J Clin Periodontol. 1982, 9, 290–96.

[174] Simion M, Baldoni M, Rossi P, Zaffe D. A comparative study of the effectiveness of e-PTFE membranes with and without early exposure during the healing period. Int J Period Restor Dent. 1994, 14, 166–80.

[175] Aaboe M, Pinholt EM, Hjørting-Hansen E. Healing of experimentally created defects: A review. Br J Oral Maxillofac Surg. 1995, 33, 312–18.

[176] Zhang Y, Zhang X, Shi B, Miron RJ. Membranes for guided tissue and bone regeneration. Ann Oral Maxillofac Surg. 2013, 1, 1–10.

[177] Buser D, Bragger U, Lang NP, Nyman S. Regeneration and enlargement of jaw bone using guided tissue regeneration. Clinl Oral Implants Res. 1990, 1, 22–32.

[178] Becker W, Dahlin C, Becker BE, Lekholm U, van Steenberghe D, Higuchi K, Kultje C. The use of e-PTFE barrier membranes for bone promotion around titanium implants placed into extraction sockets: A prospective multicenter study. Int J Oral Maxillofac Implants. 1994, 9, 31–40.

[179] Nyman S, Lang NP, Buser D, Bragger U. Bone regeneration adjacent to titanium dental implants using guided tissue regeneration: A report of two cases. Int J Oral Maxillofac Implants. 1990, 5, 9–14.

[180] Becker W, Becker BE. Guided tissue regeneration for implants placed into extraction sockets and for implant dehiscences: Surgical techniques and case report. Int J Periodont Restorat Dent. 1990, 10, 376–91.

[181] Bartee BK. Evaluation of a new polytetrafluoroethylene guided tissue regeneration membrane in healing extraction sites. Compend Cont Edu Dent. 1998, 19, 125–28.

[182] Barber HD, Lignelli J, Smith BM, Bartee BK. Using a dense PTFE membrane without primary closure to achieve bone and tissue regeneration.Journal of oral and maxillofacial surgery. J Am Assoc Oral Maxillofac Sur. 2007, 65, 748–52.

[183] Hoffmann O, Bartee BK, Beaumont C, Kasaj A, Deli G, Zafiropoulos G-G. Alveolar bone preservation in extraction sockets using non-resorbable dPTFE membranes: A retrospective non-randomized study. J Periodontol. 2008, 79, 1355–69.

[184] Crump TB, Rivera-Hidalgo F, Harrison JW, Williams FE, Guo IY. Influence of three membrane types on healing of bone defects. Oral Surg Oral Med Oral Pathol Oral Radiol Endod. 1996, 82, 365–74.

[185] Sumi Y, Miyaishi O, Tohnai I, Ueda M. Alveolar ridge augmentation with titanium mesh and autogenous bone. Oral Surg Oral Med Oral Pathol Oral Radiol Endod. 2000, 89, 268–70.

[186] Malchiodi L, Scarano A, Quaranta M, Piattelli A. Rigid fixation by means of titanium mesh in edentulous ridge expansion for horizontal ridge augmentation in the maxilla. Int J Oral Maxillofac Implants. 1998, 13, 701–05.

[187] Steflik DE, Corpe RS, Young TR, Buttle K. In vivo evaluation of the biocompatibility of implanted biomaterials morphology of the implant-tissue interactions. Implant Dent. 1998, 7, 338–50.

[188] Brunette DM. Titanium in Medicine: Material Science, Surface Science, Engineering, Biological Responses and Medical Applications. Springer, Berlin. 2001.

[189] Elgali I, Oamr O, Dahlin C, Thomsen P. Guided bone regeneration: Materials and biological mechanisms revisited. Eur J Oral Sci. 2017, 125, 315–37.

[190] Jovanovic SA, Schenk RK, Orsini M, Kenney EB. Supracrestal bone formation around dental implants an experimental dog study. Int J Oral Maxillofac Implants. 1995, 10, 23–31.

[191] Jaquiéry C, Aeppli C, Cornelius P, Palmowsky A, Kunz C, et al. Reconstruction of orbital wall defects: Critical review of 72 patients. Int J Oral Maxillofac Surg. 2007, 36, 193–99.

[192] Xie Y, Li S, Zhang T, Wang C, Cai X. Titanium mesh for bone augmentation in oral implantology: Current application and progress. Int J Oral Sci. 2020, 12; doi: 10.1038/s41368-020-00107-z.

[193] von Arx T, Hardt N, Wallkamm B. The TIME technique: A new method for localized alveolar ridge augmentation prior to placement of dental implants. Int J Oral Maxillofac Implants. 1996, 11, 387–94.

[194] Boonsirijarungradh S, Udomsom S, Manaspon C, Paengnakorn P. Development of customized biodegradable mesh membrane for dental bone graft using three-dimensional printing technique. Mater Today. 2022, 65; doi: 10.1016/j.matpr.2022.05.303.

[195] Nguyen T-D, Moon S-H, Oh TJ, Seol J-J, Lee M-H, Bae T-S. Comparison of guided bone regeneration between surface-modified and pristine titanium membranes in a rat calvarial model. Int J Oral Maxillofac Implants. 2016, 31, 581–90.

[196] Nguyen T-D, Kim Y-K, Kim S-Y, Lee M-H, Bae T-S. Osteogenesis-related gene expression and guided bone regeneration of a strontium-doped calcium-phosphate-coated titanium mesh. ACS Biomater Sci Eng. 2019, 5, 6715–24.

[197] Hirota M, Ikeda T, Tabuchi M, Ozawa T, Tohnai I, Ogawa T. Effects of ultraviolet photofunctionalization on bone augmentation and integration capabilities of titanium mesh and implants. Int J Oral Maxillofac Implants. 2017, 32, 52–62.

[198] Att W, Ogawa T. Biological aging of implant surfaces and their restoration with ultraviolet light treatment: A novel understanding of osseointegration. Int J Oral Maxillofac Implants. 2012, 27, 753–61.

[199] Okubo T, Tsukimura N, Taniyama T, Ishijima M, Nakhaei K, Rezaei NM, Hirota M, Park W, Akita D, Tateno A, Ishigami T, Ogawa T. Ultraviolet treatment restores bioactivity of titanium mesh plate degraded by contact with medical gloves. J Oral Sci. 2018, 60, 567–73.

[200] Miyazaki T, Yutani T, Murai N, Kawata A, Shimizu H, Uejima N, Miyazaki Y, Oshida Y. Early osseointegration attained by UV-photo treated implant into piezosurgery-prepared site. Report III. influence of surface treatment by hydrogen peroxide solution and determination of early loading timing. Int J Dent Oral Health. 2021, 7; dx.doi.org/10.16966/2378-7090.381.

[201] Rakhmatia YD, Ayukawa Y, Furuhashi A, Koyano K. Current barrier membranes: Titanium mesh and other membranes for guided bone regeneration in dental applications. J Prosthodont Res. 2013, 57, 3–14.

[202] Hasegawa H, Masui S, Ishihata H. New microperforated pure titanium membrane created by laser processing for guided regeneration of bone. Br J Oral Maxillofac Surg. 2018, 56, 642–43.

[203] Otawa N, Sumida T, Kitagaki H, Sasaki K, Fujibayashi S, Takemoto M, Nakamura T, Yamada T, Mori Y, Matsushita T. Custom-made titanium devices as membranes for bone augmentation in implant treatment: Modeling accuracy of titanium products constructed with selective laser melting. J Cranio-Maxillofac Surg. 2015, 43, 1289–95.

[204] Decco O, Cura A, Beltrán V, Lezcano M, Engelke W. Bone augmentation in rabbit tibia using microfixed cobalt-chromium membranes with whole blood, tricalcium phosphate and bone marrow cells. Int J Clin Exp Med. 2015, 8, 135–44.

[205] Lin W-C, Yao C, Huang T-Y, Cheng S-J, Tang C-M. Long-term in vitro degradation behavior and biocompatibility of polycaprolactone/cobalt-substituted hydroxyapatite composite for bone tissue engineering. Dent Mater. 2019, 35, 751–62.

[206] Eliaz N. Corrosion of metallic biomaterials: A review. Materials (Basel). 2019, 12, 407; doi: 10.3390/ma12030407.

[207] Harris RJ. Clinical evaluation of a composite bone graft with a calcium sulfate barrier. J Periodontol. 2004, 75, 685–92.

[208] Melo LG, Nagata MJ, Bosco AF, Ribeiro LL, Leite CM. Bone healing in surgically created defects treated with either bioactive glass particles, a calcium sulfate barrier, or a combination of both materials. A histological and histometric study in rat tibias. Clin Oral Implants Res. 2005, 16, 683–91.

[209] Anderud J, Jimbo R, Abrahamsson P, Isaksson SG, Adolfsson E, Malmstrom J, Kozai Y, Hallmer F, Wennerberg A. Guided bone augmentation using a ceramic space-maintaining device. Oral Surg Oral Med Oral Pathol Oral Radiol. 2014, 118, 532–38.

[210] Basile MA, d'Ayala GG, Malinconico M, Laurienzo P, Coudane J, Nottelet B, Ragione FD, Oliva A. Functionalized PCL/HA nanocomposites as microporous membranes for bone regeneration. Mater Sci Eng C Mater Biol Appl. 2015, 48, 457–68.

[211] Ribeiro N, Sousa SR, van Blitterswijk CA, Moroni L, Monteiro FJ. A biocomposite of collagen nanofibers and nanohydroxyapatite for bone regeneration. Biofabrication. 2014, 6; doi: 10.1088/1758-5082/6/3/035015.

[212] Shim JH, Huh JB, Park JY, Jeon YC, Kang SS, Kim JY, Rhie JW, Cho DW. Fabrication of blended polycaprolactone/poly(lactic-co-glycolic acid)/beta-tricalcium phosphate thin membrane using solid freeform fabrication technology for guided bone regeneration. Tissue Eng Part A. 2013, 19, 317–28.

[213] Mota J, Yu N, Caridade SG, Luz GM, Gomes ME, Reis RL, Jansen JA, Walboomers XF, Mano JF. Chitosan/bioactive glass nanoparticle composite membranes for periodontal regeneration. Acta Biomater. 2012, 8, 4173–80.

[214] Hutmacher D, Hurzeler MB, Schliephake H. A review of material properties of biodegradable and bioresorbable polymers and devices for GTR and GBR applications. Int J Oral Maxillofac Implants. 1996, 11, 667–78.

[215] Lekovic V, Camargo PM, Klokkevold PR, Weinlaender M, Kenny EB, Dimitrijevic B, Nedic M. Preservation of alveolar bone in extraction sockets using bioabsorbable membranes. J Periodontol. 1998, 69, 1044–49.

[216] Simon BI, Von Hagen S, Deasy MJ, Faldu M, Resnansky D. Changes in alveolar bone height and width following ridge augmentation using bone graft and membranes. J Periodontol. 2000, 71, 1774–91.

[217] Piattelli A, Scarano A, Coraggio F, Matarasso S. Early tissue reactions to polylactic acid resorbable membranes: A histological and histochemical study in rabbit. Biomaterials. 1998, 19, 889–96.

[218] Simion M, Maglione M, Iamoni F, Scarano A, Piattelli A, Salvato A. Bacterial penetration through Resolut resorbable membrane in vitro: An histological and scanning electron microscopic study. Clin Oral Implants Res. 1997, 8, 23–31.

[219] Sam G, Pillai BR. Evolution of barrier membranes in periodontal regeneration-"Are the third generation membranes really here?". J Clin Diagn Res. 2014, 8, ZE14–ZE17.

[220] Bottino MC, Pankajakshan D, Nör JE. Advanced scaffolds for dental pulp and periodontal regeneration. Dent Clin North Am. 2017, 61, 689–711.

[221] Döri F, Huszár T, Nikolidakis D, Arweiler NB, Gera I, Sculean A. Effect of platelet-rich plasma on the healing of intra-bony defects treated with a natural bone mineral and a collagen membrane. J Clin Periodontol. 2007, 34, 254–61.

[222] Hoogeveen EJ, Gielkens PF, Schortinghuis J, Ruben JL, Huysmans M-CDNJM, Stegenga B. Vivosorb as a barrier membrane in rat mandibular defects. An evaluation with transversal microradiography. Int J Oral Maxillofac Surg. 2009, 38, 870–75.

[223] Wang J, Wang L, Zhou Z, Lai H, Xu P, Liao L, Wei J. Biodegradable polymer membranes applied in guided bone/tissue regeneration: A review. Polymers. 2016, 8, 115; doi: 10.3390/polym8040115.

[224] Wang Z, Liang R, Jiang X, Xie J, Cai P, Chen H, Zhan X, Lei D, Zhao J, Zheng L. Electrospun PLGA/PCL/OCP nanofiber membranes promote osteogenic differentiation of mesenchymal stem cells (MSCs). Mater Sci Eng C Mater Biol Appl. 2019, 10, 109796; doi: 10.1016/j.msec.2019.109796.

[225] Annunziata M, Nastri L, Cecoro G, Guida L. The use of poly-D,L-lactic acid (PDLLA) devices for bone augmentation techniques: A systematic review. Molecules. 2017, 22, 2214; doi: 10.3390/molecules22122214.

[226] Dos Santos VI, Merlini C, Aragones Á, Cesca K, Fredel MC. In vitro evaluation of bilayer membranes of PLGA/hydroxyapatite/β-tricalcium phosphate for guided bone regeneration. Mater Sci Eng C Mater Biol Appl. 2020, 112; doi: 10.1016/j.msec.2020.110849.

[227] Yoshimoto I, Sasaki JI, Tsuboi R, Yamaguchi S, Kitagawaw H, Imazato S. Development of layered PLGA membranes for periodontal tissue regeneration. Dent Mater. 2018, 34, 538–50.

[228] Haghighat A, Shakeri S, Mehdikhani M, Dehnavi SS, Talebi A. Histologic, histomorphometric, and osteogenesis comparative study of a novel fabricated nanocomposite membrane versus cytoplast membrane. J Oral Maxillofac Surg. 2019, 77, 2027–39.

[229] Zhang HY, Jiang HB, Ryu JH, Kang H, Kim K-M. Kwon J-S. Comparing properties of variable pore-sized 3D-printed PLA membrane with conventional PLA membrane for guided bone/tissue regeneration. Materials. 2019, 12, 1718; doi: 10.3390/ma12101718.

[230] Abe GL, Sasaki JI, Katata C, Kohno T, Tsuboi R, Kitagawa H, Imazato S. Fabrication of novel poly(lactic acid/caprolactone) bilayer membrane for GBR application. Dent Mater. 2020, 36, 626–34.

[231] Bunyaratavej P, Wang HL. Collagen membranes: A review. J Periodontol. 2001, 72, 215–29.

[232] Wang HL, Carroll MJ. Guided bone regeneration using bone grafts and collagen membranes. Quint Int. 2001, 32, 504–15.

[233] Postlethwaite AE, Seyer JM, Kang AH. Chemotactic attraction of human fibroblasts to type I II and III collagens and collagen-derived peptides. Proc Nat Acad Sci USA. 1978, 75, 871–75.

[234] Locci P, Calvitti M, Belcastro S, Pugliese M, Guerra M, Marinucci L, Staffolani N. Phenotype expression of gingival fibroblasts cultured on membranes used in guided tissue regeneration. J Periodontol. 1997, 68, 857–63.

[235] Schlegel AK, Mohler H, Busch F, Mehl A. Preclinical and clinical studies of a collagen membrane (Bio-Gide). Biomaterials. 1997, 18, 535–38.

[236] Rothamel D, Schwarz F, Sculean A, Herten M, Scherbaum W, Becker J. Biocompatibility of various collagen membranes in cultures of human PDL fibroblasts and human osteoblast-like cells. Clin Oral Implants Res. 2004, 15, 443–49.

[237] Pitaru S, Tal H, Soldinger M, Noff M. Collagen membranes prevent apical migration of epithelium and support new connective tissue attachment during periodontal wound healing in dogs. J Periodont Res. 1989, 24, 247–53.

[238] Chung KM, Salkin LM, Stein MD, Freedman AL. Clinical evaluation of a biodegradable collagen membrane in guided tissue regeneration. J Periodontol. 1990, 61, 732–36.

[239] Blumenthal N, Steinberg J. The use of collagen membrane barriers in conjunction with combined demineralized bone-collagen gel implants in human infrabony defects. J Periodontol. 1990, 61, 319–27.

[240] Paul BF, Mellonig JT, Towle HJ III, Gray JL. Use of a collagen barrier to enhance healing in human periodontal furcation defects. Int J Periodont Restorat Dent. 1992, 12, 123–31.

[241] Van Swol RL, Ellinger R, Pfeifer J, Barton RL, Ellinger R, Pfeifer J, Barton NE, Blumenthal N. Collagen membrane barrier therapy to guide regeneration in Class II furcations in humans. J Periodontol. 1993, 64, 622–29.

[242] Blumenthal NM. A clinical comparison of collagen membranes with e-PTFE membranes in the treatment of human mandibular buccal class II furcation defects. J Periodontol. 1993, 64, 925–33.

[243] Ma S, Adayi A, Liu Z, Li M, Wu M, Xiao L, Sun Y, Cai Q, Yang X, Zhang X, Gao P. Asymmetric collagen/chitosan membrane containing minocycline-loaded chitosan nanoparticles for guided bone regeneration. Sci Rep. 2016, 6, 31822; doi: 10.1038/srep31822.

[244] Wu C, Su H, Karydis A, Anderson KM, Ghadri N, Tang S, Wang Y, Bumgardner JD. Mechanically stable surface-hydrophobilized chitosan nanofibrous barrier membranes for guided bone regeneration. Biomed Mater. 2017, 13, 015004; doi: 10.1088/1748-605X/aa853c.

[245] Zhou T, Liu X, Sui B, Liu C, Mo X, Sun J. Development of fish collagen/bioactive glass/chitosan composite nanofibers as a GTR/GBR membrane for inducing periodontal tissue regeneration. Biomed Mater. 2017, 12, 055004; doi: 10.1088/1748-605X/aa7b55.

[246] Huang D, Niu L, Li J, Du J, Wei Y, Hu Y, Lian X, Chen W, Wang K. Reinforced chitosan membranes by microspheres for guided bone regeneration. J Mech Behav Biomed Mater. 2018, 81, 195–201.

[247] Ishikawa K, Ueyama Y, Mano T, Koyama Y, Suzuki K, Matsumura T. Self-setting barrier membrane for guided tissue regeneration method: Initial evaluation of alginate membrane made with sodium alginate and calcium chloride aqueous solutions. J Biomed Mater Res. 1999, 47, 111–15.

[248] Ueyama Y, Ishikawa K, Mano T, Koyama T, Nagatsuka H, Suzuki K, Ryoke K. Usefulness as guided bone regeneration membrane of the alginate membrane. Biomaterials. 2002, 23, 2027–33.

[249] He H, Huang J, Ping F, Sun G. Chen G. Calcium alginate film used for guided bone regeneration in mandible defects in a rabbit model. Cranio. 2008, 26, 65–70.

[250] Lundgren AK, Sennerby L, Lundgren D, Taylor A, Gottlow J, Nyman S. Bone augmentation at titanium implants using autologous bone grafts and a bioresorbable barrier: An experimental study in the rabbit tibia. Clin Oral Implants Res. 1997, 8, 82–89.

[251] Simion M, Misitano U, Gionso L, Salvato A. Treatment of dehiscences and fenestrations around dental implants using resorbable and nonresorbable membranes associated with bone autografts: A comparative clinical study. Int J Oral Maxillofac Implants. 1997, 12, 159–67.

[252] Donos N, Kostopoulos L, Karring T. Alveolar ridge augmentation using a resorbable copolymer membrane and autogenous bone grafts: An experimental study in the rat. Clin Oral Implants Res. 2002, 13, 203–13.

[253] Lee J-Y, Kim Y-K. Comparative analysis of guided bone regeneration using autogenous tooth bone graft material with and without resorbable membrane. J Dental Sci. 2013, 8, 281–86.

[254] Liu Y, Zheng Y. Degradable, absorbable or resorbable – What is the best grammatical modifier for an implant that is eventually absorbed by the body? Sci China Mater. 2017, 60; doi: 10.1007/s40843-017-9023-9.

[255] Doble M. Biodegradation and Bioresorption; https://www.youtube.com/watch?v=ppJec7BNrRE

[256] Presnyakov EV, Bozo IY, Smirnov IV, Komlev VS, Popov VK, Mironov AV, Deev RV. Bioresorption and biodegradation of the 3D-printed gene-activated bone substitutes based on octacalcium phosphate. Genes Cells. 2020, 15, 66–70.

[257] LeGeros RZ. Biodegradation and bioresorption of calcium phosphate ceramics. Clin Mater. 1993, 14, 65–88.

[258] Das D, Zhang Z, Winkleer T, Mour M, Gunter C, Morlock M, Machens H-G, Schilling AF. Bioresorption and degradation of biomaterials. Adv Biochem Eng Biotechnol. 2012, 126, 317–33.

[259] Forrestal B, Case BC, Yerasi C, Musallam A, Chezar-Azerrad C, Waksman R. Bioresorbable scaffolds: Current technology and future perspectives. Ranbam Maimonides Med J. 2020, 11, e0016; doi: 10.5041/RMMJ.10402.

[260] Peng X, Qu W, Jia Y, Wang Y, Yu B, Tian J. Bioresorbable scaffolds: Contemporary status and future directions. Front Cardiovasc Med. 2020, 7, 589571; doi: 10.3389/fcvm.2020.589571.

[261] Moravej M, Mantovani D. Biodegradable metals for cardiovascular stent application: Interests and new opportunities. Int J Mol Sci. 2011, 12, 4250–70.

[262] Bartosch M, Schubert S, Berger F. Magnesium stents – Fundamentals, biological implications and applications beyond coronary arteries. BioNano Mater. 2015, 16; https://doi.org/10.1515/bnm-2015-0004.

[263] Zheng YF, Gu XN, Witte F. Biodegradable metals. Mater Sci Eng. 2014, 77, 1–34.

[264] Jahr H, Li Y, Zhou J, Zadpoor AA, Schröder K-U. Additively manufactured absorbable porous metal implants – Processing, alloying and corrosion behavior. Front Mater. 2021, 8; https://doi.org/10.3389/fmats.2021.628633.

[265] Hermawan H. Updates on the research and development of absorbable metals for biomedical applications. Prog Biomater. 2018, 7, 93–110.

[266] Erbel R, Di Mario C, Bartunek J, Bonnier J, de Bruyne B, Eberli FR, Erne P, Haude M, Heublein B, Horrigan M, Ilsley C, Böse D, Koolen J, Lüscher TF, Weissman N, Waksman R. Temporary scaffolding of coronary arteries with bioabsorbable magnesium stents: A prospective, non-randomised multicentre trial. Lancet. 2007, 369, 1869–75.

[267] González S, Pellicer E, Suriñach S, Baró MD, Sort J. Biodegradation and mechanical integrity of magnesium alloys suitable for implants. In: Biodegradation – Engineering and Technology. 2013; https://www.intechopen.com/books/biodegradation-engineering-and-technology/biodegradation-and-mechanical-integrity-of-magnesium-and-magnesium-alloys-suitable-for-implants.

[268] Sezer N, Evis Z, Kayhan SM, Tahmasebifar A, Koç M. Review of magnesium-based biomaterials and their applications. J Magnes Alloys. 2018, 6, 23–43.

[269] Oshida Y, Tuna EB, Aktören O, Gençay K. Dental Implant Systems. Int J Mol Sci. 2010, 11, 1580–678.

[270] Rahim MI, Ullah S, Mueller PP. Advances and challenges of biodegradable implant materials with a focus on magnesium-alloys and bacterial infections. Metals. 2018, 8, 532; doi: 10.3390/met8070532.

[271] Sheikh Z, Najeeb S, Khurshid Z, Verma V, Rashid H, Glogauer M. Biodegradable Materials for Bone Repair and Tissue Engineering Applications. Materials (Basel). 2015, 8, 5744–94.

[272] Middleton JC, Tipton AJ. Synthetic biodegradable polymers as orthopedic devices. Biomaterials. 2000, 21, 2335–46.

[273] Pogorielov M, Husak E, Solodivnik A, Zhdanov S. Magnesium-based biodegradable alloys: Degradation, application, and alloying elements. Interv Med Appl Sci. 2017, 9, 27–38.

[274] Sanchez AH, Luthringer BJ, Feyerabend F, Willumeit R. Mg and Mg alloys: How comparable are in vitro and in vivo corrosion rates? A review. Acta Biomater. 2015, 13, 16–31.

[275] Pierson D, Edick J, Tauscher A, Pokorney E, Bowen P, Gelbaugh J, Stinson J, Getty H, Lee CH, Drelich J, Goldman J. A simplified in vivo approach for evaluating the bioabsorbable behavior of candidate stent materials. J Biomed Mater Res B Appl Biomater. 2012, 100, 58–67.

[276] Hermawan H, Purnama A, Dube D, Couet J, Mantovani D. Fe-Mn alloys for metallic biodegradable stents: Degradation and cell viability studies. Acta Biomater. 2010, 6, 1852–60.

[277] Yang H, Wang C, Liu C, Chen H, Wu Y, Han J, Jia Z, Lin W, Zhang D, Li W, Yuan W, Guo H, Li H, Yang G, Kong D, Zhu D, Takashima K, Ruan L, Nie J, Li X, Zheng Y. Evolution of the degradation mechanism of pure zinc stent in the one-year study of rabbit abdominal aorta model. Biomaterials. 2017, 145, 92–105.

[278] Zhao D, Wang T, Nahan K, Guo X, Zhang Z, Dong Z, Chen S, Chou DT, Hong D, Kumta PN, Heineman WR. In vivo characterization of magnesium alloy biodegradation using electrochemical H_2 monitoring, ICP-MS, and XPS. Acta Biomater. 2017, 50, 556–65.

[279] Zhao D, Witte F, Lu F, Wang J, Li J, Qin L. Current status on clinical applications of magnesium-based orthopaedic implants: A review from clinical translational perspective. Biomaterials. 2017, 112, 287–302.

[280] Li X, Liu X, Wu S, Yeung KWK, Zheng Y, Chu PK. Design of magnesium alloys with controllable degradation for biomedical implants: From bulk to surface. Acta Biomater. 2016, 45, 2–30.

[281] Evis Z, Kayhan SM, Tahmasebifar A, Koç M. Review of magnesium-based biomaterials and their applications. J Magnes Alloys. 2018, 6, 23–43.

[282] Hu H, Nie X, Ma Y. Corrosion and surface treatment of magnesium alloys. Magnesium Alloys – Properties in Solid and Liquid States. 2014; https://www.intechopen.com/books/magnesium-alloys-properties-in-solid-and-liquid-states/corrosion-and-surface-treatment-of-magnesium-alloys.

[283] Agarwal S, Curtin J, Duffy B, Jaiswal S. Biodegradable magnesium alloys for orthopaedic applications: A review on corrosion, biocompatibility and surface modifications. Mater Sci Eng C. 2016, 68, 948–63.

[284] Song Y, Zhang S, Li J, Zhao C, Zhang X. Electrodeposition of Ca-P coatings on biodegradable Mg alloy: In vitro biomineralization behavior. Acta Biomater. 2010, 6, 1736–42.

[285] Matykina E, Garcia I, Arrabal R, Mohedano M, Pardo A. Role of PEO coatings in long-term biodegradation of a Mg alloy. Appl Surf Sci. 2016, 389, 810–23.

[286] Cui X-J, Lin X-Z, Liu C-H, Yang R-S, Zheng X-W, Gong M. Fabrication and corrosion resistance of a hydrophobic micro-arc oxidation coating on AZ31 Mg alloy. Corros Sci. 2015, 90, 402–12.

[287] Hench LL, Polak JM. Third-generation biomedical materials. Science. 2002, 295, 1014–17.

[288] Wei S, Ma J-X, Xu L, Gu X-S, Ma X-L. Biodegradable materials for bone defect repair. Military Med Res. 2020, 7, 54; https://doi.org/10.1186/s40779-020-00280-6.

[289] Chanlalit C, Shukla DR, Fitzsimmons JS, An KN, O'driscoll SW. Stress shielding around radial head prostheses. J Hand Surg Am. 2012, 37, 2118–25.

[290] Salgado AJ, Coutinho OP, Reis RL. Bone tissue engineering: State of the art and future trends. Macromol Biosci. 2004, 4, 743–65.

[291] Garric X, Nottelet B, Pinese C, Leroy A, Coudane J. Biodegradable synthetic polymers for the design of implantable medical devices: The ligamentoplasty case. Med Sci (Paris). 2017, 33, 39–45.

[292] Wegst UG, Bai H, Saiz E, Tomsia AP, Ritchie RO. Bioinspired structural materials. Nat Mater. 2015, 14, 23–36.

[293] Di Martino A, Sittinger M, Risbud MV. Chitosan: A versatile biopolymer for orthopaedic tissue-engineering. Biomaterials. 2005, 26, 5983–90.

[294] Noori A, Ashrafi SJ, Vaez-Ghaemi R, Hatamian-Zaremi A, Webster TJ. A review of fibrin and fibrin composites for bone tissue engineering. Int J Nanomed. 2017, 12, 4937–61.

[295] Malafaya PB, Silva GA, Reis RL. Natural-origin polymers as carriers and scaffolds for biomolecules and cell delivery in tissue engineering applications. Adv Drug Deliv Rev. 2007, 59, 207–33.

[296] Mondrinos MJ, Dembzynski R, Lu L, Byrapogu VKC, Wootton DM, Lelkes PI, Zhou J. Porogen-based solid freeform fabrication of polycaprolactone-calcium phosphate scaffolds for tissue engineering. Biomaterials. 2006, 27, 4399–408.

[297] Okazaki M, Sano Y, Mori Y, Sakao N, Yukumi S, Shigematsu H, Izutani H. Two cases of granuloma mimicking local recurrence after pulmonary segmentectomy. J Cardiothorac Surg. 2020, 15, 7; doi: 10.1186/s13019-020-1055-z.

[298] Sherwood JK, Riley SL, Palazzolo R, Brown SC, Monkhouse DC, Coates M, Griffith LG, Landeen LK, Ratcliffe A. A three-dimensional osteochondral composite scaffold for articular cartilage repair. Biomaterials. 2002, 23, 4739–51.

[299] Pietrzak WS. Principles of development and use of absorbable internal fixation. Tissue Eng. 2000, 6, 425–33.

[300] Giordano C, Sanginario V, Ambrosio L, Silvio LD, Santin M. Chemical-physical characterization and *in vitro* preliminary biological assessment of hyaluronic acid benzyl ester-hydroxyapatite composite. J Biomater Appl. 2006, 20, 237–52.

[301] Alizadeh-Osgouei M, Li Y, Wen C. A comprehensive review of biodegradable synthetic polymer-ceramic composites and their manufacture for biomedical applications. Bioact Mater. 2019, 4, 22–36.

[302] Ko CL, Chen JC, Hung CC, Wang JC, Tien YC, Chen WC. Biphasic products of dicalcium phosphate-rich cement with injectability and nondispersibility. Mater Sci Eng C Mater Biol Appl. 2014, 39, 40–46.

[303] Hench LL. The story of bioglass. J Mater Sci Mater Med. 2006, 17, 967–78.

[304] Fiume E, Barberi J, Verné E, Baino F. Bioactive glasses: From parent 45s5 composition to scaffold-assisted tissue-healing therapies. J Funct Biomater. 2018, 9, 24; https://doi.org/10.3390/jfb9010024.

[305] Hench LL, Jones JR. Bioactive glasses: Frontiers and challenges. Front Bioeng Biotechnol. 2015, 3, 2015; https://doi.org/10.3389/fbioe.2015.00194.

[306] Brauer DS. Bioactive glasses – Structure and properties. J German Chem Soc. 2015; https://doi.org/10.1002/anie.201405310.

[307] Seitz JM, Durisin M, Goldman J, Drelich JW. Recent advances in biodegradable metals for medical sutures: A critical review. Adv Healthc Mater. 2015, 4, 1915–36.

[308] Myrissa A, Agha NA, Lu Y, Martinelli E, Eichler J, Szakács G, Kleinhans C, Willumeit-Römer R, Schäfer U, Weinberg A-M. In vitro and in vivo comparison of binary mg alloys and pure mg. Mater Sci Eng C Mater Biol Appl. 2016, 61, 865–74.

[309] Wang H, Zheng Y, Liu J, Jiang C, Li Y. In vitro corrosion properties and cytocompatibility of Fe-Ga alloys as potential biodegradable metallic materials. Mater Sci Eng C Mater Biol Appl. 2017, 71, 60–66.

[310] Drelich AJ, Zhao S, Guillory RJ 2nd, Drelich JW, Goldman J. Long-term surveillance of zinc implant in murine artery: Surprisingly steady biocorrosion rate. Acta Biomater. 2017, 58, 539–49.

[311] Tan L, Yu X, Peng W, Ke Y. Biodegradable materials for bone repairs: A review. J Mater Sci Technol. 2013, 29, 503–13.

[312] Kubota K, Mabuchi M, Higashi K. Review processing and mechanical properties of fine-grained magnesium alloys. J Mater Sci. 1999, 34, 2255–62.

[313] Erbel R, Di Mario C, Bartunek J, Bonnier J, de Bruyne B, Eberli FR, Erne P, Haude M, Heublein B, Horrigan M, Ilsley C, Böse D, Koolen J, Lüscher TF, Weissman N, Waksman R. Temporary scaffolding of coronary arteries with bioabsorbable magnesium stents: A prospective, non-randomised multicentre trial. Lancet. 2007, 369, 1869–75.

[314] He J, He FL, Li DW, Liu YL, Liu YY, Ye YJ, Yin D-C. Advances in Fe-based biodegradable metallic materials. Rsc Adv. 2016; https://pubs.rsc.org/en/content/articlelanding/2016/ra/c6ra20594a.

[315] Kraus T, Moszner F, Fischerauer S, Fiedler M, Martinelli E, Eichler J, Witte F, WIllbold E, Schinhammer M, Meischel M, Uggowitzer PJ, Löffler JF, Weinberg A. Biodegradable Fe-based alloys for use in osteosynthesis: Outcome of an in vivo study after 52 weeks. Acta Biomater. 2014, 10, 3346–53.

[316] Peuster M, Hesse C, Schloo T, Fink C, Beerbaum P, Von Schnakenburg C. Long-term biocompatibility of a corrodible peripheral iron stent in the porcine descending aorta. Biomaterials. 2006, 27, 4955–62.

[317] Zhu S, Nan H, Li X, Yu Z, Liu H, Lei Y, Sun H, Yao Y. Biocompatibility of Fe–O films synthesized by plasma immersion ion implantation and deposition. Surf Coat Technol. 2009, 203, 1523–29.

[318] Hermawan H, Dubé D, Mantovani D. Degradable metallic biomaterials: Design and development of Fe-Mn alloys for stents. J Biomed Mater Res A. 2010, 93, 1–11.

[319] Huang T, Cheng J, Bian D, Zheng Y. Fe-Au and Fe-Ag composites as candidates for biodegradable stent materials. J Biomed Mater Res B Appl Biomater. 2016, 104, 225–40.

[320] van Hengel IAJ, Riool M, Fratila-Apachitei LE, Witte-Bouma J, Farrell E, Zadpoor AA, Zaat SAJ, Apachitei I. Selective laser melting porous metallic implants with immobilized silver nanoparticles kill and prevent biofilm formation by methicillin-resistant Staphylococcus aureus. Biomaterials. 2017, 140, 1–15.

[321] Li Y, Jahr H, Lietaert K, Pavanram P, Yilmaz A, Fockaert LI, Leeflang MA, Pouran B, Gonzalez-Garcia Y, Weinans H, Mol JMC, Zhou J, Zadpoor AA. Additively manufactured biodegradable porous iron. Acta Biomater. 2018, 77, 380–93.

[322] Li Y, Jahr H, Pavanram P, Bobbert FSL, Paggi U, Zhang XY, Pouran B, Leeflang MA, Weinans H, Zhou J, Zadpoor AA. Additively manufactured functionally graded biodegradable porous iron. Acta Biomater. 2019, 96, 646–61.

[323] Vojtěch D, Kubásek J, Serák J, Novák P. Mechanical and corrosion properties of newly developed biodegradable Zn-based alloys for bone fixation. Acta Biomater. 2011, 7, 3515–22.

[324] Mccall KA, Huang C, Fierke CA. Function and mechanism of zinc metalloenzymes. J Nutr. 2000, 130, 1437s–46s.

[325] Yusa K, Yamamoto O, Iino M, Takano H, Fukuda M, Qiao Z, Sugiyama T. Eluted zinc ions stimulate osteoblast differentiation and mineralization in human dental pulp stem cells for bone tissue engineering. Arch Oral Biol. 2016, 71, 1629.

[326] An S, Gong Q, Huang Y. Promotive effect of zinc ions on the vitality, migration, and osteogenic differentiation of human dental pulp cells. Biol Trace Elem Res. 2017, 175, 112–21.

[327] Bowen PK, Drelich J, Goldman J. Zinc exhibits ideal physiological corrosion behavior for bioabsorbable stents. Adv Mater. 2013, 25, 2577–82.

[328] Li HF, Xie XH, Zheng YF, Cong Y, Zhou FY, Qiu KJ, Wang X, Chen SH, Huang L, Tian L, Qin L. Development of biodegradable Zn-1X binary alloys with nutrient alloying elements Mg, Ca and Sr. Sci Rep. 2015, 5; doi: 10.1038/srep10719.

[329] Tiffany AS, Gray DL, Woods TJ, Subedi K, Harley BAC. The inclusion of zinc into mineralized collagen scaffolds for craniofacial bone repair applications. Acta Biomater. 2019, 93, 86–96.

[330] Zhang N, Zhao D, Liu N, Wu Y, Yang J, Wang Y, Xie J, Wang Y, Yan J. Assessment of the degradation rates and effectiveness of different coated mg-Zn-Ca alloy scaffolds for in vivo repair of critical-size bone defects. J Mater Sci Mater Med. 2018, 29, 138; https://doi.org/10.1007/s10856-018-6145-2.

7 Cell reaction of bone-grafting materials

Orthopedic implant treatments as well as implant dentistry involve the traumatized bone structure and its healing processes, which are composed of mainly four wound healing phases that every wound goes through – hemostasis, inflammation, proliferation, and maturation. For each stage of the healing process, interfacial reaction between the vital cell and the surface of foreign material (either implant body, bonegrafting material and/or scaffold, mesh/membrane materials) play a crucial role in controlling the successful outcome of the entire treatment. In this chapter, cell reactions with foreign materials, in particular, bine grafting materials, will be reviewed.

7.1 Healing processes and cell reactions

Materials are needed to stimulate the body's own regenerative mechanisms, and restore the diseased or damaged tissue to its original state and function [1]. Taking bone as an example – after trauma and pathological conditions, the bone can repair itself (e.g., fracture healing) if the defect is small; however, above a critical size, bone defects will not always heal spontaneously [2, 3], and a template, or "scaffold", is required to guide bone repair, which could serve as a supporting device during implant treatment procedures.

Schukarev et al. [4] mentioned that the mechanisms by which biomaterials support bone formation are largely unknown but will certainly be determined by cell–surface interactions and the interface between the biomaterial and cells. Bone-graft materials may regulate cell adhesion, and functions through chemical and physical properties, dissolution and precipitation reactions, surface charge, protein adsorption, and surface topography. In the case of a bone-graft substitute, it occurs at the interface of the biomaterial, which is in equilibrium with the surrounding body fluid. When the bone-graft substitute has been implanted in the patient, the biomaterial and the surrounding tissue fluids will start interacting, within nanoseconds. After minutes to hours, depending on the material and the solution's properties, equilibrium will be reached, and organic components will be attracted. The development of a biomaterial-tissue interface can initiate and this interface and facilitate pre-osteoblast attachment. A possible attachment mechanism could be the integrin-RGD (Arg-Gly-Asp amino acid) sequence binding, leading to the activation of intracellular signaling and change of cellular functions. Activated and differentiated osteoblasts will then produce the bone matrix (osteoid), which needs to be mineralized to become the bone. The mature mineralized bone will undergo bone remodeling, involving both osteoclast and osteoblast activity, in response to physiological changes in the environment [4].

The phenomena at the interface of the biomaterial and biological media is crucial in maintaining optimal cellular functions, initiating biomineralization, and bone tis-

https://doi.org/10.1515/9783111136691-007

sue regeneration [4]. In contrast to "inert" implants substituting hard and soft tissues, bioactive ceramics used as bone-graft substitutes are known to possess a reactive surface with respect to body liquids and similar model solutions [5]. It is at the interface that the approaching cells "decide" to build or not to build a new bone. At interfacial equilibrium, an electrical double layer (EDL) is formed at the interface. The key feature that determines the EDL structure is the surface charge developed due to chemical interactions between the bone-graft substitute material and the biological media. Three scenarios are possible, depending on the biomaterial surface chemistry, composition of the media, and the media pH value. Positively or negatively charged surfaces encourage electrostatic (long-range forces) attraction of oppositely charged species, for example, ions, protonated/deprotonated aminoacids, peptides, proteins, and probably even, cells. For particles with zero surface charge, short-range (chemical binding) interactions are preferable. Since body fluids have relatively constant composition and stable buffered pH, it is mainly the bulk and the surface chemistry of the biomaterial that affect the surface charge development. Electrostatic attraction of inorganic cations and anions from biological media is particularly important to initiate biomineral surface nucleation. The surface charge of the biomaterial plays a pivotal role in the adsorption of organic species from the media. The adsorption can change the hydrophobic/hydrophilic properties of the surface if preferential accumulation of lipids, proteins, and vitamins from one side, or sugars, small organic acids, and aminoacids from another side occurs. The adsorbed organic molecules will determine if the cells can attach to the bioactive surface with functional groups on the cell membrane. Moreover, the surface charge will influence the interface pH and bone-graft substitute dissolution/re-precipitation processes. The extent of biomaterial dissolution is very important, considering the rate of bone regeneration and biomaterial substitution by new bone. Another important factor that can significantly improve the initial adhesion of the cell of the bone-graft substitute materials is the surface morphology. Considering the cell size (μm) and the interface (EDL) thickness (nm), it should be the surface morphology that certainly maintains comfortable hosting for the cell to adhere, fine tune its microenvironment, and communicate with other cells. The main parameters to consider are the particles size, shape, porosity, surface (macro and micro) roughness, and availability of natural and artificial scaffolds. Although the initial morphology is not kept during the formation of new bone, it can contribute at the first stage of the tissue regeneration process. "Biomaterial in biological media" system can be separated into three different parts: solid phase (macro), solution phase (micro), and solid-solution interface (nano). Continuous phases are easy to investigate, but the most important information about the interface is almost inaccessible. The formation of new bone can be experimentally observed at the micro level using histological sections of biopsies from patients; changes in composition of biological media are routinely monitored by standard chemical tools. However, specific characteristics necessary for a potential interfacial "nano-reactor" and key elements that initiate and control cellular activity are still unknown. To understand these phenomena at a molecular level, we need supplementary physicochemical information about the surface of bone-graft substitute material and the immediate interface between the material and the biological media. This information will allow material scientists and bone biologists to improve existing tissue substitutes or create new tissue substitutes with attractive interface properties, leading the cell toward the decision – "to build or not to build" the bone" [4].

Implant treatments in both dental and orthopedic implants should involve healing processes of a traumatized bone. Integrating several information available on the wound healing process, there are four wound healing phases that every wound goes through: hemostasis, inflammation, proliferation, and maturation [6–8]. For each stage of the healing process, interfacial reaction between the vital cell and the surface of foreign material (either implant body, bone-grafting material, and/or scaffold, mesh/membrane materials) play a curial role to control the successful outcome of the entire treatment.

7.2 Hemostasis phase

Hemostasis (bleeding) stage is a short stage (lasting up to 24 h, depending on the extent of soft tissue injury) that occurs directly after an injury, which focuses on protection and reaction so that the initial injury can be managed, in order to simply stop any bleeding. To do so, your body activates its blood clotting system. There are many different cells that come rushing in to prevent and stop the bleeding. Thrombocytes are cells that help form clots and trigger vasoconstriction, whereas platelets assist in clot formation, resulting in activation and aggregation. An enzyme called thrombin is at the center, and it initiates the formation of a fibrin mesh, which strengthens the platelet clumps into a stable clot. Lastly, leukocytes are a type of white blood cells, protecting the body against foreign bodies and helping fight any infection present. In this stage, a clot is formed, while leukocytes begin to remove foreign particles.

7.3 Inflammation phase

Inflammation (as key defense mechanism) stage generally takes up to six days and should go away. It focuses on protection and early repair, and overlaps with the proliferation stage. Typically, inflammation presents with swelling, bruising, increased temperature, pain, and loss in function at the injury site. The main function of this stage is to have a type of white blood cells called neutrophils to enter the wound to destroy bacteria and remove debris with white and other blood cells. Inflammation ensures that the wound is clean and ready for the new tissue to start growing. As the white blood cells leave, specialized cells called macrophages arrive to continue clearing the debris. These cells also secrete growth factors and proteins that attract immune system cells to the wound to facilitate tissue repair. The inflammatory cells stay around so that when the proliferation starts, there is an overlapping zone.

An understanding of the foreign body reaction is important as the foreign body reaction may impact the biocompatibility of the medical device, prosthesis, or implanted biomaterial and may significantly impact short- and long-term tissue responses with tissue-engineered constructs containing proteins, cells, and other biological components for use in tissue engineering and regenerative medicine [9]. The foreign body reaction, composed of macrophages and foreign body giant cells, is the end-stage response of the inflammatory and wound healing responses, following the implantation of a medical device, prosthesis, or biomaterial. These events in the foreign body reaction include protein adsorption, monocyte/macrophage adhesion, macrophage fusion to form foreign body giant cells, manage consequences of the foreign body response on biomaterials, and cross talk between macrophages/foreign body giant cells and inflammatory and/or wound healing cells [10, 11].

Host hard/soft tissue's response to biomaterials should be considered first as an inflammatory reaction. Any biomaterial implanted into the body will be perceived as a threat. The host defenses will attempt to eliminate it. This will not happen if the biomaterial is inert and cannot be degraded. Eventually, inert biomaterials will be integrated into the tissue or will become walled off [12]. Biodegradable biomaterials are slowly resorbed over time and are eventually eliminated. Some biomaterials have chemical reactivity and will continue to stimulate over long periods of time. The inflammatory processes should include four main stages: initial events (redness, swelling, and pain), cellular invasion (white cells invade tissues), tissue remodeling, repair (being orchestrated by macrophages; occurs differently in different tissues; bone may completely remodel; and usually completed within three to four weeks), and resolution (extrusion, resorption, integration, and encapsulation). Furthermore, cellular invasion has neutrophil action (main function is phagocytes; die at tissue site; and release further inflammatory medicators, prostaglandins, and leukotrienes) and macrophages (phagocytes; removal of dead cells; numbers and activity depend on the pressure of a particulate material). Moreover, resolution is an attempt to return to the original condition: extrusion and resorption of implant material for reestablishment of homeostasis, and integration and encapsulation with a layer of fibrous tissue to establish a steady state [13–15].

Carnicer-Lambarte et al. [16] described that the implantation of any foreign material (including implant main body, occasionally accompanied and supported by scaffold and mesh/membrane) into the body leads to the development of an inflammatory and fibrotic process – the foreign body reaction (FBR). Upon implantation into a tissue, cells of the immune system become attracted to the foreign material and attempt to degrade it. If this degradation fails, fibroblasts envelop the material and form a physical barrier to isolate it from the rest of the body. Long-term implantation of medical devices faces a great challenge presented by FBR, as the cellular response disrupts the interface between the implant and its target tissue. This is particularly true for nerve neuroprosthetic implants – devices implanted into nerves to address conditions such as sensory loss, muscle paralysis, chronic pain, and epilepsy. Nerve neuroprosthetics rely on tight interfacing between the nerve tissue and the electrodes to detect the tiny electrical signals carried by axons, and/or electrically stimulate small subsets of axons within a nerve. Moreover, as advances in microfabrication drive the field to increasingly miniaturized nerve implants, the need for a stable, intimate implant-tissue interface is likely to quickly become a limiting factor for the development of new neuroprosthetic implant technologies.

Biomaterials are widely used in GBR and GTR. After application, there is an interaction between the host immune system and the implanted biomaterial, leading to a biomaterial-specific cellular reaction. Al-Maawi et al. [17] mentioned that two types of cellular reactions were observed. There was a physiological reaction with the induction of only mononuclear cells and a pathological reaction with the induction of multinucleated giant cells (MNGCs). Attention was directed to the frequently observed MNGCs and the consequences of their appearance within the implantation region. MNGCs have different subtypes. The different morphological phenotypes observed within the biomaterial implantation bed was reviewed and the critical role of MNGCs, their subtypes, and their precursors were described. The characteristics and differences between biomaterial-related MNGCs and osteoclasts were also compared. Polymeric biomaterials that only induced mononuclear cells underwent integration and maintained their integrity, while polymeric biomaterials that induced MNGCs underwent disintegration, with material breakdown and loss of integrity. Hence, there is a question whether our attention should be directed to alternative biological concepts, in combination with biomaterials, which induce a physiological mononuclear cellular reaction to optimize biomaterial-based tissue regeneration. Currently, a wide range of different biomaterials is available to support hard and soft tissue regeneration following the principles of guided bone and guided tissue regeneration (GBR/GTR). In this context, biomaterials are used as scaffolds to hold a place for delayed tissue regeneration in bone defects as well as to prevent premature soft tissue ingrowth into the defect area [18]. After biomaterial implantation, an interaction between the host immune system and the implanted biomaterial occurs, resulting in a biomaterial-specific tissue response during a complex biological process [9]. Two types of cellular reactions toward biomaterials have been observed. They are a cellular reaction, based on physiologically existing mononuclear cells, such as macrophages, lymphocytes and fibroblasts, and a foreign body reaction, based on the additional presence of multinucleated giant cells [19]. The inflammatory pattern induced by biomaterials includes an early innate immune response from macrophages, whereas lymphocytes, as a part of the adaptive immune system, play a crucial role in the foreign body reaction and formation of foreign body multinucleated giant cells (MNGCs) [20].

The host response to foreign objects has plagued researchers in the development of therapeutics and devices. The body's recognition of self and the aim of containing or destroying the nonself is the basis of the foreign body reaction (FBR) [21]. The FBR involves many complex molecular and cellular players but can be broadly categorized into five sequential phases: (1) blood-biomaterial interaction, (2) acute inflammation, (3) chronic inflammation, (4) foreign body giant cell formation and (5) encapsulation [22, 23]. The initial exposure to body fluids spurs a cascade of biochemical events that modify the surface of implanted materials. This concept is critically important in nanomedicine and in nanostructured materials, in which the material surface has been engineered to choreograph a specific cellular outcome. Since a requisite condition of nano engineered materials/devices is direct interaction with cells, consideration must be given to factors influencing protein adsorption and the provisional matrix. These early events ultimately dictate cellular responses that determine the fate of the implant. Addressing the FBR holistically will be a significant challenge and can only be accomplished via a thorough understanding of cell–biomaterial interactions. To this end, integration of the early inflammatory reaction (biomolecules, cells) into in vitro models may be a promising step along a more efficient path in product development.

The use of temporary, functional porous materials in regenerative medicine has great potential to reduce the need for replacement of damaged tissue. Several new nanostructured macroporous materials are being designed to mimic tissue structure and properties, and to stimulate new tissue growth while degrading in the body. However, cell response studies must be carried out in vitro before in vivo tests and clinical trials can be undertaken. Here, some techniques for characterizing these materials and for investigating their cell response are discussed. The use of conventional two-dimensional cell culture assays is not always possible on bulk porous materials. There is a need to move away from tissue replacement with "bioinert" materials, such as metal alloy hip replacements, because they only tend to survive for 15~25 years, which is unsatisfactory for a patient having an implant at the age of 60. Materials are needed that can stimulate the body's own regenerative mechanisms, restoring diseased or damaged tissue to its original state and function1. Taking bone as an example – after trauma, the bone can repair itself if the defect is small; however, above a critical size, a template or a "scaffold", is required to guide bone repair [24]. Besides muscle growth and increased endurance, peptides provide many more benefits than one can imagine. One of these includes rapid healing, and tissue repair and regeneration. Most healing peptides work by increasing angiogenesis, i.e., forming new blood vessels. This provides a direct pathway for the rejuvenating tissues to get nutrients and enough oxygen to heal. Besides this, peptides often boost our immune system and act as a chemokine for anti-inflammatory cells. This allows our body's protective agent to clear the debris and fight off infection at the earliest so that wound healing can take place. Interactions between peptides and scaffolds can result in a completely different surface chemistry, topography, surface energy, and charge [25]. They can also lead to conformational changes in the peptide structure, which is usually undesirable. Proteins are usually adsorbed or bonded onto material surfaces in solution by immersing the material in phosphate-buffered saline (PBS) containing the protein.

Conventional cell biology studies involve the culture of cells in polystyrene wall plates, often on thin tissue culture plastic slides, which are optically transparent and termed two-dimensional cultures. This allows the use of inverted optical microscopes to visualize cell behavior with respect to changes in the cell culture medium by the addition of drugs or growth factors. However, when cells are grown on a three-dimensional material, inverted optical microscopes cannot be used as the materials are not transparent. Cell function in three dimensions has been found to be very different to two dimensions because of changes in the precise distribution of proteins, so it is vital to be able to determine how introducing macropores or a three-dimensional hydrogel matrix affects the cell response [26–28]. At the present time, there is no technique that can be used to observe cell interactions at the nanoscale in three-dimensional structures. Therefore, it is necessary to create the scaffold material as a disc, a bulk porous scaffold, and determine the cell responses to each [24].

7.4 Proliferation phase

Proliferation (meaning reproduce rapidly) can last for four days to almost a month, depending on the surface area of the wound. The phagocytes will start to leave and therefore inflammation will reduce. This stage is when cells that are commonly referred to as "blasts" cells will come to the area to change the fibrin-like matrix into a more structured matrix. "Blasts" means to build new tissue. Here, the fibrin matrix from the bleeding and inflammation state would be replaced by "immature" disorganized tissue. In addition, new blood cells will start to form at the injured site, i.e., if one broke a bone, the blood vessels may have been broken at this time and therefore new vessels will form in the healing process. The blasts cells will lay down new tissue (whether it is bone, muscle, cartilage, etc.) which is "immature" – meaning that it cannot withstand the final forces yet. This stage can be broken down into three semiphases, including: (1) filling the wound with new connective tissue and blood vessels, (2) contracting the edges of the wound – this will feel like the wound is tightening towards the center, and (3) covering the wound – epithelial cells (cells that create a protective barrier between the inside and outside of your body; epithelialization) flood in and multiply to close the wound completely.

Wound healing is a normal, multiphase, physiological process and is of obvious importance after surgery. Growth factors play critical roles in all three phases of wound healing: (1) the inflammatory phase, (2) the proliferative phase, and (3) the remodeling phase. Platelets are the principal cell type for the inflammatory phase because they release platelet-derived growth factor (PDGF) and transforming growth factor (TGF)-beta, both of which are growth factors that attract macrophages and neutrophils. Neutrophils are white blood cells that can kill bacteria and thereby prevent sepsis at the wound site. Macrophages secrete growth factors that both attract fibroblasts and also play a key role in the second phase of wound healing, the proliferative phase. TGF-beta is also secreted by macrophages and plays a role in wound fibrosis. Both macrophages and endothelial cells can secrete VEGF, and this growth factor promotes angiogenesis (growth of new blood vessels). Macrophages also secrete epidermal growth factor (EGF), and it stimulates fibroblasts to secrete collagenase, which is essential during the remodeling phase. There is evidence suggesting that the complex and changing mixture of growth factors occurring during wound healing modulates the macrophage functions [29–31]. Growth factors can act on specific cell surface receptors that subsequently transmit their growth signals to other intracellular components and eventually result in altered gene expression. The general process of transmitting an external molecular signal to a cell to evoke a cellular response is called signal transduction [32–34].

With improved understanding of fracture healing and bone regeneration at the molecular level [35, 36], a number of key molecules that regulate this complex physiological process have been identified, and are already in clinical use or under investigation to enhance bone repair. Of these molecules, bone morphogenetic proteins (BMPs) have been the most extensively studied to recognize as potent osteoinductive factors. They induce the mitogenesis of mesenchymal stem cells (MSCs) and other osteoprogenitors, and their differentiation towards osteoblasts. Since the discovery of BMPs, a number of experimental and clinical trials have supported the safety and efficacy of their use as osteoinductive bone-graft substitutes for bone regeneration. BMP-2 and BMP-7 (FDA approved) molecules have been used in a variety of clinical conditions, including nonunion, open fractures, joint fusions, aseptic bone necrosis, and critical bone defects [37]. Extensive research is ongoing to develop injectable formulations for minimally invasive application, and/or novel carriers for prolonged and targeted local delivery [38]. Other growth factors, besides BMPs, which have been implicated during bone regeneration, with different functions in terms of cell proliferation, chemotaxis and angiogenesis, are also being investigated or are currently being used to augment bone repair [39, 40], including platelet-derived growth factor, transforming growth factor (TGF)-β, insulin-like growth factor-1, vascular endothelial growth factor (EGF), and fibroblast growth factor, among others [36]. These have been used either alone or in combinations in a number of in vitro and in vivo studies, with debatable results [39, 40]. One current approach to enhance bone regeneration and soft-tissue healing by local application of growth factors is the use of platelet-rich plasma, a volume of the plasma fraction of autologous blood with platelet concentrations above baseline that is rich in many of the aforementioned molecules [41].

A growth factor is a naturally occurring substance capable of stimulating cell proliferation, wound, and occasionally, cellular differentiation. Usually it is a secreted protein or a steroid hormone. Growth factors typically act as signaling molecules between cells. Growth factors are important for regulating a variety of cellular processes. Growth factors (GFs) are soluble, secreted signaling polypeptides capable of instructing specific cellular responses (cell survival, control over migration, differentiation, or proliferation) in a biological environment. Growth factor is a protein molecule (which is much larger than peptides) made by the body; it functions to regulate cell division and cell survival.

Growth factors can also be produced by genetic engineering in the laboratory and used in biological therapy. Healing promotive factors such as growth factors have been extensively used to treat bony defects and for osteoinduction [42, 43]. Some growth factors such as vascular endothelial growth factor (VEGF), TGF-β, PDGF, and BMPs such as BMP-2, BMP-7, and IGF are present in the healthy bone matrix, and are expressed during bone healing [44–46]. These factors can regulate vascularization and induce proliferation and differentiation of the osteoblasts and their precursors, so they can be useful in improving the healing processes [44]. They often promote cell differentiation and maturation, which varies between growth factors. For example, epidermal growth factor (EGF) enhances osteogenic differentiation (osteogenesis,; aka bone-formation mechanism) while fibroblast growth factors and vascular endothelial growth factors stimulate blood vessel differentiation (angiogenesis) [47].

Growth factors (GFs) such as BMPs, platelet-derived growth factors (PDGFs) and insulin-like growth factors (IGFs) have been found to possess osteoinductive properties, allowing for accelerated bone regeneration in bony defects [48–51]. In the dental field, the first use of bioactivated materials with growth factors is in the use of plasma rich in growth factors (PRGF), platelet-rich plasma (PRP) and plasma-rich in fibrin (PRF), to accelerate bone healing in patients with bisphosphonate-related osteonecrosis of the jaw [52, 53]. Although the use of GFBSs has presented a promising area for the introduction of new bone substitutes, novel bioactivated products with GFs have generally not progressed beyond the animal study stage in recent years. These products are generally used in combination with a structural scaffold or carriers such as allograft material, collagen sponges, titanium mesh, or β-TCP/HA [53, 54].

Sticky bone is another recently developed concept, which utilizes a bone-graft matrix enriched with growth factors using autologous fibrin glue [55, 56]. The use of sticky bone is to stabilize bone-graft material in bony defects, allowing for accelerated bone regeneration and minimizing bone loss. The advantages of this material include good moldability, good structural stability, selectivity for osteogenic progenitor through prevention of soft tissue cell migration via fibrin interconnections; and fibrin network, allowing for rapid cell adhesion and accelerated healing [56]. When used in combination with a concentrated growth factor (CGF) membrane or a titanium mesh, grafting with sticky bone in an atrophic alveolar ridge resulted in favorable three-dimensional ridge augmentation over a 4-month period [55].

7.5 Maturation phase

Maturation (or remodeling) phase is crucial to allow the "immature" tissue that was laid down to be converted to a more mature, organized, and structured matrix. It lasts anywhere from 21 days to two years, or sometimes takes over a year to fully repair. The cells take the disorganized immature tissue from the proliferation stage and align the tissue in an organized and linear manner to construct the final tissue type.

During the maturation phase, the new tissue slowly gains strength and flexibility. Here, collagen fibers reorganize – the tissue remodels and matures and there is an overall increase in tensile strength (though maximum strength is limited to 80% of the pre-injured strength). Healing process is remarkable and complex, and it is also susceptible to interruption due to local and systemic factors, including moisture, infection, and maceration (local); and age, nutritional status, and body type (systemic). When the right healing environment is established, the body works in wondrous ways to heal and replace the devitalized tissue [6–8].

References

[1]   Hench LL, Polak JM. Third-generation biomedical materials. Science. 2002, 295, 1014–17.

[2]   Feng X, McDonald JM. Disorders of bone remodeling. Annu Rev Pathol. 2011, 6, 121–45.

[3]   Nakahama K. Cellular communications in bone homeostasis and repair. Cell Mol Life Sci. 2010, 67, 4001–09.

[4]   Shchukarev A, Ransjö M, Mladenović Ž. To build or not to build: The interface of bone graft substitute materials in biological media from the view point of the cells. In: Pignatello R (Ed.)., Biomaterials Science and Engineering. 2011; doi: 10.5772/24578.

[5]   Hench LL, Wilson J. An Introduction to Bioceramics – Advanced Series in Bioceramics. Vol. 1, World Scientific Publishing. 1993.

[6]   Four stages of healing process. 2016; https://www.woundsource.com/blog/four-stages-wound-healing.

[7]   What are the Phases of Wound Healing? https://www.elastoplast.com.au/first-aid/wound-care/what-are-phases-of-wound-healing.

[8]   Maynard J. How Wounds Heal: The 4 Main Phases of Wound Healing. 2015; http://www.shieldhealthcare.com/community/popular/2015/12/18/how-wounds-heal-the-4-main-phases-of-wound-healing/.

[9]   Anderson JM, Rodriguez A, Chang DT. Foreign body reaction to biomaterials. Semin Immunol. 2008, 20, 86–100.

[10] Sheikh Z, Brooks PJ, Barzilay O, Fine N, Glogauer M. Macrophages, foreign body giant cells and their response to implantable biomaterials. Materials (Basel). 2015, 8, 5671–701.

[11] Klopfleisch R. Macrophage reaction against biomaterials in the mouse model – Phenotypes, functions and markers. Acta Biomater. 2016, 43, 3–13.

[12] Oshida Y. Bioscience and Bioengineering of Titanium Materials. Elsevier. 2007.

[13] Chen L, Deng H, Cui H, Fang J, Zuo Z, Deng J, Li Y, Wang X, Zhao L. Inflammatory responses and inflammation-associated diseases in organs. Oncotarget. 2018, 9, 7204–18.

[14] Van LS, Miteva K, Tschöpe C. Crosstalk between fibroblasts and inflammatory cells. Cardiovasc Res. 2014, 102, 258–69.

[15] Robb CT, Regan KH, Dorward DA, Rossi AG. Key mechanisms governing resolution of lung inflammation. Semin Immunopathol. 2016, 38, 425–48.

[16] Carnicer-Lombarte A, Chen S-T, Malliaras GG, Barone DG. Foreign body reaction to implanted biomaterials and Its Impact in Nerve Neuroprosthetics. Front Bioeng Biotechnol. 2021, 9, 2021; https://doi.org/10.3389/fbioe.2021.622524.

[17] Al-Maawi S, Orlowska A, Sader R, Kirkpatrick CJ, Ghnaaati S. In vivo cellular reactions to different biomaterials – Physiological and pathological aspects and their consequences. Semin Immunol. 2017, 29, 49–61.

[18] Hämmerle CH, Karring T. Guided bone regeneration at oral implant sites. Periodontol. 2000, 17, 151–75.

[19] Ghanaati S. Non-cross-linked porcine-based collagen I-III membranes do not require high vascularization rates for their integration within the implantation bed: A paradigm shift. Acta Biomater. 2012, 8, 3061–72.

[20] McNally AK, Anderson JM. Phenotypic expression in human monocyte-derived interleukin-4-induced foreign body giant cells and macrophages in vitro: Dependence on material surface properties. J Biomed Mater Res. 2015, 103, 1380–90.

[21] Li J. Biomaterial and cell interactions – The foreign body response as an obstacle in nanomedicine. J Nanomed Res. 2017, 5, 00135; doi: 10.15406/jnmr.2017.05.00135.

[22] Williams DF. On the nature of biomaterials. Biomaterials. 2009, 30, 5897–909.

[23] Klopfleisch R, Jung F. The pathology of the foreign body reaction against biomaterials. J Biomed Mater Res. 2017, 105, 927–40.

[24] Jones JR. Observing cell response to biomaterials. Mater Today. 2006, 9, 34–43.

[25] Hole BB, Schwarz JA, Gilbert JL, Atkinson BL. A study of biologically active peptide sequences (P-15) on the surface of an ABM scaffold (PepGen P-15™) using AFM and FTIR. J Biomed Mater Res. 2005, 74A, 712–21.

[26] Cukierman E, Pankov R, Stevens DR, Yamada KM. Taking cell-matrix adhesions to the third dimension. Science. 2001, 294, 1708–12.

[27] Gumbiner BM. Cell adhesion: The molecular basis of tissue architecture and morphogenesis. Cell. 1996, 84, 345–57.

[28] Discher DE, Lanmey P, Wang Y-L. Tissue cells feel and respond to the stiffness of their substrate. Science. 2005, 310, 1139–43.

[29] Stone WL, Leavitt L, Varacallo M. Physiology, Growth Factor. 2022; https://www.ncbi.nlm.nih.gov/books/NBK442024/.

[30] Dao DT, Anez-Bustillos L, Adam RM, Puder M, Bielenberg DR. Heparin-binding epidermal growth factor-like growth factor as a critical mediator of tissue repair and regeneration. Am J Pathol. 2018, 188, 2446–56.

[31] Popescu M, Bogdan C, Pintea A, Rugină D, Ionescu C. Antiangiogenic cytokines as potential new therapeutic targets for resveratrol in diabetic retinopathy. Drug Des Devel Ther. 2018, 12, 1985–96.

[32] Naderali E, Khaki AA, Rad JS, Ali-Hemmati A, Rahmati M, Charoudeh HN. Regulation and modulation of PTEN activity. Mol Biol Rep. 2018, 45, 2869–81.

[33] Eddy AC, Bidwell GL, George EM. Pro-angiogenic therapeutics for preeclampsia. Biol Sex Differ. 2018, 25, 36; doi: 10.1186/s13293-018-0195-5.

[34] Sharma D, Jaggi AS, Bali A. Clinical evidence and mechanisms of growth factors in idiopathic and diabetes-induced carpal tunnel syndrome. Eur J Pharmacol. 2018, 837, 156–63.

[35] Dimitriou R, Jones E, McGinagle D, Giannoudis PV. Bone regeneration: Current concepts and future directions. BMC Med. 2011, 9, https://doi.org/10.1186/1741-7015-9-66.

[36] Dimitriou R, Tsiridis E, Giannoudis PV. Current concepts of molecular aspects of bone healing. Injury. 2005, 36, 1392–404.

[37] Giannoudis PV, Einhorn TA. Bone morphogenetic proteins in musculoskeletal medicine. Injury. 2009, 40, S1–3.

[38] Blokhuis TJ. Formulations and delivery vehicles for bone morphogenetic proteins: Latest advances and future directions. Injury. 2009, 40, S8–11.

[39] Nauth A, Giannoudis PV, Einhorn TA, Hankenson KD, Friedlaender GE, Li R, Schemitsch EH. Growth factors: Beyond bone morphogenetic proteins. J Orthop Trauma. 2010, 24, 543–46.

[40] Simpson AH, Mills L, Noble B. The role of growth factors and related agents in accelerating fracture healing. J Bone Joint Surg Br. 2006, 88, 701–05.

[41] Alsousou J, Thompson M, Hulley P, Noble A, Willett K. The biology of platelet-rich plasma and its application in trauma and orthopaedic surgery: A review of the literature. J Bone Joint Surg Br. 2009, 91, 987–96.

[42] Kumar P, Vinitha B, Fathima G. Bone grafts in dentistry. J Pharm Bioallied Sci. 2013, 5, S125–7.

[43] Oryan A, Alidadi S, Moshiri A, Maffulli N. Bone regenerative medicine: Classic options, novel strategies, and future directions. J Orthop Surg Res. 2014, 9, 18; https://josr-online.biomedcentral.com/articles/10.1186/1749-799X-9-18.

[44] Zimmermann G, Moghaddam A. Allograft bone matrix versus synthetic bone graft substitutes. Injury. 2011, 42, S16–21.

[45] Janicki P, Schmidmaier G. What should be the characteristics of the ideal bone graft substitute? Combining scaffolds with growth factors and/or stem cells. Injury. 2011, 42, S77–81.

[46] Springer IN, Açil Y, Kuchenbecker S, Bolte H, Warnke PH, Abboud M, Wiltfang J, Terheyden H. Bone graft versus BMP-7 in a critical size defect–cranioplasty in a growing infant model. Bone. 2005, 37, 563–69.

[47] Del Angel-Mosqueda C, Gutiérrez-Puente Y, López-Lozano AP, Romero-Zavaleta RE, Mendiola-Jiménez A, Medina-De la Garza CE, Márquez MM, De la Garza-Ramos MA. Epidermal growth factor enhances osteogenic differentiation of dental pulp stem cells in vitro. Head Face Medicine. 2015, 11, 29; doi: 10.1186/s13005-015-0086-5.

[48] Zhao R, Yang R, Cooper PR, Khurshid Z, Shavandi A, Ratnayake J. Bone grafts and substitutes in dentistry: A review of current trends and developments. Molecules. 2021, 26, 3007; doi: 10.3390/molecules26103007.

[49] Oliveira É, Nie L, Podstawczyk D, Allahbakhsh A, Ratnayake J, Brasil D, Shavandi A. Advances in growth factor delivery for bone Tissue. Int J Mol Sci. 2021, 22, 903; doi: 10.3390/ijms22020903.

[50] Kim J. The Use of Biologic Growth Factors (PRF, CGF, AFG) to Enhance Clinical Predictability with Minimally Invasive Surgery. 2017; https://www.aaid.com/education/Biologics-Protein-Growth-Factors-Clinical-Clarity.html.

[51] Marx RE, Carlson ER, Eichstaedt RM, Schommele SR, Strauss JE, Georgeff KR. Platelet-rich plasma: Growth factor enhancement for bone grafts. Oral Surg Oral Med Oral Pathol Oral Radiol Endod. 1998, 85, 638–46.

[52] Albanese A, Licata ME, Polizzi B, Campisi G. Platelet-rich plasma (PRP) in dental and oral surgery: From the wound healing to bone regeneration. Immun Age. 2013, 10, 1–23.

[53] Cicciù M. Growth factor applied to oral and regenerative surgery. Int J Mol Sci. 2020, 21, 7752; doi: 10.3390/ijms21207752.

[54] Sallent I, Capella-Monsonís H, Procter P, Bozo IY, Deev RV, Zubov D, Vasyliev R, Perale G, Pertici G, Baker J, Gingras P, Bayon Y, Zeugolis DI. The few who made it: Commercially and clinically successful innovative bone grafts. Front Bioeng Biotechnol. 2020, 8; doi: 10.3389/fbioe.2020.00952.

[55] Sohn D-S, Huang B, Kim J, Park WE, Park CC. Utilization of autologous concentrated growth factors (CGF) enriched bone graft matrix (Sticky bone) and CGF-enriched fibrin membrane in Implant Dentistry. J Implant Adv Clin Dent. 2015, 7, 11–18.

[56] Whitman DH, Berry RL, Green DM. Platelet gel: An autologous alternative to fibrin glue with applications in oral and maxillofacial surgery. J Oral Maxillofac Surg. 1997, 55, 1294–99.

8 Technique-sensitive bone-grafting method

In dental clinic cases, there are several reasons causing remarkable bone volume loss, including tooth extraction prior to implant placement, periodontitis, or traumatization on hard and soft tissue [1–3]. Furthermore, the volumetric alteration in the maxillary and mandibular bone is a critical consequence of tooth loss, limiting the rehabilitation with dental implants [4, 5]. Tan et al. [6] reported that the reduction in volume is described between 29% and 63% horizontally and between 11% and 22% vertically after six months of tooth loss. It was also mentioned that, after tooth extraction, an average alveolar bone loss of 1.5 ~ 2 mm (vertical) and 40% ~ 50% (horizontal) occurs within six months [7, 8]. Schropp et al. [9] described that most alveolar dimensional changes occur during the first three months. If no treatment to restore the dentition is provided, then continued bone loss occurs and up to 40%–60% of ridge volume is lost in the first three years [10, 11]. The loss of vertical bone height leads to great challenges to dental implant placement due to surgical difficulties and anatomical limitations [2]. Figure 8.1 depicts Siebert classification of bone loss scheme [3]. In Siebert class I ridge defects there is horizontal bone loss with adequate height, which leads to insufficient bone volume for successful placement of regular diameter implants. In class II there is vertical bone loss with adequate width, which leads to insufficient bone volume for proper positioning of regular length implants in correct prosthetic corono-apical position. In class III there is vertical and horizontal bone loss that prevents placement of successful implants in all spatial dimensions.

It is well documented that such loss of sufficient bone volume and/or height will jeopardize the treatment outcome with respect to implant success and survival [12, 13]. To this end, various surgical and augmentation techniques have been proposed and bone-grafting materials have been developed as well. For horizontal gain, autogenous and allogeneic block grafts, autogenous, xenogenous, and alloplastic particulate grafts, expansion of the alveolar crest, and guided bone regeneration (GBR) have been proposed [4, 14]. As an alternative for vertical loss, short implants (smaller than 7 mm), lateralization of the lower alveolar nerve, autogenous block graft, osteogenic distraction, use of growth factors and tissue engineering, and GBR have been developed and used [5, 15, 16]. Some of methods are considered to be technique-sensitive.

8.1 Vertical vs. horizontal bone augmentation

Vertical bone augmentation is aimed at regenerating bone extra-skeletally (outside the skeletal envelope) in order to increase bone height [17–19]. Vertical bone augmentation aims to restore the previous levels of bone height and is one of the most challenging surgical procedures in dentistry as it requires the formation and maintenance of extra-skeletal bone (i.e., outside the newly established skeletal envelope) [18, 19].

https://doi.org/10.1515/9783111136691-008

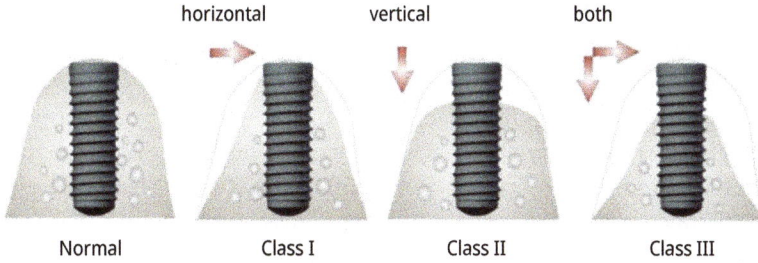

Figure 8.1: Siebert classification of bone [3].

Several existing techniques aimed at vertical bone augmentation, such as autologous bone block graft, distraction osteogenesis, and GBR combined with particulate grafts, have shown varying levels of success and are generally considered to be technique-sensitive and clinically unpredictable [11, 19, 20]. Ewers et al. [21] stated that the ridge augmentation for implant site development can be achieved through several techniques based on basic principles of bone regeneration, availability of tissue at recipient site, and desired clinical outcomes. Among various methods, it was further mentioned that although osteoperiosteal flaps (micro-anastomosed free bone-grafts or interpositional bone-grafts) and distraction osteogenesis can achieve high bone volume gain, these techniques require adequate local tissues at recipient site and are highly technique-sensitive.

Horizontal bone loss occurs at a faster rate and to a greater extent compared to vertical bone loss; leading to the development of several horizontal bone augmentation techniques, such as guided bone regeneration, ridge expansion, distraction osteogenesis, and block grafts [22, 23] and these proposed augmentation techniques aim to place the implant in an ideal 3D position for successful restorative therapy. It was mentioned that horizontal bone augmentation is fairly predictable if certain criteria are fulfilled [23, 24]; however, with numerous techniques and materials currently available, it is difficult to choose the most suitable treatment modality [22]. Mendoza-Azpur et al. [25] evaluated dimensional bone alterations following horizontal ridge augmentation using GBR with or without autogenous block graft (ABG) for the rehabilitation of atrophic jaws with dental implants. Forty-two patients, with 42 severe horizontal bone atrophy sites in the maxilla or mandible were randomly assigned to two groups: ABG or GBR. The ABG group received a combination of ABG with particulate xenograft, covered by a collagen membrane, while the GBR group received particulate xenograft alone, covered by a collagen membrane. After 6 ~ 9 months of healing, implants were inserted. All implants were definitively restored six months after implant placement. It was found that (i) both groups developed enough bone ridge width for implant placement, (ii) a total of 65 implants were placed (implant survival rate was 100% in both groups), (iii) mean increases in HBW (horizontal bone width) amounted to 5.6 ± 1.35 mm in GBR sites and 4.8 ± 0.79 mm in ABG sites, and (iv) the

ABG group had a statistically significant higher prevalence of sensory disturbances and hematomas compared to the GBR group; indicating that (v) either GBR with or without ABG is an effective approach in augmenting resorbed horizontal deficient ridges prior to implant placement; however, more complications may be seen with the use of ABG related to the donor sites. A split bone block technique has been used to graft a horizontal mandibular defect for latter dental implant-supported prosthesis rehabilitation [26]. Due to the bone defect, a horizontal bone augmentation was performed previously to implantation using split bone block technique on a 50-year-old female patient. The split bone block technique has been proven to be a suitable and predictable technique for osseous augmentation due the use of the gold standard grafting material. Wang et al. [27] introduced a digitally guided in situ autogenous onlay grafting technique and compared its effectiveness with the conventional (ex situ) onlay technique in augmenting horizontal bone defects of the anterior maxilla. It was concluded that in situ onlay grafting combined with GBR was an effective and reliable approach for horizontal bone augmentation in the anterior maxilla and appeared to demonstrate better stability in vertical bone remodeling. Ding et al. [28] evaluated the use of bone ring technique with xenogeneic bone-grafts in treating horizontal alveolar bone defects. In total, 11 patients in need of horizontal bone augmentation treatment before implant placement were included. All patients received simultaneous bone augmentation surgery and implant placement with xenogeneic bone ring grafts. The postoperative efficacy of the bone ring technique with xenogeneic bone-grafts using radiographical and clinical parameters were evaluated. Survival rates of implants were 100%. Cone-beam computed tomography revealed that the xenogeneic bone ring graft had significantly sufficient horizontal bone augmentation below the implant neck platform to 0 mm, 1 mm, 2 mm, and 3 mm. It could also provide an excellent peri-implant tissue condition during the 1-year follow-up after loading. It was, accordingly, concluded that the bone ring technique with xenogeneic bone ring graft could increase and maintain horizontal bone mass in the region of the implant neck platforms in serious horizontal bone defects [28].

The use of GBR for vertical and horizontal bone gain is a predictable approach to correct bone defects before implant installation; however, the use of different protocols is associated with different clinical results. It is suggested that platelet-rich fibrin (PRF) could improve the outcomes of regenerative procedures [5]. Based on this background, Valladão et al. [5] investigated the bone gain associated with GBR procedures combining membranes, bone-grafts, and PRF for vertical and horizontal bone augmentation. Eighteen patients who needed vertical or horizontal bone regeneration before installing dental implants were included. The horizontal bone defects were treated with a GBR protocol that includes the use of a mixture of particulate autogenous and xenogenous grafts in the proportion of 1:1, injectable form of PRF (i-PRF) to agglutinate the graft, an absorbable collagen membrane covering the regenerated region, and leukocyte PRF (L-PRF) membrane covering the GBR membrane. The vertical bone defects were treated with the same grafted mixture protected by a titanium-reinforced non-resorbable high-density polytetrafluoroethylene (d-PTFE-Ti) membrane and covered by L-PRF. The bone gain was measured using a cone-beam computed tomography at baseline and after a period of 7.5 months. It was found that (i) all patients underwent surgery to install implants after this regenerative protocol, (ii) the GBR produces an increase in bone thickness and height after treatment, with a bone gain of 5.9 ± 2.4 for horizontal defects and 5.6 ± 2.6 for vertical defects, (iii) in horizontal defects, the gain was higher in the maxilla than in mandible and in anterior than the posterior region, and (iv) no differences related to GBR location were observed in vertical defects suggesting that (v) GBR associated with a mixture of particulate autogenous and xenogenous grafts and i-PRF was evaluated to be effective for vertical and horizontal bone augmentation in maxillary and mandibular regions, permitting sufficient bone gain to future implant placement.

8.2 Guided bone regeneration

Guided bone regeneration (GBR) is also known by the term "membrane-protected bone regeneration." GBR has been broadly documented for the reconstruction of alveolar ridge defects simultaneously with or staged to implant placement [11, 29]. The term implies the use of barrier membranes with the goal of fulfilling the principle by the term "compartmentalization" [20]. The concept of GBR implies the use of cell-occlusive membranes for space provision over a vertical or horizontal defect, promoting the ingrowth of osteogenic cells while preventing migration of undesired cells from the overlying soft tissue [30–32]. It also effectively stabilizes the blood coagulum and thereby allows for faster healing to occur. This technique can be used before or at the same time as implant placement [30]. Initially, it was advocated for the repair of the periodontium [20], although it was later used for implant site development [33]. In other words, the barrier membrane aims to promote bone formation while acting as a passive barrier to preclude soft tissue ingrowth. Moreover, the effect of the barrier membrane has been further shown to promote bone formation, as it induces molecular and cellular events [11]. The principles of GBR were initially applied for implant site development in atrophic jaws [34]. The expansion of GBR to a large variety of bone defect types led to the widespread use of this technique in clinical practice [7, 35, 36]. In some cases, the use of barrier membranes is not warranted and the graft material can be used alone to fill the defect [37]. GBR is a technique that works on the principle of separating particulate graft material from surrounding soft tissue to allow for bone regeneration, which occurs at a slower rate compared to soft tissues [3, 38, 39]. Resorbable (e.g., collagen-based material) or non-resorbable (such as e-PTFE-based material) membranes are frequently used to stabilize the graft material, limit graft resorption, and act as an occlusive barrier toward the surrounding soft tissue regeneration and infiltration [3, 39].

Alkudmani et al. [40] conducted a systematic review on the effect and type of bone-graft and GBR around immediate implants on hard and soft tissue changes, and concluded that (i) the application of guided bone regeneration techniques aids in soft tissue preservation and prevents resorption of the buccal plate of the immediately placed implant, despite the type of membrane used and (ii) in terms of comparing different bone-grafting materials, autogenous bone-graft proved to be superior to synthetic polylactic polyglycolic acid polymer alloplast. On the other hand, it was also described that a combination of autogenous bone with polylactic polyglycolic acid polymer alloplast further improved the vertical defect fill.

8.3 Onlay bone block technique

To increase the vertical height of mandibular and maxillary edentulous ridges, Isaksson et al. [41] firstly employed onlay bone block grafting technique. It was followed by numerous clinical reports involving the use of an autologous bone block fixed to the recipient ridge with osteosynthesis screws or dental implants [3, 42–45]. After performing recipient site corticotomy to encourage blood clot formation and bone marrow osteoblast precursor migration into the graft, the latter is laid over the defective recipient bed devoid of soft tissues and immobilized [3]. Several autologous bone-grafting techniques have been used for the treatment of severely resorbed edentulous mandible and maxilla [46]. In particular, the use of barrier membranes for block grafts seems to significantly improve the clinical outcome [e.g., [47–49]].

8.4 Bone manipulation techniques

For implant treatments, if inadequate bone exists, several surgical techniques may be used to reconstruct the deficient ridge for implant placement [50]. Success of dental implants depends largely on the quality and quantity of the available bone in the recipient site. This, however, may be compromised or unavailable due to tumor, trauma, periodontal disease, etc., which in turn necessitates additional bone manipulation [50]. Bone manipulation techniques are capable of manipulating the patient's bones to alter their density to make them extremely durable and strong. These techniques mobilize vital bone with plastic bending, shaping, or condensation of tissue as a bone flap or bone-periosteal flap [51]. These result in contour or dimensional changes, while preserving bone integrity and viability. The concept is to manipulate the residual bone to create an intrabony cavity with a wider base or taller roof that heals like an extraction site, with access to mesenchymal stem cells and the normal wound healing mechanisms. The morphology of bony defect is an important consideration in the selection of a method for ridge manipulation. The fewer the number of remaining bony walls, the greater is the need for osteopromotive techniques. Current bone manipulation techniques include inlay and onlay grafting, GBR, bone expansion, bone-splitting osteotomy, and different fixation devices such as bone screws, pins, titanium mesh, different augmentation materials, and different barrier membranes [52]. Bone is a biologically privileged tissue in that it has the capacity to undergo regeneration as a part of repair process [53]. Adequate bone volume prerequisites implant therapy and proper esthetic result. Inadequate alveolar bone height and width often require bone manipulation procedures performed before, at the time of, or after the implant surgery. There are various techniques in the bone manipulation, which can be categorized [31, 54].

8.4.1 Osteoperiosteal flap technique

The osteoperiosteal flap (OPF) or "bone flap" commonly used in segmental orthognathic surgery is a bone fragment moved in space without detachment of the investing periosteum [55]. The prerequisite for simultaneous implant placement in a vertical repositioning bone flap is adequate width within the transport segment. It is always a fine balance between allowing enough exposure to place the fixation device and not significantly compromising periosteal vascular input into the bone segment [56]. As it is well documented both clinically and experientially, full thickness mucoperiosteal releases will cause some degree of bone resorption at the labial plate [55].

OPF treatment is achieved through a vascularized segmental osteotomy performed on alveolar bone. The biologic principles of this technique are based on vascularization studies and experience with Le Fort I techniques in craniomaxillofacial surgery. Alveolar bone receives blood supply form both bone marrow and periosteum, the latter becoming more significant with age when atrophy of the ridge is associated with decreased bone marrow blood flow. This technique depends on maintenance of vascularization in bone fragments from periosteum. OPFs through segmental osteotomies are used in combination with interpositional grafts in the gap generated by transposition of the flap in the desired position to achieve vertical ridge gain [3].

The micro-anastomosed free bone flaps such as fibular grafts are used in craniomaxillofacial surgery to correct severe bone deficiencies. These grafts offer bone gain of the highest quality (native bone) combined with greatest bone volumetric stability due to continuous blood flow achieved through venular and arterial microsurgical anastomosis [57]. However, free bone-grafts are extremely technique-sensitive and are associated with significant morbidity at the donor site compared to other techniques used for implant site development. Therefore, free bone flaps are reserved for reconstruction of severe mandibular deficiencies due to trauma or cancer of dysplasia. Numerous studies reported successful and stable results with fibular free grafts used for implant site development [58, 59]. OPF combined with interpositional (inlay) grafts are increasingly being used more for implant site development in ridges with height deficiencies. The main advantage of osteotomy-based techniques is the preservation of the attached gingiva and even the papillae in some cases [60, 61].

8.4.2 Distraction osteogenesis

Distraction osteogenesis uses the long-standing biologic phenomenon that new bone fills in the gap defect created when two pieces of bone are separated slowly under tension [31, 62, 63]. Distraction osteogenesis is a technique used in craniomaxillofacial surgery to achieve high bone volume gain in all spatial dimensions [3]. Distraction of the segment can be achieved in a vertical and/or a horizontal direction [64]. The basic principles involved in distraction osteogenesis technique includes three phases: (i) the latency phase of seven days, when soft tissues heal around the surgical site where the distractor is placed; (ii) the distraction phase, when the two bone fragments are separated incrementally at a rate of $0.5 \sim 1$ mm/day; and (iii) the consolidation phase, when the newly formed bone mineralizes and matures [62, 65–71].

The promising outcome of the distraction osteogenesis technique as a significant vertical bone augmentation technique has been extensively reported. Hidding et al. [72] introduced a new technique for the treatment of alveolar ridge atrophy called "vertical distraction osteogenesis (VDO)" to move dentulous and edentulous segments of the alveolar process vertically with a device in microplate design. It was reported that (i) VDO was completed successfully in nine patients with segments length ranging from 6.5 to 43 mm with an average of 23.7 mm, (ii) the vertical distraction rate was 9.9 mm on average in seven mandibular and two maxillary segments, and (iii) in all cases good stability and the predicted movement of the segments were recognized. It was furthermore mentioned that main advantages of vertical distraction osteogenesis are: (1) no bone harvesting, (2) decreased resorption tendency, (3) lower morbidity compared with conventional techniques, (4) lower infection rate, (5) feasibility to insert dental implants 12 weeks after distraction procedure, and (6) gain of soft tissue.

Chiapasco et al. [73] examined the opportunities offered by intraoral distraction osteogenesis to vertically elongate insufficient alveolar ridges and thereby improve local anatomy for ideal implant placement. Eight patients presenting with vertically deficient edentulous ridges were treated by means of the distraction osteogenesis principle with an intraoral alveolar distractor. Two to three months after consolidation of the distracted segments, 26 implants were placed in the distracted areas. Four to six months later, abutments were connected and prosthetic loading of the implants was started. The mean follow-up after initial prosthetic loading was 14 months. It was found that (i) in all patients, the desired bone gain was reached at the end of distraction (mean vertical bone gain of 8.5 mm), (ii) the cumulative success rate of implants was 100%, and (iii) radiographic examinations 12 months after functional loading of implants showed a significant increase in the density of the newly generated bone in the distracted areas. Based on these findings, it was suggested that (iv) this technique seems to be reliable, and the regenerated bone has withstood the functional demands of implant loading, (v) success rates of implants, periodontal indices of peri-implant soft tissues, and (vi) Periotest values were consistent with those reported in the literature regarding implants placed in native bone. Gaggl et al. [74] treated 35 patients with distraction implants for the correction of alveolar ridge deficiency. In 10 patients with atrophy of the mandible or maxilla, 16 patients with severe defects of the alveolar process after trauma, and 9 patients with localized alveolar ridge defects after single tooth loss, alveolar ridge distraction was carried out with the aid of 62 distraction implants. It was reported that (i) for 5% of the implants, pathologic probing depth of more than 3 mm and sulcus bleeding were registered prior to prosthetic treatment and these observations decreased during the next nine months, and (ii) Periotest values were normal before the start of prosthetic treatment and there was a decrease in the Periotest values, thus an increase in implant stability, during the following nine months. It was then concluded that (iii) alveolar ridge distraction using distraction implants can be a successful technique for alveolar ridge augmentation with a low rate of complication and (iv) acceptable esthetic and functional results can be achieved by this atraumatic technique of surgery and distraction.

Urbani et al. [75] presented a surgical technique for vertical ridge augmentation. The procedure, performed in a 30-year-old woman with an atrophied alveolar ridge in the anterior portion of the mandible, is based on the biologic concept of osteogenesis distraction previously introduced in orthopedic and maxillofacial surgery. After elevation of a full-thickness flap a horizontal osteotomy was performed 7–8 mm from the top of the ridge. Two vertical osteotomies were prepared with drills of increasing diameter (2, 2.8, and 3.25 mm), tapping was performed for the first 5–6 mm, and two distractor base plugs were placed at the base of the osteotomies with a repositioning tool. An intraosseous distraction implant was then inserted and two inward vertical cuts were made in the bone to allow proper distraction to take place. It was reported that histologic evaluation of the biopsy specimen showed woven bone formation approximately 75 days after the initial procedure. McAllister [76] evaluated the placement of 10 consecutive distractors in seven patients. The surgical technique, latency period, distraction rate, and consolidation period were also reviewed. It was obtained that (i) the technique of distraction osteogenesis resulted in an average vertical augmentation of 7 mm, with a range of 5– mm, (ii) there were no complications affecting the outcome of the distraction procedure, and (iii) no failures have occurred to date in the 16 implants that were placed and loaded following distraction; hence it was concluded that (iv) clinical, histologic, and radiographic evidence of consistent vertical bone augmentation was found with this technique of distraction osteogenesis for vertical ridge augmentation.

8.4.3 Bone expansion technique

Ridge expansion refers to the surgical, lateral widening of the residual ridge (i.e., buccolingually) using chisels and/or osteotomes for creating enough room for the patient to receive a bone-graft and/or a dental implant. Once the ridge expansion process is complete, the implants can be placed. Screws are inserted and the surgeon gradually increases their thickness. While this process may take a bit longer, a good clinical outcome is likely. Ridge splitting is an alternative to the various techniques described for horizontal ridge augmentation, including distraction osteogenesis; it has a similar healing pattern and end result [31, 77, 78]. Many times, implant placement involves an inadequate buccolingual width of the edentulous ridge [31, 54]. This technique involves the manipulation of the bone to form a receptor site for an implant without the removal of any bone from the patient [79]. This technique maintains the existing soft bone by pushing the buccal bony plates of the residual ridge laterally with minimal trauma. This technique takes an advantage of the softer bone quality found in Types III and IV maxillary bone by relocating the alveolar bone rather than losing the precious bone by drilling [80]. The most common anatomic area in which ridge expansion is performed is in the narrow anterior maxilla, followed by posterior maxilla and then the anterior and posterior mandible, respectively. With a narrow ridge, splitting the alveolar bone longitudinally, using chisels, osteotomes, or piezosurgical devices [81], can be performed to increase the horizontal ridge width, provided the buccal and lingual cortical plates are not fused and some intervening cancellous bone is present [31].

Scipioni et al. [78] presented clinical results of a surgical technique that expands a narrow ridge when its orofacial width precludes the placement of dental implants. In 170 people, 329 implants were placed in sites needing ridge enlargement using the edentulous ridge expansion procedure. This technique involves a partial-thickness flap, crestal and vertical intraosseous incisions into the ridge, and buccal displacement of the buccal cortical plate, including a portion of the underlying spongiosa. Implants were placed in the expanded ridge and allowed to heal for 4– 5 months. It was claimed that the results yielded a success rate of 98.8%. Similarly, it was mentioned that, with adequate vascularity and stabilization of the mobile bone segment, together with sufficient interpositional bone-grafting and soft tissue protection, a comparable result to alternate techniques can be obtained [78]. Sethi et al. [82], with the technique of maxillary ridge expansion, placed 449 implants in 150 patients and observed over a period of up to 93 months. Thin maxillary ridges of adequate height and comprising two separate cortical plates with intervening cancellous bone were selected for maxillary ridge expansion and simultaneous implant placement. Two-stage implants were used and allowed to heal in a closed environment for six months prior to loading. It was reported that (i) single and multiple teeth were replaced using this technique and (ii) an estimated mean survival rate better than 97% after a 5-year observation period was calculated and good esthetic and functional outcomes were observed.

Engelke et al. [83] reported a surgical technique for reconstruction of the buccolingually reduced alveolar process, involving the preparation of an artificial socket with immediate implant placement, which reduces total treatment time compared with two-stage procedures. Alveolar preparation comprises lamellar cortical splitting of the alveolus, interlamellar implant placement, and primary stabilization based on a microfixation technique. It was used for a wide range of indications involving single and multiple alveoli related to the partially dentate and the edentulous alveolar process.

The results of 24 Brånemark standard implants and 97 ITI implants with 44 consecutively treated patients have been reviewed with a mean observation time of 34.3 months (range 6–68 months). The main indicator for alveolar reconstruction was the narrow anterior maxillary arch. It was described that (i) the 5-year cumulated success rate was 86.2%, (ii) 12 implants failed during the observation period, (iii) the mean marginal bone loss was 1.7 mm, and (iv) there was a low infection rate compared with membrane-based GBR techniques. Simion et al. [52] selected five patients with sufficient vertical bone height but insufficient bone width for implant placement for treatment with a split-crest technique combined with guided tissue regeneration. The surgical technique involved splitting the alveolar ridge longitudinally in two parts, provoking a greenstick fracture. A chisel was then used to make a fine cut and spread apart the two cortical plates. Implants were then placed. Implants and defects were covered with expanded polytetrafluoroethylene (ePTFE) membranes. It was found that (i) biometrical examination showed a gain in bone width, varying between 1 mm and 4 mm; maxillary sites showed greater ridge enlargement, (ii) histologic examination showed regeneration of bone tissue between the two portions of the split crest, and (iii) this membrane technique could be effective and predictable for horizontal ridge augmentation associated with immediate implant placement. Vercellotti [81] presented a new surgical technique that, thanks to the use of modulated-frequency piezoelectric energy scalpels, permits the expansion of the ridge and the placement of implants in single-stage surgery in positions that were not previously possible with any other method. The technique involves the separation of the vestibular osseous flap from the palatal flap and the immediate positioning of the implant between the two cortical walls. The case report illustrates the ridge expansion and positioning of implants step-by-step in bone of quality 1–2 with only 2–3 mm of thickness that is maintained for its entire height. To obtain rapid healing, the expansion space that was created for the positioning of the implant was filled, following the concepts of tissue engineering, with bioactive glass synthetic bone-graft material as an osteoconductive factor and autogenous platelet-rich plasma as an osteoinductive factor. The site was covered with a platelet-rich plasma membrane. It was reported that a careful evaluation of the site when reopened after three months revealed that the ridge was mineralized and stabilized at a thickness of 5 mm and the implants were osseointegrated.

8.4.4 Sandwich osteotomy

Sandwich techniques are similar to distraction osteogenesis in terms of surgical approach and in having similar healing patterns and end results [77, 78]. Sandwich osteotomy is a technique for vertical augmentation based on the principle of a graft being placed between two pedicled native bones. The inherent vascularization helps in graft consolidation. The aim is to review the bone height gained, implant survival, and pitfalls of sandwich osteotomy. As a replacement to callus distraction, a gap is formed by the placement of the bone fragment in final position and stabilized and fixed with either osteosynthesis screws or the dental implant itself [84, 85].

The continuing resorption of the alveolar ridge will eventually result in insufficient bone height superior to the inferior alveolar nerve (IAN), making dental implant placement impossible. The augmentation procedure above the IAN in terms of height provides sufficient bone for implant placement and allows long-term successful restoration of missing teeth with implant-supported prosthesis [86]. Mavriqi et al. [86] introduced a case report on a 49-year-old female patient with a bilaterally atrophic mandible and a need for implant therapy. A thorough radiographic examination using cone-beam tomography revealed mandibular ridges that were not suitable for immediate implant placement in terms of height (6.2 mm on the left side and 7.2 on the right side). It was concluded that (i) moderate to severe posterior mandibular atrophy was successfully treated with interpositional sandwich osteotomy bone-grafts, leading to the successful placement of implants and fixed prosthetic implant restorations, thus allowing ever more patients to be considered for implant treatment, (ii) the placement of implants of 10 mm in height was made possible, and (iii) the technique, which has been recently revisited, permits dental rehabilitation in terms of raising the bone above the nerve, reshaping the alveolar crest, and normalizing the interocclusal distance and the crown-implant ratio.

Sandwich osteotomy is a technique for vertical augmentation based on the principle of a graft being placed between two pedicled native bones. The inherent vascularization helps in graft consolidation. Bera et al. [87] reviewed the bone height gained, implant survival, and pitfalls of sandwich osteotomy. It was reported that (i) the overall implant survival rate ranged from 90% to 100% and prosthetic survival rate from 87% to 95%, (ii) an overall 6 ~ 10 mm of bone can be gained in the anterior mandible and 4–8 mm in the posterior mandible, (iii) a total of 1,030 implants were placed of which 988 implants survived after the mean follow-up periods, (iv) implant survival is independent of the graft being used, and (v) vertical augmentation in the posterior mandible is limited compared to anterior owing to the presence of IAN and the keratinized tissue deficiency. It was hence concluded that (vi) sandwich osteotomy is a successful procedure for vertical bone augmentation in the mandible, (vii) overall a bone gain of 6 ~ 10 mm in the anterior region and 4 ~ 8 mm in the posterior region can be obtained with this procedure, and (viii) the overall implant survival rate ranged from 90% to 100% and prosthetic survival rate from 87% to 95% with very low heterogeneity.

Bormann et al. [88] introduced the treatment outcome after alveolar ridge augmentation in the atrophic posterior mandible by segmental sandwich osteotomy combined with an interpositional autograft prior to placement of endosseous implants, involving 13 consecutive patients (five males, mean age 48 years, and eight females, mean age 61 years). The postoperative course was uneventful in six patients. Sensory disturbances in the mental nerve were found in five patients, all of them with hypoesthesia. It was found that (i) none of these patients complained of permanent sensory disturbances; vertical gain ranged from 2.0 to 7.8 mm (mean value 4.61 mm), (ii) horizontal gain ranged from 2.0 to 6.3 (mean value 3.42 mm), and (iii) a total of 41 implants were placed in 22 surgical sites, 12 weeks after bone reconstruction; concluding that segmental mandibular sandwich osteotomy is recommended to meet the dimensional requirements of pre-implant bone augmentation in atrophic posterior mandible.

Ferreira et al. [89] reported the clinical case of a patient with a vertical defect of the alveolar ridge, which prevented the installation of dental implants without first treating the defect in question – a 32-year-old female patient with a height defect of approximately 6 mm in the region of the missing absent teeth. The patient was treated using the sandwich osteotomy technique with the interposition of a block bone-graft of bovine origin. It was mentioned that (i) no complications were reported in the post-operative period, (ii) after seven months, two dental implants were installed in the relevant region, (iii) the bovine bone-graft was incorporated into the relevant area, and (iv) the bovine bone block graft used in this clinical case was shown to be a viable option for interposition between bone segments that have been osteotomized via sandwich osteotomy.

A three-dimensionally favorable mandibular bone crest is desirable to success-fully implant placement to meet the aesthetic and functional criteria in the implant-prosthetic rehabilitation. Several surgical procedures have been advocated for bone augmentation of the atrophic mandible, and the sandwich osteotomy is one of these techniques. Santagata et al. [90] presented a case report to assess the suitability of seg-mental mandibular sandwich osteotomy combined with a tunnel technique of soft tis-sue. A 59-year-old woman with a severely atrophied right mandible was treated with the sandwich osteotomy technique filled with autologous bone-graft harvested by a cortical bone collector from the ramus. Clinical examination revealed that the mandi-ble was edentulous bilaterally from the first molar to the second molar region. Radio-graphically, atrophy of the mandibular alveolar ridge in the same teeth site was observed. A horizontal osteotomy of the edentulous mandibular bone was then made with a piezoelectric device after tunnel technique of the soft tissue. The segmental mandibular sandwich osteotomy was completed by two (mesial and distal) slightly di-vergent vertical osteotomies. The entire bone fragment was displaced cranially, and the desirable position was obtained. The gap was filled completely with autologous bone chips harvested from the mandibular ramus through a cortical bone collector. No barrier membranes were used to protect the grafts. The vertical incisions were closing with interruptive suturing of the flaps with a resorbable material. In this way, the suture will not fall on the osteotomy line of the jaw; the result will be a better predictability of soft and hard tissue healing. Based on the presented evidences, it was concluded that (i) nobody described before the sandwich osteotomy with tunnel technique to improve the healing of the wound and meet the dimensional require-ments of pre-implant bone augmentation in cases of a severely atrophic mandible, and (ii) segmental mandibular sandwich osteotomy is an easy and safety technique that could be performed in an atrophic posterior mandible.

8.4.5 Alveolar ridge splitting technique

Most changes around the alveolus occur as a result of extraction, which leads to alteration in the width and height of the ridge. This usually happens due to the normal remodeling of bone following extraction, traumatic extraction, periodontal disease, surgical resection, prolonged denture wear, or disuse atrophy. The loss of a tooth can cause significant ridge resorption in all three planes, the most prominent being in the horizontal direction [91]. Such narrow, atrophic alveolar ridges pose a challenge and greater difficulty for successful restoratively driven implant placement as a tooth replacement option. We can place dental implants only if there is adequate bone to stabilize them, and there should be a minimum of 1–1.5 mm of bone all around the implant for successful osseointegration [92]. Several methods have been implicated to augment the narrow alveolar ridge, such as guided bone regeneration (GBR) using various graft materials (autograft, allograft, xenograft, and alloplast), autogenous onlay block grafts harvested intra-orally or extra-orally, distraction osteogenesis, ridge expansion osteotomy, and ridge splitting [93].

This technique is performed by gaining access to a ridge that is more than 3 mm wide by splitting the buccal and palatal bone flaps with a scalpel first by separating two cortices through its cancellous bone. This technique is employed in cases where there is an insufficient width to use round osteotomes. This procedure provides a quicker method wherein an atrophic ridge can be predictably expanded and grafted with bone allografts, eliminating the need for a second donor site and a second stage surgery [94]. Ideal sites to perform this procedure include clinical scenarios where there is a knife-edge ridge that widens further apically, and that consists of adequate cortical thickness but with some degree of interpositional lamellar bone. This is particularly true in the anterior region of the maxilla.

Pandey et al. [95] present a case report on a 47-year-old female patient who had partial edentulism on the lower left jaw region associated with a narrow alveolar ridge, using the ridge split technique. A piezosurgical unit was used for splitting the ridge, followed by simultaneous implant placement. This alveolar ridge split technique (ARST) is considered to be more predictable, reliable, and successful as compared to other techniques such as autogenous onlay bone-graft and guided bone regeneration. It was indicated that (i) the ridge-splitting technique provides the advantage of ridge expansion and simultaneous implant placement for the management of a narrow alveolar ridge, (ii) proper patient evaluation and case selection are crucial for achieving successful surgical and prosthetic outcomes, and (iii) digital impression and CAD/CAM design prostheses provide more successful outcomes.

Moro et al. [96] introduced some new tips that have been specifically designed for the treatment of atrophic ridges with transversal bone deficit. A two-step piezosurgical split technique is also described, based on specific osteotomies of the vestibular cortex and the use of a mandibular ramus graft as interpositional graft. A total of 15 patients were treated with the proposed new tips by our department. All the expanded areas were successful in providing an adequate width and height to insert implants according to the prosthetic plan and the proposed tips allowed obtaining the most from the alveolar ridge split technique and piezosurgery. It was mentioned that (i) these tips had made ARST simple, safe, and effective for the treatment of horizontal and vertical bone defects, and (ii) the proposed piezosurgical split technique allows obtaining horizontal and vertical bone augmentation. The proposed tips help the surgeon to obtain the most from the alveolar ridge split technique and piezosurgery. The main advantages offered are the protection of the delicate anatomical structures, the ability to modulate the depth of the cut, and the precision of the incision, which permits their usage even for the expansion of very thick alveolar ridge. These tips have made ARST simple, safe, and effective for the treatment of horizontal and vertical bone defects. The proposed tips have also proved to be useful in other surgical procedures such as bone harvesting, sinus lift, dentoalveolar surgery, and orthognathic and craniofacial surgery. Furthermore the use of the described piezosurgical split technique is an effective procedure to obtain ideal positioning of implants and both horizontal and vertical bone augmentation.

ARST is a surgical procedure performed for horizontal ridge augmentation in cases of narrow crestal ridges. It is a biologically oriented technique, taking advantage of the osteogenic and osteoconductive dynamic of the native bone. Recently, the utilization of devices such as thin diamond disks or piezoelectric cutting devices has enhanced ARST success rate and reduced surgical time and patient's morbidity, regardless of bone quality. Oikonomou et al. [97] presented and discussed clinical cases where ARST was performed for horizontal bone augmentation with simultaneous implant placement, using classical or piezoelectric-assisted surgical procedures. ARST was performed in six patients with 10 sites of horizontal defects. Under local anesthesia, a crestal incision was performed and a full thickness mucoperiosteal flap was reflected. Initial osteotomy was done using the classical technique in six cases (diamond disk, rotary burs), while a piezoelectric device was used in the remaining four cases. Osteotomes and chisels with gradually increasing dimensions were used to expand the alveolar ridge to facilitate the immediate placement of 18 implants, approximately 2 mm subcrestally in order to prevent marginal bone loss. The remaining space between bony plates was filled with allograft or xenograft and covered with collagen membrane in nine cases, while in one case no bone-graft or membrane were used. Primary closure without tension was achieved with periosteal releasing incisions of the buccal flap. Postoperative instructions included administration of antibiotic and analgesic, chlorhexidine solution, smoke cessation, and thorough oral hygiene. It was found that (i) one patient did not comply with the postoperative instructions, applying poor oral hygiene and continuing smoking as before, which resulted in exposure of the collagen-membrane the second week, leading to secondary wound closure compromising the augmentation, (ii) a new CBCT (cone-beam computed tomography) was performed four months after the split, which revealed successful bone augmentation and implant osseointegration without any significant bone loss around the implants, and (iii) oral rehabilitation with either fixed or removable prosthetics was accomplished and the patients remained under systematic review. Based on these clinical findings, it was concluded that (iv) ARST is a predictable and reliable procedure characterized by its low invasiveness and morbidity, (v) the major advantage is the simultaneous bone augmentation with implant placement, which provides reduced treatment time and cost, (vi) piezoelectric devices make the split technique easier and safer, decreasing the risk of complications in the treatment of extremely atrophic crests, and (vii) ARST's success and efficacy depends greatly on patient selection in order for the basic anatomical criteria to be fulfilled.

8.5 Minimally invasive approaches

It is generally stated that surgery is conducted using small incisions (cuts) and few stitches. During minimally invasive surgery, one or more small incisions may be made in the body. A laparoscope (thin, tube-like instrument with a light and a lens for viewing) is inserted through one opening to guide the surgery. Tiny surgical instruments are inserted through other openings to do the surgery. Minimally invasive surgery (MIS) may cause less pain, scarring, and damage to healthy tissue, and the patient may have a faster recovery than with traditional surgery [98]. Common MIS practices can include the following: (1) Laparoscopy involves placing a lighted telescope (called a laparoscope) with a camera into the belly through a very small incision in the belly button. This allows the surgeon to see inside the belly and pelvis. (2) Robotic laparoscopy uses laparoscopic instruments controlled by a surgeon seated at a console. (3) Hysteroscopy uses a lighted telescope (called a hysteroscope) that is inserted through the vagina and cervix (neck of the womb) to see the inside of the uterus (womb). MIS possesses several benefits, including: (i) Fewer scars on the outside; scars from minimally invasive surgery are much smaller than in traditional open surgery. Laparoscopy involves one incision in the belly button and 1–3 others in the lower belly. These incisions are usually 1/4 ~ 1/2 inch long. Hysteroscopy leaves no scar because the instrument goes through the natural opening (neck of the womb) from the vagina into the uterus. (ii) Fewer scars on the inside; in general, all surgery can cause adhesions or scar tissue inside your lower belly abdomen. These scars can cause pain, problems with getting pregnant, or bowel blockage. Minimally invasive surgery may cause less scarring. (iii) Quicker recovery: MIS does not usually require a woman to stay overnight in the hospital, compared to 2–4 days after open surgery. This reduces the risk of problems such as blood clots in the legs or infection.. (iv) Less pain, less medication: as incisions are smaller, MIS is less painful than open surgery. This means that women need less pain medication and recover more quickly. On the other hand, following disadvantages are recognized: (i) it is not suitable for everyone: some MIS is riskier for women who have had previous "open" surgery in the upper or lower part of their belly, or women with other medical problems; the surgeon may have other reasons to choose open, and not MIS; not all surgeries can be done with minimally invasive techniques; and (ii) special training and equipment: surgeons need special training before they can perform MIS. Not all doctors are qualified to do these types of procedures and not all hospitals have the special equipment necessary to do some or all of these kinds of surgeries [99].

The concept of MIS was devised to achieve vertical bone regeneration and to prevent postoperative complications and graft exposure [3, 100–103]. Since HA (hydroxyapatite) particles were unstable and diffused into adjacent tissues causing the formation of a fibrous capsule that prevented bone formation [101, 104], newer graft materials with optimized viscosity and an improved surgical technique continued to offer potential for this method but results are controversial [100, 101, 104, 105]. Calcium phosphate-based biomaterials such as brushite cement pastes have been evaluated in various in vivo studies as injectable pastes with controlled viscosity and additives to achieve minimally invasive vertical bone augmentation [106–109].

8.5.1 Minimally invasive tunnel technique

It is considered to be a safe and simple patient compliance method to augment bone. The subperiosteal tunneling approach is a minimally invasive procedure, which allows the surgeon to allocate the graft in a space that is obtained between the soft tissues and the underlying bone through an access represented by a single incision on the mesial limit of the bone defect [100, 110]. This approach is believed to warrant minimal discomfort to the patient in the immediate postoperative phase in addition to ensuring a steady coverage of the graft during the healing time, with minimal risk of exposure, infection, and failure [111, 112]. The subperiosteal tunneling technique is considered superior to other bone-grafting techniques since it uses autologous bone, which is still considered the gold standard for bone regeneration [113]. Flap necrosis and wound dehiscence are the two major problems in bone-grafting surgery. They both contribute to uncovering of the graft with subsequent infection of the surgical site and failure of the surgical procedure. Compared to the conventional bone-grafting techniques, this technique has several advantages. It is relatively less morbid and less technique-sensitive, and it does not require flap elevation. Since there is no flap elevation, it ensures better preservation of keratinized gingiva, and adequate overfilling is also possible since primary coverage is not directly required. Furthermore, it ensures minimal implant exposure and infection that result in good mechanical stability of bone-grafting material and prevents postoperative complications.

The subperiosteal tunneling technique involved a small surgical incision made in the alveolar ridge to elevate the periosteum and inject a low-viscosity paste of hydroxyapatite particles [114]. There is still insufficient comparable quantitative data to assess the clinical usefulness of this technique. However, some studies demonstrated that tunneling combined with screw or membrane-mediated stabilization of the grafts can be a predictable vertical augmentation technique [100–104, 114–117].

Williams et al. [101] mentioned that (i) the use of this technique has presented several prosthodontic problems, such as diffusion of material into adjacent tissues, incorrect positioning of the material on the ridge, and inadequate ridge height, and (ii) use of the transpositional flap technique described by Lew [118] and modified by Williams has been used to overcome these problems. Twenty-five patients were treated with this method during a 30-month period. The patients had worn a maxillary complete denture for at least five years and had opposing natural mandibular anterior teeth with unsupported posterior occlusion. It was then concluded that the use of the transpositional flap technique was shown to overcome the common prosthodontic problems seen with the tunneling procedure for anterior maxilla augmentation with hydroxyapatite. Kfir et al. [102] reported a new minimally invasive GBR (which is indicated when there is a volume deficiency of the residual ridge that prohibits implantation or optimal implant installation for esthetic and functional needs) and its clinical application in several patients. A vertical incision is made mesial to the augmentation zone. The periosteum is initially elevated with a miniature chisel and then through a series of sequential balloon inflations, yielding a tunnel with adequate space for membrane insertion, decortication, and grafting with substitute bone and platelet rich fibrin (PRF) filling. Primary closure is obtained by two or three simple interrupted sutures. Vertical and horizontal gains were measured on computerized tomography obtained before and 5–6 months after the procedure. Eleven patients were treated with this procedure. There were no significant adverse events. The range of vertical gain was 2.4–5.1 mm, while horizontal gain measured 1.3–3.9 mm. Implants were successfully placed in six patients.

Hasson [100] evaluated the use of subperiosteal tunneling (pocket technique) combined with bovine bone-graft and resorbable collagen membrane insertion as an alternative surgical approach for augmenting deficient alveolar lateral ridge. Seven patients were included in the study. All had lateral alveolar ridge augmentation using the subperiosteal tunneling approach with bovine bone as graft material combined with the insertion of collagen membrane. Alveolar ridge augmentations were performed in the anterior maxilla (3 patients) and in the posterior area of the mandible (4 patients). It was found that (i) 11 implants ranging in diameter from 3.75 to 4.7 mm were inserted after 4 months of graft maturation, and (ii) all implants had good primary stability at the time of insertion and were stable at uncovering. Based on these results, it was concluded that alveolar ridge augmentation using the subperiosteal tunneling dissection is a minimally invasive procedure that does not require a secondary site for bone harvesting and should be considered when performing augmentation of deficient lateral alveolar ridge.

Mazzocco et al. [119] reported on the tunnel technique, an approach to alveolar ridge augmentation, in partially edentulous patients that uses bone blocks immobilized with titanium screws prior to implant placement. Twenty patients (7 men and 13 women) between the ages of 35 years and 65 years were treated during a two-year period with the tunnel technique. The technique consists of creating the tunnel, exposing the crestal defect, harvesting the graft, and final adaptation and stabilization of the graft in the defect site. Nineteen of the 20 patients treated had an adequate level of bone postoperatively to place implants 3.75 mm or 4 mm in diameter and at least 10 mm in length. None of the patients reported temporary or permanent lower lip paresthesia. There were also no infections reported in the donor sites. It was indicated that this method eliminates the need for a membrane because the integrity of the periosteum is preserved, and it greatly reduces patient discomfort since only one surgical field is needed. Li et al. [103] evaluated histological and clinical results in nine patients who underwent a subperiosteal tunneling procedure with a Bio-Oss block onlay graft in an atrophic area of the mandible. Nine months after grafting, at the time of dental implantation, biopsy samples were taken from the grafted areas of nine patients and were analyzed histologically. It was found that (i) new bone formation through the bovine bone block was observed consistently in the nine cases and (ii) there was direct deposition of bone on the surface of the graft material. This indicated that (iii) ridge augmentation using a subperiosteal tunneling procedure with Bio-Oss bone blocks might be useful for implant placement in the atrophic alveolar ridges.

8.5.2 GBR pocket technique

Following tooth removal, the normal healing process takes place over approximately 40 days, starting with a clot formation and culminating in a socket filled with bone covered by connective tissue and epithelium [120]. However, loss of a tooth causes a resorption of bone crest. Ashman [121] mentioned that (i) ridge preservations is the prevention of the 40% ~ 60% jawbone atrophy that normally takes place two to three years post extraction, (ii) it is the new paradigm in patient care that is effective, simple, and beneficial for a multitude of reasons, (iii) it is achieved by the immediate grafting of the extraction socket with or without the use of an immediate implant, and (iv) the ability to not only preserve the alveolar ridge for future restorative dentistry but also to preserve and restore esthetics and prevent postoperative pain and bleeding makes this modality important for all practicing dentists.

GBR is a technique that allows restoration of a defective alveolar ridge after extractions. It can be used before or in combination with implant placement. It may also be employed after extractions to reduce crestal bone resorption [19]. In the GBR technique, a resorbable or non-resorbable membrane is used, the purpose of which is to cover the blood clot, preventing soft tissue cells from invading the underlying space so that osteogenic cells are allowed to colonize the site of augmentation. The fundamental characteristics of barrier membranes in regenerative therapy include biocompatibility, integration by host tissues, clinical manageability, and the ability to create space. These requirements can be satisfied by polytetrafluoroethylene (PTFE) [48]. Nowadays, GBR is a predictable technique in vertical bone defects treatment. A tension-free wound closure is a key factor to achieve success in bone-grafting because it minimizes flap dehiscence and graft exposure [19, 121, 122]. The traditional GBR technique usually requires the exposure of a large skeletal area and therefore an extended crestal incision with large vertical release incisions. Moreover, it often requires a second surgical access to take a portion of autologous bone. An extended wound, however, involves long healing times, high number of sutures, and an extremely cooperative patient following postoperative hygiene instructions. This GBR pocket technique is a new surgical approach that generates from the need to reduce the invasiveness of the traditional procedures through [119]: (1) a minimal surgical wound, thanks to a single incision and a shorter healing time; (2) a single area of grafting and bone harvesting through the use of a tunnel surgical scraper; and (3) a contraction of the risk of failure of soft tissue healing, preserving blood circulation.

Scavia et al. [120] compared the clinical outcomes of GBR pocket technique one-site tunneling with the results reported in literature of vertical bone augmentation and crestal bone remodeling after one-year follow-up. Twenty-eight patients received 28 GBR procedures in the posterior region due to vertical and horizontal defects. A 50/50 mixture of autologous bone component and heterologous bone of equine origin was then made with the use of a bone scraper tunnel with internal reservoir. A d-PTFE membrane with titanium reinforcement was then fixed to the residual bone structure with screws in order to maintain the graft in place. Radiographic checks were made before graft procedures and implants insertion, then six months later and one year after implants placement. It was found that (i) the average bone augmentation after surgery seems to be aligned, or even better, than the average reported in literature with alternative surgical approach and (ii) the mean crestal remodeling after one year and the rate of complications are aligned with other previous surgical techniques with a vertical bone augmentation of 8.78 mm and a bone remodeling after one year of 0.59 mm. Accordingly, it was concluded that (iii) the advantages of this technique are preservation of blood circulation and consequently risk of flap necrosis, dehiscence, and graft exposure, (iv) this technique also reduces mucosal healing times even if it takes longer surgical time, and (v) GBR pocket technique is the use of a minimally invasive surgical wound to reduce patient morbidity and compliance.

Clinical and radiographic outcomes of the PPF (periosteal pocket flap) technique with the use of xenogeneic and autologous bone and plasma rich in growth factors (PRGF) were evaluated, in comparison with conventional GBR procedures [123]. Nine patients were enrolled in the study (seven women and two men, mean age: 53 ± 2.74 years) and allocated to PPF or GBR. In both groups implant placement was performed simultaneously to bone regeneration. Preoperative CBCT scans were performed for each patient. Surgical time and postoperative pain were recorded, as well as tissue healing. Horizontal bone gain (mm), graft surface area (mm^2) and graft volume (mm^3) were evaluated. It was found that (i) nine surgeries were performed: six PPF and three GBR, (ii) regarding clinical outcomes, operative time was significantly greater in GBR group than in PPF group (51.67 min vs. 37 min), (iii) postoperative pain was higher in GBR compared to PPF, and (iv) regarding radiographical results, there were not significant differences in horizontal bone gain (PPF: 9.43 mm; GBR: 9.28 mm), surface area (PPF: 693.33 mm^2; GBR: 655.61 mm^2), and volume (PPF: 394.97 mm^3; GBR: 261.66 mm^3)

between groups. Based on these results, it was indicated that the combination of auto-graft/xenograft and PRGF in PPF technique is a simpler, cheaper, and faster technique than GBR technique for achieving moderate lateral bone augmentation in implant treatment.

8.5.3 Sausage technique

The sausage technique is a term used in implant dentistry to describe a specific technique used for bone regeneration. Created by Hungarian periodontist Dr. Istvan Urbán, the sausage technique is much less invasive than its predecessors. Before this technique was developed, more autogenous bone had to be harvested, which typically resorbs over time. Now, periodontists attempting to regenerate bone prior to a dental implant can use 50% autogenous bone and 50% xenogenic bone. Instead of using only one material or the other, both materials are used and much less autogenous bone is necessary, which results in a less invasive harvesting procedure. The sausage technique receives its name from the way the native collagen membrane looks when it is stretched out like a skin with small tacks to keep the bone-graft from moving. The membrane allows for improved blood flow during healing and bone regeneration, and the host bone is typically reabsorbed by six weeks [124, 125].

Briefly, the technique is similar to conventional GBR in that the membrane is fixed with pins; however, the graft material is filled inside the fixed membrane in a sufficient quantity to show a balloon effect and to push the graft material in the crestal direction to create tension on the membrane [126]. In this technique, two types of collagen membranes can be used: a cross-linked synthetic collagen membrane or a native collagen membrane. When a cross-linked synthetic collagen membrane is exposed, approximately 48.5% of the grafted bone may be lost [127]. An inflammatory reaction may also occur during membrane degradation [128]. On the other hand, due to rapid degradation of a natural collagen membrane, epithelialization proceeds rapidly in case of exposure and the risk of infection is relatively low [129, 130]. A cross-linked synthetic membrane does not allow capillaries to pass through it, but a natural collagen membrane is vascularized within the first weeks of healing [128]. These advantages can reduce patient discomfort and the surgeon's burden in terms of complications. It is known that resorbable membranes are less commonly used for GBR than non-resorbable membranes [131, 132]. However, according to the PASS (primary closure, angiogenesis, space maintenance, stability of wound) principle for successful GBR reported by Wang et al. [133], the sausage technique is theoretically considered a predictable procedure, as it meets most of the principles. Stability of the wound and space maintenance can be enhanced by a combination of titanium pins, bone-grafts, and membranes. The balloon effect of the membrane can be achieved by using a sufficient amount of bone-graft with a slow resorption rate and pin fixation [19]. Therefore, the author decided to proceed with the sausage technique while focusing on these advantages.

Various techniques have been introduced for the guided bone regeneration procedure. Among these, the sausage technique using a natural collagen membrane has many advantages. Lee et al. [127] described two cases wherein the sausage technique was employed in the maxillary anterior area using deproteinized bovine bone mineral (DBBM) and native bilayer collagen membrane (NBCM) without autogenous bone-graft. In both cases, sufficient horizontal bone regeneration for implant placement was achieved. In addition, the bone and implant conditions were maintained without any complications maintained for more than 15 months. It was indicated that the sausage technique using NBCM with DBBM was able to achieve satisfactory horizontal alveolar bone augmentation for implant placement in the maxillary anterior area.

8.6 Summary

There are a wide range of surgical procedures that can be used to correct deficient edentulous ridges. Based on the existing literature, it is challenging to conclude that one surgical procedure offers a better outcome than another, as far as predictability of the augmentation and survival/success rates of implants placed in the augmented sites are concerned. Every surgical technique poses its own benefits and hindrances, which must be carefully evaluated before the surgical intervention. It is still unclear as to whether some surgical procedures that are widely used in clinical practice, such as sinus grafting procedures in the case of limited/moderate sinus pneumatization or reconstruction of atrophic edentulous mandibles with onlay autogenous bone-grafts, are really useful for improving the long-term survival of implants. Based on the literature, it is advocated that in order to understand when bone augmentation procedures are needed and which are the most effective techniques for the specific clinical indications, larger, well-designed, long-term trials are needed. It is difficult to provide clear indications with respect to which procedures are actually needed. Priority should be given to procedures that are simpler and less invasive, involve less risk of complications, and reach their goals within the shortest time frame [30]. McAllister et al. [31] stated many techniques exist for effective bone augmentation. The approach is largely dependent on the extent of the defect and specific procedures to be performed for the implant reconstruction. It is most appropriate to use an evidenced-based approach when a treatment plan is being developed for bone augmentation cases.

As final words, referring to statements made by Mittal et al. [134], on the basis of available data, it is difficult or impossible to conclude that one surgical procedure offers a better outcome than another, as far as predictability of the augmentation and survival/success rates of implants placed in the augmented sites are concerned. Every surgical procedure presents advantages and disadvantages, which must be carefully evaluated before surgery. Moreover, it is not yet known if some surgical procedures that are widely used in clinical practice, such as sinus grafting procedures in the case of limited/moderate sinus pneumatization or reconstruction of atrophic edentulous mandibles with an onlay autogenous bone-grafts, are useful for improving the long-term survival of implants. The predictable outcome of these procedures depends on several biologic principles that must be followed. Diagnosis, treatment planning, careful execution of the surgical treatment, postoperative follow-up, and appropriate implant loading are all important factors in achieving success.

References

[1] Khoury F, Buchmann R. Surgical therapy of peri-implant disease: A 3-year follow-up study of cases treated with 3 different techniques of bone regeneration. J Periodontol. 2001, 72, 1498–508.
[2] Rocchietta I, Fontana F, Simion M. Clinical outcomes of vertical bone augmentation to enable dental implant placement: A systematic review. J Clin Periodontol. 2008, 35, 203–15.
[3] Sheikh Z, Sima C, Glogauer M. Bone replacement materials and techniques used for achieving vertical alveolar bone augmentation. Materials (Basel). 2015, 8, 2953–93.
[4] Chiapasco M, Casentini P, Zaniboni M. Bone augmentation procedures in implant dentistry. Int J Oral Maxillofac Implant. 2009, 24, 237–59.
[5] Valladão CAA, Monteiro MF, Joly JC. Guided bone regeneration in staged vertical and horizontal bone augmentation using platelet-rich fibrin associated with bone grafts: A retrospective clinical study. Int J Implant Dent. 2020, 6, 72; https://doi.org/10.1186/s40729-020-00266-y.
[6] Tan WL, Wong TLT, Wong MCM, Lang NP. A systematic review of post-extractional alveolar hard and soft tissue dimensional changes in humans. Clin Oral Implants Res. 2012, 23, 1–21.
[7] Liu J, Kerns DG. Mechanisms of guided bone regeneration: A review. Open Dent J. 2014, 8, 56–65.
[8] Van der Weijden F, Dell'Acqua F, Slot DE. Alveolar bone dimensional changes of post-extraction sockets in humans: A systematic review. J Clin Peirodontol. 2009, 36, 1048–58.
[9] Schropp L, Wenzel A, Kostopoulos L, Karring T. Bone healing and soft tissue contour changes following single-tooth extraction: A clinical and radiographic 12-month prospective study. Int J Periodontics Restor Dent. 2003, 23, 313–23.
[10] Tallgren A. The continuing reduction of the residual alveolar ridges in complete denture wearers: A mixed-longitudinal study covering 25 years. J Prosthet Dent. 2003, 89, 427–35.
[11] Bernstein S, Cooke J, Fotek P, Wang HL. Vertical bone augmentation: Where are we now?. Implant Dent. 2006, 15, 219–28.
[12] Tolman DE. Advanced residual ridge resorption: Surgical management. Int J Prosthodont. 1993, 6, 118–25.
[13] Oshida Y, Miyazaki T. Biomaterials and Engineering for Implantology. De Gruyter, 2022.
[14] Milinkovic I, Cordaro L. Are there specific indications for the different alveolar bone augmentation procedures for implant placement? A systematic review. Int J Oral Maxillofac Surg. 2014, 43, 606–25.
[15] Saletta JM, Garcia JJ, Caramês JMM, Schliephake H, da Silva Marques DN. Quality assessment of systematic reviews on vertical bone regeneration. Int J Oral Maxillofac Surg. 2019, 48, 364–72.
[16] Nisand D, Picard N, Rocchietta I. Short implants compared to implants in vertically augmented bone: A systematic review. Clin Oral Implants Res. 2015, 26, 170–79.

[17] Vaquette C, Mitchell J, Ivanovski S. Recent advances in vertical alveolar bone augmentation using additive manufacturing technologies. Front Bioeng Biotechnol. 2022, 9, 798393; doi: 10.3389/fbioe.2021.798393.

[18] Esposito M, Grusovin MG, Felice P, Karatzopoulos G, Worthington HV, Coulthard P. The Efficacy of Horizontal and Vertical Bone Augmentation Procedures for Dental Implants – A Cochrane Systematic Review. Eur J Oral Implantol. 2009, 2, 167–84.

[19] Urban IA, Montero E, Monje A, Sanz-Sánchez I. Effectiveness of vertical ridge augmentation interventions: A systematic review and meta-analysis. J Clin Periodontol. 2019, 46, 319–39.

[20] Urban IA, Montero E, Amerio E, Palombo D, Monje A. Techniques on vertical ridge augmentation: Indications and effectiveness. Periodont. 2023, https://doi.org/10.1111/prd.12471.

[21] Ewers RTB, Ghali G, Jensen O. The Osteoperiosteal Flap: A Simplified Approach to Alveolar Bone Reconstruction. Quintessence Publishing, Chicago, IL, USA, 2010. A new biologic classification of bone augmentation.

[22] Fu J-H, Wang H-L. Horizontal bone augmentation: The decision tree. Int J Periodontics Restorative Dent. 2011, 31, 429–36.

[23] Benic GI, Hämmerle CHF. Horizontal bone augmentation by means of guided bone regeneration. Periodontology. 2014, https://doi.org/10.1111/prd.12039.

[24] Urban IA, Nagursky H, Lozada JL. Horizontal ridge augmentation with a resorbable membrane and particulated autogenous bone with or without anorganic bovine bone-derived mineral: A prospective case series in 22 patients. Int J Oral Maxillofac Implants. 2011, 26, 404–14.

[25] Mendoza-Azpur G, De la Fuente A, Chavez E, Valdivia E, Khouly I. Horizontal ridge augmentation with guided bone regeneration using particulate xenogenic bone substitutes with or without autogenous block grafts: A randomized controlled trial. Clin Implant Dent Rel Res. 2019, https://doi.org/10.1111/cid.12740.

[26] Skiba THI. Horizontal mandibular augmentation with split bone block technique: A case report. J Den Oral Sci. 2022, 4; https://doi.org/10.37191/Mapsci-2582-3736-4(2)-126.

[27] Wang J, Luo Y, Qu Y, Man Y. Horizontal ridge augmentation in the anterior maxilla with in situ onlay bone grafting: A retrospective cohort study. Clin Oral Invest. 2022, 26, 5893–908.

[28] Ding Y, Wang L, Su K, Gao J, Li X, Cheng G. Horizontal bone augmentation and simultaneous implant placement using xenogeneic bone rings technique: A retrospective clinical study. Sci Rep. 2021, 11, 4947; https://doi.org/10.1038/s41598-021-84401-8.

[29] Melcher AH. On the repair potential of periodontal tissues. J Periodontol. 1976, 47, 256–60.

[30] Vathare AS, Bhanot R, Khader AB, Kuntamukkula VKS. Novel bone grafting techniques in implant dentistry. Saudi J Oral Dent Res. 2019, https://saudijournals.com/media/articles/SJODR_411_767-769_FT.pdf.

[31] McAllister BS, Haghighat K. Bone augmentation techniques. J Periodont. 2007, 78, 377–96.

[32] Rodriguez R, Hartmann N, Weingart D. Current concepts of bone regeneration in implant dentistry. J Surgery. 2015, 10, 263–65.

[33] Dahlin C, Sennerby L, Lekholm U, Linde A, Nyman S. Generation of new bone around titanium implants using a membrane technique: An experimental study in rabbits. Int J Oral Maxillofac Implants. 1989, 4, 19–25.

[34] Simion M, Jovanovic SA, Trisi P, Scarano A, Piattelli A. Vertical ridge augmentation around dental implants using a membrane technique and autogenous bone or allografts in humans. Int J Periodontics Restor Dent. 1998, 18, 8–23.

[35] Bhola M, Kinaia BM, Chahine K. Guided bone regeneration using an allograft material: Review and case presentations. Pract Proced Aesthet Dent PPAD. 2008, 20, 551–57.

[36] Chiapasco M, Zaniboni M. Clinical outcomes of GBR procedures to correct peri-implant dehiscences and fenestrations: A systematic review. Clin Oral Implant Res. 2009, 20, 113–23.

[37] Clarizio LF. Successful implant restoration without the use of membrane barriers. J Oral Maxillofac Surg. 1999, 57, 1117–21.

[38] Dahlin C, Linde A, Gottlow J, Nyman S. Healing of bone defects by guided tissue regeneration. Plast Reconstr Surg. 1988, 81, 672–76.

[39] Buser D, Dula K, Hess D, Hirt HP, Belser UC. Localized ridge augmentation with autografts and barrier membranes. Periodontology. 1999, 19, 151–63.

[40] Alkudmani H, Al Jasser R, Andreana S. Is bone graft or guided bone regeneration needed when placing immediate dental implants? A systematic review. Implant Dent. 2017, 26, 936–44.

[41] Isaksson S, Alberius P. Maxillary alveolar ridge augmentation with onlay bone-grafts and immediate endosseous implants. J Cranio Maxillo Fac Surg. 1992, 20, 2–7.

[42] Chiapasco M, Zaniboni M, Rimondini L. Autogenous onlay bone grafts *vs.* Alveolar distraction osteogenesis for the correction of vertically deficient edentulous ridges: A 2–4-year prospective study on humans. Clin Oral Implant Res. 2007, 18, 432–40.

[43] Barone A, Covani U. Maxillary alveolar ridge reconstruction with nonvascularized autogenous block bone: Clinical results. J Oral Maxillofac Surg. 2007, 65, 2039–46.

[44] Bahat O, Fontanessi RV. Implant placement in three-dimensional grafts in the anterior jaw. Int J Periodontics Restor Dent. 2001, 21, 357–65.

[45] Turker N, Basa S, Vural G. Evaluation of osseous regeneration in alveolar distraction osteogenesis with histological and radiological aspects. J Oral Maxillofac Surg. 2007, 65, 608–14.

[46] Keller EE, Tolman DE, Eckert S. Surgical-prosthodontic reconstruction of advanced maxillary bone compromise with autogenous onlay block bone grafts and osseointegrated endosseous implants: A 12-year study of 32 consecutive patients. Int J Oral Maxillofac Implant. 1999, 14, 197–209.

[47] Jardini MA, De Marco AC, Lima LA. Early healing pattern of autogenous bone grafts with and without e-PTFE membranes: A histomorphometric study in rats. Oral Surg Oral Med Oral Pathol Oral Radiol Endod. 2005, 100, 666–73.

[48] Ronda M, Rebaudi A, Torelli L, Stacchi C. Expanded vs. Dense polytetrafluoroethylene membranes in vertical ridge augmentation around dental implants: A prospective randomized controlled clinical trial. Clin Oral Implant Res. 2014, 25, 859–66.

[49] Urban IA, Lozada JL, Jovanovic SA, Nagursky H, Nagy K. Vertical ridge augmentation with titanium-reinforced, dense-PTFE membranes and a combination of particulated autogenous bone and anorganic bovine bone-derived mineral: A prospective case series in 19 patients. Int J Oral Maxillofac Implant. 2014, 29, 185–93.

[50] Mittal Y, Jindal G, Garg S. Bone manipulation procedures in dental implants. Indian J Dent. 2016, 7, 86–94.

[51] Cullum D. Advances in bone manipulation: Part 2. Osteomobilization for horizontal and vertical implant site development. SORMS. 2010, 18, 5, 1–44.

[52] Simion M, Baldoni M, Zaffe D. Jawbone enlargement using immediate implant placement associated with a split-crest technique and guided tissue regeneration. Int J Periodontics Restorative Dent. 1992, 12, 462–73.

[53] Basa S, Varol A, Turker N. Alternative bone expansion technique for immediate placement of implants in the edentulous posterior mandibular ridge: A clinical report. Int J Oral Maxillofac Implants. 2004, 19, 554–58.

[54] Vathare AS, Bhanot R, Khader AB, Kuntamukkula VKS. Novel bone grafting techniques in implant dentistry. Saudi J Oral Dent Res. 2019, https://saudijournals.com/media/articles/SJODR_411_767-769_FT.pdf.

[55] Jensen OT, Cullum DR, Baer D. Marginal bone stability using three different flap approaches for alveolar expansion for dental implants – A one year clinical study. J Oral Maxillofac Surg. 2009, 67, 19–21.

[56] Tsegga T, Wright T. Maxillary segmental osteoperiosteal flap with simultaneous placement of dental implants: Case report of a novel technique. Int J Implant Dent. 2017, 3; doi: 10.1186/s40729-017-0067-5.

[57] Malizos KN, Zalavras CG, Soucacos PN, Beris AE, Urbaniak JR. Free vascularized fibular grafts for reconstruction of skeletal defects. J Am Acad Orthop Surg. 2004, 12, 360–69.

[58] Kramer FJ, Dempf R, Bremer B. Efficacy of dental implants placed into fibula-free flaps for orofacial reconstruction. Clin Oral Implant Res. 2005, 16, 80–88.

[59] Raoul G, Ruhin B, Briki S, Lauwers L, Patou GH, Capet JP, Maes JM, Ferri J. Microsurgical reconstruction of the jaw with fibular grafts and implants. J Craniofac Surg. 2009, 20, 2105–17.

[60] Jensen OT, Kuhlke KL. Maxillary full-arch alveolar split osteotomy with island osteoperiosteal flaps and sinus grafting using bone morphogenetic protein-2 and retrofitting for immediate loading with a provisional: Surgical and prosthetic procedures and case report. Int J Oral Maxillofac Implant. 2013, 28, e260–71.

[61] Kilic E, Alkan A, Ulu M, Zortuk M, Gumus HO. Vertical ridge augmentation using sandwich osteotomy: 2 case reports. Gen Dent. 2013, 61, e22–5.

[62] Block MS, Almerico B, Crawford C, Gardiner D, Chang A. Bone response to functioning implants in dog mandibular alveolar ridges augmented with distraction osteogenesis. Int J Oral Maxillofac Implants. 1998, 13, 342–51.

[63] Oda T, Sawaki Y, Ueda M. Experimental alveolar ridge augmentation by distraction osteogenesis using a simple device that permits secondary implant placement. Int J Oral Maxillofac Implants. 2000, 15, 95–102.

[64] Takahashi T, Funaki K, Shintani H, Haruoka T. Use of horizontal alveolar distraction osteogenesis for implant placement in a narrow alveolar ridge: A case report. Int J Oral Maxillofac Implants. 2004, 19, 291–94.

[65] Ilizarov GA. The tension-stress effect on the genesis and growth of tissues: Part II. The influence of the rate and frequency of distraction. Clin Orthop. 1989, 239, 263–85.

[66] Ilizarov GA. Basic principles of transosseous compression and distraction osteosynthesis. Ortop Travmatol Protez. 1971, 32, 7–15.

[67] Davies J, Turner S, Sandy JR. Distraction osteogenesis – A review. Br Dent J. 1998, 185, 462–67.

[68] Maffuli N, Fixsen JA. Distraction osteogenesis in congenital limb length discrepancy: A review. J R Coll Surg Edinb. 1996, 41, 258–64.

[69] Rachmiel A, Srouji S, Peled M. Alveolar ridge augmentation by distraction osteogenesis. Intl J Oral Maxillofac Surg. 2001, 30, 510–17.

[70] Saulačić N, Martín MS, Camacho MDLAL, García AG. Complications in alveolar distraction osteogenesis: A clinical investigation. Intl J Oral Maxillofac Surg. 2007, 65, 267–74.

[71] Mohanty R, Kumar NN, Ravindran C. Vertical alveolar ridge augmentation by distraction osteogenesis. J Clin Diagn Res. 2015, 9, ZC43–6.

[72] Hidding J, Lazar F, Zöller JE. Initial outcome of vertical distraction osteogenesis of the atrophic alveolar ridge. Mund Kiefer Gesichtschirurgie MKG. 1999, 3, S79–S83.

[73] Chiapasco M, Romeo E, Vogel G. Vertical distraction osteogenesis of edentulous ridges for improvement of oral implant positioning: A clinical report of preliminary results. Int J Oral Maxillofac Implants. 2001, 16, 43–51.

[74] Gaggl A, Schultes G, Kärcher H. Vertical alveolar ridge distraction with prosthetic treatable distractors: A clinical investigation. Int J Oral Maxillofac Implants. 2000, 15, 701–10.

[75] Urbani G, Lombardo G, Santi E, Consolo U. Distraction osteogenesis to achieve mandibular vertical bone regeneration: A case report. Int J Periodontics Restor Dent. 1999, 19, 321–31.

[76] McAllister BS. Histologic and radiographic evidence of vertical ridge augmentation utilizing distraction osteogenesis: 10 consecutively placed distractors. J Periodontol. 2001, 72, 1767–79.

[77] Duncan JM, Westwood RM. Ridge widening for the thin maxilla: A clinical report. Int J Oral Maxillofac Implants. 1997, 12, 224–27.

[78] Scipioni A, Bruschi GB, Calesini G. The edentulous ridge expansion technique: A five-year study. Int J Periodontics Restorative Dent. 1994, 14, 451–59.

[79] Borgner RA, Kirkos LT, Gougaloff R, Cullen MT, Delk PL. Computerized tomography scan interpretation of a bone expansion technique. J Oral Implantol. 1999, 25, 102–08.

[80] Siddiqui AA, Sosovicka M. Lateral bone condensing and expansion for placement of endosseous dental implants: A new technique. J Oral Implantol. 2006, 32, 87–94.

[81] Vercellotti T. Piezoelectric surgery in implantology: A case report – A new piezoelectric ridge expansion technique. Int J Periodontics Restorative Dent. 2000, 20, 358–65.

[82] Sethi A, Kaus T. Maxillary ridge expansion with simultaneous implant placement: 5-year results of an ongoing clinical study. Int J Oral Maxillofac Implants. 2000, 15, 491–99.

[83] Engelke WG, Diederichs CG, Jacobs HG, Deckwer I. Alveolar reconstruction with splitting osteotomy and microfixation of implants. Int J Oral Maxillofac Implants. 1997, 12, 310–18.

[84] Yoshioka I, Tanaka T, Khanal A, Habu M, Kito S, Kodama M, Oda M, Wakasugi-Sato N, Matsumoto-Takeda S, Seta Y, Tominaga K, Sakoda S, Morimoto Y. Correlation of mandibular bone quality with neurosensory disturbance after sagittal split ramus osteotomy. Br J Oral Maxillofac Surg. 2011, 49, 552–56.

[85] Jensen OT, Kuhlke KL. Maxillary full-arch alveolar split osteotomy with island osteoperiosteal flaps and sinus grafting using bone morphogenetic protein-2 and retrofitting for immediate loading with a provisional: Surgical and prosthetic procedures and case report. Int. J Oral Maxillofac Implant. 2013, 28, e260–71.

[86] Mavriqi L, Baca E, Demiraj A. Sandwich osteotomy of the atrophic posterior mandible prior to implant placement. Clin Case Rep. 2015, 3, 610–14.

[87] Bera RN, Tandon S, Singh AK, Bhattacharjee B, Pandey S, Chirakkattu T. Sandwich osteotomy with interpositional grafts for vertical augmentation of the mandible: A meta-analysis. Natl J Maxillofac Surg. 2022, 13, 347–56.

[88] Bormann K-H, Suarez-Cunquwiro MM, von See C, Hokemüller H, Schumann P, Gellrich N-C. Sandwich osteotomy for vertical and transversal augmentation of the posterior mandible. Intl J Oral Maxillofac Surg. 2010, 39, 554–60.

[89] Ferreira C, Ortega-Lopes R, Martins B, Ferreira C, Coelho F, Olate S. Sandwich osteotomy with interposition of a bovine block bone graft for vertical ridge augmentation. Intl J Med Surg Sci. 2015, 2, 475–79.

[90] Santagata M, Sgaramella N, Ferrieri I, Corvo G, Tartaro G, D'Amato S. Segmental sandwich osteotomy and tunnel technique for three-dimensional reconstruction of the jaw atrophy: A case report. Intl J Implant Dent. 2017, 3, 14; https://doi.org/10.1186/s40729-017-0077-3.

[91] Ferrigno N, Laureti M. Surgical advantages with ITI TE implants placement in conjunction with split crest technique: 18-month results of an ongoing prospective study. Clin Oral Implants Res. 2005, 16, 147–55.

[92] Nedir R, Bischof M, Briaux JM, Beyer S, Szmukler-Moncler S, Bernard JP. A 7-year life table analysis from a prospective study on ITI implants with special emphasis on the use of short implants. Clin Oral Implants Res. 2004, 15, 150–57.

[93] Sanz-Sánchez I, Sanz-Martín I, Ortiz-Vigón A, Molina A, Sanz M. Complications in bone-grafting procedures: Classification and management. Periodontol. 2022, 88, 86–102.

[94] Demetriades N, Park JI, Laskarides C. Alternative bone expansion technique for implant placement in atrophic edentulous maxilla and mandible. J Oral Implantol. 2011, 37, 463–71.

[95] Pandey KP, Kherdekar RS, Advani H, Dixit S, Dixit A. Mandibular alveolar ridge split with simultaneous implant placement: A case report. Cureus. 2022, 14, e31156; doi: 10.7759/cureus.31156.

[96] Moro A, Gasparini G, Foresta E, Saponaro G, Falchi M, Cardarelli L, De Angelis P, Forcione M, Garagiola U, D'Amato G, Pelo S. Alveolar ridge split technique using piezosurgery with specially designed tips. BioMed Res Intl. 2017, Article ID 4530378, https://doi.org/10.1155/2017/4530378.

[97] Oikonomou ME, Chliaoutakis A, Samanidis K, Papachristodima A, Ntagiantis G, Giannouslis G, Chatzichalepli C, Choipis K. Alveolar ridge split technique and immediate implant placement in atrophic maxilla. Case series. Clin Oral Implants Res. 2020, 31, 297; https://doi.org/10.1111/clr.235_13644.

[98] Minimally invasive procedure; https://en.wikipedia.org/wiki/Minimally_invasive_procedure.

[99] Minimally invasive surgery. 2014; https://www.reproductivefacts.org/news-and-publications/patient-fact-sheets-and-booklets/documents/fact-sheets-and-info-booklets/minimally-invasive-surgery/.

[100] Hasson O. Augmentation of deficient lateral alveolar ridge using the subperiosteal tunneling dissection approach. Oral Surg Oral Med Oral Pathol Oral Radiol Endod. 2007, 103, e14–9.

[101] Williams CW, Meyers JF, Robinson RR. Hydroxyapatite augmentation of the anterior portion of the maxilla with a modified transpositional flap technique. Oral Surg Oral Med Oral Pathol. 1991, 72, 395–99.

[102] Kfir E, Kfir V, Eliav E, Kaluski E. Minimally invasive guided bone regeneration. J Oral Implantol. 2007, 33, 205–10.

[103] Li J, Xuan F, Choi BH, Jeong SM. Minimally invasive ridge augmentation using xenogenous bone blocks in an atrophied posterior mandible: A clinical and histological study. Implant Dent. 2013, 22, 112–16.

[104] Marshall SG. The combined use of endosseous dental implants and collagen/hydroxylapatite augmentation procedures for reconstruction/augmentation of the edentulous and atrophic mandible: A preliminary report. Oral Surg Oral Med Oral Pathol. 1989, 68, 517–25.

[105] Rothstein SS, Paris DA, Zacek MP. Use of hydroxylapatite for the augmentation of deficient alveolar ridges. J Oral Maxillofac Surg. 1984, 42, 224–30.

[106] Kuemmerle JM, Oberle A, Oechslin C, Bohner M, Frei C, Boecken I, Von Rechenberg B. Assessment of the suitability of a new brushite calcium phosphate cement for cranioplasty – An experimental study in sheep. J Cranio Maxillofac Surg. 2005, 33, 37–44.

[107] Apelt D, Theiss F, El-Warrak AO, Zlinszky K, Bettschart-Wolfisberger R, Bohner M, Matter S, Auer JA, von Rechenberg B. In vivo behavior of three different injectable hydraulic calcium phosphate cements. Biomaterials. 2004, 25, 1439–51.

[108] Flautre B, Lemaitre J, Maynou C, Van Landuyt P, Hardouin P. Influence of polymeric additives on the biological properties of brushite cements: An experimental study in rabbit. J Biomed Mater Res. 2003, 66A, 214–23.

[109] Lu JX, About I, Stephan G, Van Landuyt P, Dejou J, Fiocchi M, Lemaitre J, Proust JP. Histological and biomechanical studies of two bone colonizable cements in rabbits. Bone. 1999, 25, 41S–5S.

[110] Borgonovo AE, Marchetti A, Vavassori V, Censi R, Boninsegna R, Re D. Treatment of the atrophic upper jaw: Rehabilitation of two complex cases. Case Reports in Dent. 2013, 2013; https://doi.org/10.1155/2013/154795.

[111] Maestre-Ferrín L, Boronat-López A, Peñarrochadiago M, Peñarrocha-Diago M. Augmentation procedures for deficient edentulous ridges, using onlay autologous grafts: An update. Med Oral Patol Oral Cir Bucal. 2009, 14, e402–7.

[112] De Macedo LGS, De Macedo NL, Monteiro ADSF. Fresh-frozen human bone graft for repair of defect after adenomatoid odontogenic tumour removal. Cell Tissue Banking. 2009, 10, 221–26.

[113] Khan SN, Cammisa FP Jr, Sandhu HS, Diwan AD, Girardi FP, Lane JM. The biology of bone grafting. J Amer Acad Orthopaedic Surg. 2005, 13, 7–86.

[114] Kent JN, Quinn JH, Zide MF, Finger IM, Jarcho M, Rothstein SS. Correction of alveolar ridge deficiencies with nonresorbable hydroxylapatite. J Am Dent Assoc. 1982, 105, 993–1001.

[115] Smiler D, Soltan M, Lee JW. A histomorphogenic analysis of bone grafts augmented with adult stem cells. Implant Dent. 2007, 16, 42–53.

[116] Vanassche BJ, Stoelinga PJ, de Koomen HA, Blijdorp PA, Schoenaers JH. Reconstruction of the severely resorbed mandible with interposed bone grafts and hydroxylapatite. A 2–3 year follow-up. Int J Oral Maxillofac Surg. 1988, 17, 157–60.

[117] Block MS. Horizontal ridge augmentation using particulate bone. Atlas Oral Maxillofac Surg Clin North Am. 2006, 14, 27–38.

[118] Lew D. A method for augmenting the severely atrophie maxilla using hydroxylapatite. J Oral and Maxillofac Surg. 1985, 43, 57–60.

[119] Mazzocco C, Buda S, De Paoli S. The tunnel technique: A different approach to block grafting procedures. Int J Periodontics Restor Dent. 2008, 28, 45–53.

[120] Scavia S, Roncuccu R, Bianco E, Mirabelli L, Bader A, Maddalone M. Vertical bone augmentation with GBR pocket technique: Surgical procedure and preliminary results. J Comtemp Dent Prac. 2021; doi: 10.5005/jp-journals-10024-3243.

[121] Ashman A. Ridge preservation: Important buzzwords in dentistry. Gen Dent. 2000, 48, 304–12.

[122] Simion M, Fontana F, Raperini G, Maiorana C. Vertical ridge augmentation by expanded-polytetrafluoroethylene membrane and a combination of intraoral autogenous graft and deproteinized anorganic bovine bone (Bio Oss). Clin Oral Impl Res. 2007, 18, 620–29.

[123] Maddalone M, Mirabelli L, Scavia S, Roncucci R. Vertical bone augmentation with GBR pocket technique: Surgical procedure and preliminary results. J Contemp Dent Prac. 2022, 22, 1370–76.

[124] Iglesias-Velázquez Ó, Zamora RS, López-Pintor RM, Tresguerres FGF, Berrocal IL, García CM, Tresguerres IF, García-Denche JT. Periosteal pocket flap technique for lateral ridge augmentation. A comparative pilot study versus guide bone regeneration. Annals of Anatomy – Anatomischer Anzeiger. 2022, 243, 151950; https://doi.org/10.1016/j.aanat.2022.151950.

[125] Sausage Technique; https://www.icoi.org/glossary/sausage-technique/#:~:text=The%20sausage%20technique%20receives%20its,typically%20reabsorbed%20by%206%20weeks.

[126] Urbán I. Sausage Technique and Gain of Vestibular Depth; https://go.geistlich-na.com/egen/081016/Urban_clinical_sausage.pdf.

[127] Lee S-J. The sausage technique using collagen membrane without autogenous bone graft: A case report. J Implantol Applied Sci. 2021, 74–83; https://doi.org/10.32542/implantology.2021008.

[128] Park SH, Lee KW, Oh TJ, Misch CE, Shotwell J, Wang HL. Effect of absorbable membranes on sandwich bone augmentation. Clin Oral Implants Res. 2008, 19, 32–41.

[129] Rothamel D, Schwarz F, Sager M, Herten M, Sculean A, Becker J. Biodegradation of differently cross-linked collagen membranes: An experimental study in the rat. Clin Oral Implants Res. 2005, 16, 369–78.

[130] Tatakis DN, Promsudthi A, Wikesjo UM. Devices for periodontal regeneration. Periodontol. 1999, 19, 59–73.

[131] Kim SH, Kim DY, Kim KH, Ku Y, Rhyu IC, Lee YM. The efficacy of a double-layer collagen membrane technique for overlaying block grafts in a rabbit calvarium model. Clin Oral Implants Res. 2009, 20, 1124–32.

[132] Hämmerle C, Jung R. Bone augmentation by means of barrier membranes. Periodontol. 2003, 33, 36–53.

[133] Wang HL, Boyapati L. "PASS" principles for predictable bone regeneration. Implant Dent. 2006, 15, 8–17.

[134] Mittal Y, Jindal G, Garg S. Bone manipulation procedures in dental implants. Indian J Dent. 2016, 7, 86–94.

9 Future perspectives in bone-grafting

Delayed union, malunion, and nonunion are challenges in the treatment of orthopedic defects [1] and cranial and maxillofacial defects [2], and bone is the second most transplanted tissue in the human body [3]. There are many cases in which bone-grafts are needed in large quantity such as for reconstruction of large bone defects caused by trauma, tumors, infections, and congenital defects, and also in cases where the regeneration is compromised (osteoporosis, necrosis, and atrophic nonunions) [4, 5]. In general, four elements of bone-grafts are needed for bone regeneration: osteoconduction, osteoinduction, osteointegration, and osteogenesis [6, 7]. There are three main types of bone-grafts: autografts, allografts, and bone-graft substitutes [3]. For developing and searching the most appropriate type of bone-grafting material and their proper use in terms of surgical and clinical applications along with supportive devices including mesh/membranes and scaffold design, numerous materials and techniques have been developed and applied. Despite these efforts in research and development, as Kheirallah et al. [3] pointed out, there is no single ideal graft material to choose in clinical practice; therefore there are still windows for R&D in all relevant fields to establish modern bone regeneration protocols that may lead to the innovation of ideal graft substitutes.

Due to several drawbacks associated with the conventional biomaterials including impaired cell attachment and proliferation, or potential toxicity, advanced nanostructured biomaterials offer a crucial role in the development in treatment plans such as replacement of tissues and organs, and repair and regeneration. At the same time, nanotechnology opens new avenues to bone tissue engineering by forming new assemblies similar in size and shape to the existing hierarchical bone structure. Organic and inorganic nanobiomaterials are increasingly used for bone tissue engineering applications because they may allow overcoming some of the current restrictions put forth by bone regeneration methods. Based on these current activities, Kumar et al. [8] reviewed 550 articles extensively to cover the applications of different organic and inorganic nanobiomaterials in the field of hard tissue engineering. Several important points were addressed as follows. (1) Biomaterials science is a highly multidisciplinary area. Developments in life science and nanotechnology have enabled scientists and engineers to conceive new designs and improve the existing bone structure. For example, advances in nanotechnology allowed the development of novel methods for fabricating new nanostructured scaffolds possessing higher efficiency in tissue regeneration. (2) Nanomaterials represent an excellent tool for research and therapeutic approaches in bone tissue engineering. Organic nanomaterials are more biocompatible, nontoxic, and help more with cell regeneration than inorganic nanomaterials; on the other hand, inorganic nanomaterials provide better mechanical strength and inertness to chemical agents. It appears that nanoparticles, graphene, and nanocomposites are the most diffused types of nanostructures used for hard tissue applications. An important research trend that

https://doi.org/10.1515/9783111136691-009

has resulted in a rapidly growing number of published articles is the development of new composite nanobiomaterials especially for scaffold applications, as described in previous chapters. (3) Interactions between bone cells and nanomaterials depend on the composition of nanoparticles. Proper selection of nanoparticles may result in faster bone regeneration and recovery. Besides composition, the overall performance of a nanobiomaterial depends on porosity, microstructure, mechanical properties, and functionality. Nanomaterials-based scaffolds also play a major role in three-dimensional tissue growth. Nanostructural modifications provide a favorable environment for bone regeneration. And it was pointed out that (4) tissue engineering supports: (i) application of engineering design methods to functionally engineered tissues, (ii) development of novel biomaterials for constructing scaffolds that mimic extracellular matrix, and (iii) creation of artificial microenvironments. Overall, it was concluded that nanobiomaterials represent an excellent tool for research and therapeutic approaches in bone tissue engineering; however, further investigations should be aimed at producing advanced nanobiomaterials suitable for hard tissue engineering that can fill the gap between biomaterial fabrication and clinical implementation.

9.1 Materials

9.1.1 Nonmetallic materials

The requirements for an ideal biomaterial bone-graft are adequate mechanical properties, biocompatibility, controlled bioresorbability, and bioactivity, which lead to formation of a bond between the host tissue and the implant material (aka, osseointegration). Polymeric materials allow for precise control of properties such as architecture, mechanical properties, and degradation rate by altering composition and fabrication technique. These properties can be exploited to design implants with ideal properties for specific applications. Polymer composites retain the tunability of polymeric scaffolds while incorporating the properties of another material, which can be used to increase mechanical properties or bioactivity. Biodegradable or bioabsorbable polymers have attracted attention because as they degrade, tissue can be deposited in their place, but this is an important design criterion for bone regeneration, as the structural integrity must not be compromised as the material degrades.

Bone is a nanocomposite composed of organic (mainly collagen) and inorganic (nanocrystalline hydroxyapatite) components, with a hierarchical structure ranging from nano- to macro-scale. Its functions include providing mechanical support and transmitting physio-chemical and mechanochemical cues. Nanomaterials and nanocomposites are promising platforms to recapitulate the organization of natural extracellular matrix for the fabrication of functional bone tissues because nanostructure provides a closer approximation to native bone architecture (namely, achieving the biomechanical compatibility [9, 10]). Nanostructured scaffolds provide structural support for the cells and regulate cell proliferation, differentiation, and migration, which results in the formation of functional tissues. Unique properties of nanomaterials, such as increased wettability and surface area, lead to increased protein adsorption when compared with conventional biomaterials. Cell-scaffold interactions at the cell-material interface may be mediated by integrin-triggered signaling pathways that affect cell behavior. The materials selection and processing techniques can affect the chemical, physical, mechanical, and cellular recognition properties of biomaterials [9, 10]. Reviewing various fabrication techniques for nanomaterials and nanocomposites, McMahon et al. [11] described that novel strategies should be integrated from various compositions and nanoscale properties for providing optimal feature size and chemistry for cell adhesion, proliferation, differentiation, and protein adsorption for bone regeneration. In additions to the structural elements in native bones, BMPs (bone morphogenetic proteins) are important osteoinductive growth factors [12]. Furthermore, neovascularization adds a fourth dimension in biomimetic effort for bone-graft engineering [13, 14]. In clinical application, long-term graft survival and function after surgery is an important aspect in orthopedic surgery. All these along with these other aspects could be developed for clinical applications in the near future [11]. Nanoscale materials and composites with various chemical compositions provide optimal feature size and chemistry for cell interaction and bone regeneration. When compared with conventional materials, nanomaterials enhance protein adsorption, cell adhesion, proliferation, and differentiation. There is no one method that produces an ideal scaffold for bone tissue engineering as each technique has advantages, such as ease of fabrication, ability to incorporate bioactive factors, and control over surface or bulk properties or architecture, and drawbacks, including use of toxic solvents, lack of uniform architecture, and poor strength. Therefore, the selection of polymer and processing techniques is a matter of determining the most important scaffold properties for success of the particular tissue [11]. Due to advanced technologies in bone tissue engineering, the use of autografts as the gold standard in clinical practice has been constantly declining [15–18]; at the same time, artificial scaffolds have been extensively employed [17, 18]. However, new products are far from optimal as low fusion rates and adverse effects have been reported [15, 17]. To overcome these limitations, nature-inspired biohybrid bone-grafts have been developed, including calcium phosphate/poly-ε-caprolactone particles [19], silicon carbide collagen scaffolds: BioSiC [20], poly(N-acryloyl 2-glycine)/methacrylated gelatin hydrogels [21]. These materials combine the mechanical properties of tailored synthetic polymers and the bioactive elements of natural polymers or minerals [15].

The superior strength and partial elasticity of biological calcified tissues (e.g., bones) are due to the presence of bioorganic polymers (mainly, collagen [22]) rather than to a natural ceramic (mainly, biological apatite) phase [23]. A decalcified bone becomes very flexible and is easily twisted, whereas a bone without collagen is very brittle [24]; thus, the inorganic, nano-sized crystals of biological apatite provide hardness and stiffness, while the bioorganic fibers are responsible for the elasticity and toughness [25]. In bones, both types of materials integrate with each other on a nanometric scale in such a way that the crystallite size, fiber orientation, short-range order between the components, etc. determine its nanostructure and, therefore, the function and mechanical properties of the entire composite [22, 26, 27]. Bone is also an anisotropic material, because its properties are directionally dependent [28, 29]. It remains a great challenge to design the ideal bone-graft – one that emulates nature's own structures or functions. Certainly, the success design requires an appreciation of the structure of bone [24].

There are several requirements for the ideal bone-graft.

(1) It should be benign, be available in a variety of forms and sizes, all with sufficient mechanical properties for use in load-bearing sites, form a chemical bond at the bone/implant interface, as well as be osteogenic, osteoinductive, osteoconductive, biocompatible, completely biodegradable, at the expense of bone growth, and moldable to fill and restore bone defects [30, 31].

(2) Further, it should resemble the chemical composition of bones (thus, the presence of calcium orthophosphates is mandatory), exhibit contiguous porosity to encourage invasion by the live host tissue, as well as possess both viscoelastic and semi-brittle behavior, as bones do [32, 33].

(3) The degradation kinetics of the ideal implant should be adjusted to the healing rate of the human tissue, with absence of any chemical or biological irritation and/or toxicity caused by substances that are released due to corrosion or degradation. Ideally, the combined mechanical strength of the implant and the ingrowing bone should remain constant throughout the regenerative process [24].

(4) The substitution implant material should not significantly disturb the stress environment of the surrounding living tissue [34].

(5) There is the opinion that, in the case of serious trauma, the bone should fracture rather than the implant [35].

(6) Good sterilizability, storability, and processability, as well as a relatively low cost are also of a great importance to permit clinical application [24].

Dorozhkin [24] reviewed the state-of-the-art biocomposites and hybrid biomaterials based on calcium orthophosphates, for suitability in biomedical applications. It was suggested that although many different formulations (in terms of the material constituents, fabrication technologies, structural and bioactive properties, as well as both in vitro and in vivo characteristics) have already been proposed, among the others, the nanostructurally controlled biocomposites, those containing nanodimensional compounds, biomimetically fabricated formulations with collagen, chitin, and/or gelatin as well as various functionally graded structures seem to be the most promising candidates for clinical applications.

Today, the structure of bone is regularly influenced by medication [36]. The number of elderly patients that use bisphosphonates as protection against osteoporosis, is ever rising. Additionally, bisphosphonates are commonly used as effective therapeutic drugs for other bone diseases like the Paget's disease or metastatic bone lesions. Bisphosphonates are used to inhibit the mineralization of the bone substance as well as the bone resorption by suppressing the osteoclast activity [37]. It was indicated that bisphosphonates may be able to enhance fracture healing and also improve common bone substitutes [38], in line with reports on calcium phosphate/bisphosphonate composites for orthopedic application [39, 40]. Based on this background, Schlickewei et al. [37] developed a new injectable bone-graft substitute by combining the features of calcium phosphate and bisphosphonate as a composite bone-graft to support bone healing and to evaluate the effect of alendronate in the bone healing process in an animal model. It was found that (i) radiologically, the bone-graft materials were equally absorbed, and no fracture was documented, (ii) after two weeks, the histological analysis showed an increased new bone formation for both materials; the osteoid volume per bone volume was significantly higher for the calcium phosphate group, while after four weeks, the results were almost equal; the trabecular thickness increased in comparison to week 2 in both groups with a slight advantage for the calcium phosphate group, (iii) the total mass of the bone-graft and the bone-graft substitute surface density were consistently decreasing; after 12 weeks, the new bone volume per tissue volume was still constantly growing, (iv) both bone-grafts showed good integration; new bone was formed on the surface of both bone-grafts, and (v) the calcium phosphate as well as the calcium phosphate alendronate paste had been enclosed by the bone; the trabecular thickness was higher in both groups compared to the first time point. Based on these findings, it was concluded that (i) calcium phosphate proved its good potential as a bone-graft substitute, (ii) the composite graft induced a good and constant new bone formation, (iii) not only the graft was incorporated into the bone but also a new bone was formed on its surface, (iv) both implants proved their function as a bone-graft substitute, but the bisphosphonate alendronate does not support the bone healing process sufficiently that the known properties of calcium phosphate as a bone-graft substitute were improved in the sense of a composite graft, and (v) bisphosphonate alendronate used as a bone-graft in a healthy bony environment did not influence the bone healing process in a positive or negative way [37].

9.1.2 Metallic materials

In recent years, significant progress has been made toward the development of titanium foams or porous titanium scaffolds, for orthopedic and dental applications. Ti foam possesses various characteristics (which are interrelated) including topographic features, compositions, compositions, surface properties, and various pore characteristics. The stability of the implant-bone interface, strength and rate of osseointegration, and the long-term success of dental implants are achieved by appropriate design and proper material selection. Although there is no specific implant design that meets all of the requirements in dental applications, implants can be engineered to maximize strength, interfacial stability, and load transfer [41]. All implantable materials intended for dental and orthopedic applications must contain complex topographical features. Surface topography and roughness are of the utmost importance for bone bonding and biological fixation of dental and orthopedic implants. Most surface topographical features provide initial stabilization until bone can grow and attach to the implant surface to establish further improvement in implant bonding [42]. Surface modification and coating of dental implants are aimed at creating 3D nano- and micro-scale porous structures for accelerating and improving bone and implant integration. A variety of methods have been developed to produce a porous coating on implants, with the specific aim of enhancing osseointegration. The fabrication of porous-surfaced Ti and Ti-6Al-4V implants normally involves either plasma-spraying or sintering powders onto a solid substrate [43, 44]. Story et al. [45] investigated cylindrical dental implants coated with cancellous structured titanium (CSTi) in canine mandibles and femorals. CSTi-2-coated and HA-coated implants were placed. The porosity of the CSTi-2 coating was 9% less than that of the previously studied CSTi-1 [46], resulting in greatly improved mechanical strength and cosmetic appearance. It was also mentioned that (i) a slightly lower level of bone ingrowth was observed for CSTi-2 than for CSTi-1; however, the in vivo attachment strength of the CSTi-2 coating was comparable both to CSTi-1 and to an HA-coated control after eight weeks; (ii) more bone ingrowth for the higher porosity coatings was observed at all time points (14 weeks) in the femorals and at the initial time points (two and four weeks) in the mandibles; and (iii) measured porosity is technique-dependent; digital analysis of in vitro samples yielded higher porosity values than in vivo histology cross sections.

The interest in using porous scaffolds for dental applications has been gaining momentum and continues to be the hot spot of clinical research. The focus, thereby, has mainly been on applications that involve bone ingrowth into the porous Ti scaffolds with a fully porous structure or with a solid core and porous surface/coating [41]. For load-bearing dental implants, Ti and its alloys have so far shown the greatest potential as the basis for such scaffolds, due to their excellent corrosion resistance, biocompatibility, and high degree of strength-to-weight ratio (in other words, specific strength). Significant progress has been made in fabricating various porous Ti scaffolds by using different techniques. Some of these techniques allow for greater control of pore architecture and porosity within scaffolds. The macro- and microstructure of porous scaffolds, such as pore size, pore shape, porosity, and pore interconnectivity, affect osteogenic cell adhesion, migration, ingrowth, and the production of a mineralized matrix. However, engineering porous scaffolds with ideal pore characteristics for hard tissue engineering applications remains a complex and challenging process. In addition, most of the available techniques for the fabrication of porous scaffolds are unable to serve as reliable methods for the further development of reproducible porous scaffolds. Ongoing and future research will continue to develop new techniques to control the pore characteristics of scaffolds and generate dental implants containing gradients of pore size and porosity for the regeneration of craniofacial bone. It is only recently that researchers have begun to understand the combination of parameters that need to be addressed in the successful implementation of porous scaffolds in vivo. Numerous studies have been performed to hasten osseointegration in porous Ti scaffolds and to improve their mechanical properties under occlusal and masticatory forces by altering the surface chemistry, introducing new biocompatible alloys, designs, and pore characteristics, or impregnating surface coatings with bone growth factors such as BMPs. It is thus envisaged that the breakthroughs and developments in manufacturing methods, micro- and nanoengineering, metallurgy, and biology will lay the foundations for more advanced and functional Ti foam scaffolds for dental applications [41].

Figure 9.1 shows the trabecular metallic dental implant with a structure and elasticity similar to cancellous bone – consisting of a titanium cervical and internal core section (upper section) covered by a trabecular metal sleeve (middle section) and joined by a titanium apical section (lower section).

Figure 9.1: Trabecular metal dental implant with a structure and elasticity similar to cancellous bone [modified after 41].

Magnesium has been suggested as a revolutionary biodegradable metal for use as an orthopedic material. As a biocompatible and degradable metal, it has several advantages over the permanent metallic materials currently in use, including eliminating the effects of stress shielding, improving biocompatibility concerns in vivo, and improving degradation properties, removing the requirement of a second surgery for implant removal. The rapid degradation of magnesium, however, is a double-edged sword as it is necessary to control the corrosion rates of the materials to match the rates of bone healing. In response, calcium phosphate coatings have been suggested as a means to control these corrosion rates. The potential calcium phosphate phases and their coating techniques on substrates are numerous and can provide several different properties for different applications. The reactivity and low melting point of magnesium, however, require specific parameters for calcium phosphate coatings to be successful. Within this review, an overview of the different calcium phosphate phases, their properties and their behavior in vitro and in vivo has been provided, followed by the current coating techniques used for calcium phosphates that may be or may have been adapted for magnesium substrates [46].

Among various Mg alloys, the following alloys have been investigated as potential implant materials, including AZ31, AZ91, WE43, WE54, Mg-Y-Zn, Mg-Zn, Mg-Ca, Mg-Zn-Ca, and Mg-Nd-Zn-Zr [48]. A series of studies have confirmed that magnesium alloys have good biocompatibility, may promote osteocyte growth, and induce production of osteoblasts and osteocytes [49–51]. Unfortunately, a high corrosion rate in the physiologic environment resulting in the generation of generous hydrogen gas and rapid loss of mechanical strength is the major obstacle to the clinical use of magnesium alloys [49, 50, 52–55]. As described above, Mg and Mg-based alloys have been proposed as degradable replacements to commonly used orthopedic biomaterials such as titanium alloys and stainless steel; however, the corrosion of Mg in a physiological environment remains a difficult characteristic to accurately assess with in vitro methods. Walker et al. [56] identified a simple in vitro immersion test that could provide corrosion rates similar to those observed in vivo. Pure Mg and five alloys (AZ31, Mg-0.8Ca, Mg-1Zn, Mg-1Mn, Mg-1.34Ca-3Zn) were immersed in either Earle's balanced salt solution (EBSS), minimum essential medium (MEM), or MEM-containing 40 g/L bovine serum albumin (MEMp) for 7, 14, or 21 days before removal and assessment of corrosion by weight loss. This in vitro data was compared to in vivo corrosion rates of the same materials implanted in a subcutaneous environment in Lewis rats for equivalent time points. It was found that (i) for the alloys investigated, the EBSS buffered with sodium bicarbonate provides a rate of degradation comparable to those observed in vivo; in contrast (ii) the addition of components such as (4-(2-hydroxyethyl)-1-piperazineethanesulfonic acid) (HEPES), vitamins, amino acids, and albumin significantly increased corrosion rates. Based on these findings, it is proposed that with this in vitro protocol, immersion of Mg alloys in EBSS can be used as a predictor of in vivo corrosion. Hence, if magnesium alloys are to be used successfully in orthopedic applications, the corrosion rates must be effectively controlled so as not to exceed the healing rates of the affected tissues. Thus, the magnesium alloys should remain present in the body and maintain mechanical integrity until the affected tissues have healed. As biomaterials for orthopedic applications, the biocompatibility and surface bioactivity of magnesium alloys must be taken into consideration. To further improve the corrosion resistance and surface biocompatibility of AZ31 alloy, a calcium phosphate (Ca–P) coating was fabricated on the surface of the AZ31 alloy by a chemical deposition process. However, the Ca–P-coated AZ31 alloy has not been systematically studied as a degradable biomaterial for orthopedic applications in vitro or in vivo. In the current study, we focused on the influence of the Ca–P coating on the corrosion behavior and biocompatibility of the AZ31 alloy. The aim of the present study was to determine the feasibility of the Ca–P coating in controlling the corrosion rate and improving the surface biocompatibility of the AZ31 alloy under in vitro and in vivo conditions [47].

Wang et al. [57] fabricated a calcium phosphate (Ca–P) coating on the surface of an AZ31 alloy by a chemical deposition process, and the in vitro and in vivo studies were carried out on a Ca–P-coated and uncoated AZ31 alloy to determine the effect of Ca–P coating on the corrosion behavior and biocompatibility of the AZ31 alloy. It was found that (i) the Ca–P coating reduced the in vitro and in vivo corrosion rates of the AZ31 alloy, (ii) cell experiments showed significantly good adherence and high proliferation on the Ca–P-coated AZ31 alloy than those on the uncoated AZ31 alloy, (iii) the blood cell aggregation tests showed that the Ca–P-coated AZ31 alloy had decreased the blood cell aggregation compared to the uncoated AZ31 alloy, and (iv) the animal experiments showed that the uncoated AZ31 alloy degraded more rapidly than the Ca–P-coated AZ31 alloy and the Ca–P coating provided significantly good biocompatibility; suggesting that (v) the Ca–P coating not only slowed down the corrosion rate of the AZ31 alloy, but also improved its biocompatibility, and (vi) the Ca–P-coated AZ31 alloy can be considered as a promising biomaterial for orthopedic applications.

9.2 Technologies and strategy

Reviewing new developments in bone tissue engineering, specifically focusing on the promising role of nanotechnology and future avenues of research, Saiz et al. [58] particularly emphasized on the need to fabricate scaffolds with multidimensional hierarchies for improved mechanical integrity, and new advances to promote bioactivity by manipulating the nano-level internal surfaces of scaffolds are examined followed by an evaluation of techniques using scaffolds as a vehicle for local drug delivery to promote bone regeneration/integration and methods of seeding cells into the scaffold. It was described that (i) the development of scaffolds for bone regeneration requires a material that is able to promote rapid bone formation while possessing sufficient strength to prevent fracture under physiological loads; (ii) success in simultaneously achieving mechanical integrity and sufficient bioactivity with a single material has been limited; however, (iii) the use of new tools to manipulate and characterize matter down to the nanoscale may enable a new generation of bone scaffolds that will surpass the performance of autologous bone implants.

Bone-grafts represent one of the most common tissue transplants, with over 2.2 million performed annually worldwide. While autologous bone-grafting for the reconstruction of skeletal defects is the current gold standard, this technique is hindered by variable resorption, limited supply, donor site morbidity, and high failure rates (up to 50%) in certain sites. These limitations lead to the development of synthetic biomaterials for the replacement of bone tissue. However, these synthetic materials are hindered/limited by their potential for both foreign-body reactions and infection. In recent years, nano-engineered particles and porous 3D scaffolds that facilitate growth of new bone have garnered significant attention. Nanotechnology represents a major frontier with potential to significantly advance the field of bone tissue engineering. Current limitations in regenerative strategies include impaired cellular proliferation and differentiation, insufficient mechanical strength of scaffolds, and inadequate production of extrinsic factors necessary for efficient osteogenesis. Based on this background, Walmsley et al. [59] reviewed several major areas of research in nanotechnology with potential implications in bone regeneration: 1) nanoparticle-based methods for delivery of bioactive molecules, growth factors, and genetic material, 2) nanoparticle-mediated cell labeling and targeting, and 3) nano-based scaffold construction and modification to enhance physicochemical interactions, biocompatibility, mechanical stability, and cellular attachment/survival. As these technologies continue to evolve, ultimate translation to the clinical environment may allow for improved therapeutic outcomes in patients with large bone deficits and osteodegenerative diseases. The review study concluded that (i) as a template for three-dimensional tissue growth, scaffolds must emulate native extracellular matrix. Nanoparticle modifications of scaffolds enhance this capacity to mimic complex properties of the natural bone environment and provide a more favorable milieu for cellular attachment, ingrowth, and bone formation. Whether through promotion of cellular survival, osteoblastic differentiation, or modulation of immunological response, nanostructural changes to polymer surfaces may provide a more favorable environment for bone regenerative strategies; and (ii) notwithstanding the formidable advances within the field, the issue of better understanding the interactions between nanoscale surface topography and the biological system into which it is introduced persists. As Bruinink et al. [60] described, this interface plays a significant role in osseointegration of implants, but evidence-based nanoscale-surfaced design is an elusive technique. Walmsley et al. [59] also mentioned that (iii) further understanding of surface characteristics such as wettability, surface energy and roughness, surface curvature and nanoscale features, organic and inorganic coatings effect cell signaling, proliferation, integration, and viability will be required; and (iv) in the foreseeable future, nanotechnology may allow for specifically tailored treatment of various disease states with complex skeletal defects at distinct anatomical sites.

9.2.1 Tissue engineering

Tissue engineering involves using relevant scaffolds, introducing appropriate growth factors and cells, and, more recently, using stem cells [61, 62]. Using tissue engineering techniques, it is possible to design new scaffolds and tissue grafts aiming to decrease the disadvantages of traditional grafts and improve graft incorporation, osteogenicity, osteoconductivity, and osteoinductivity [62, 63]. Tissue engineering (TE) has limitations, including use of a wide variety of materials in producing tissue-engineered grafts or scaffolds. Consequently, translational investigations testing each material are limited, reducing their clinical applicability. Therefore, some important aspects of host graft interaction and immune response to these implants, scaffolds, and viable grafts are still not clear [61, 62]. With advances in tissue engineering, the ability to repair or regenerate bone tissue is developing, and its applications are expanding.

After reviewing and analyzing the literature on bone-grafts and tissue engineering as a strategy in a field of orthopedic surgery, Oryan et al. [62] concluded that (i) tissue engineering is a new and developing option that had been introduced to reduce limitations of bone-grafts and improve the healing processes of the bone fractures and defects and (ii) the combined use of scaffolds, healing promoting factors, together with gene therapy, and, more recently, three-dimensional printing of tissue-engineered constructs may open new insights in the near future. It was further foreseen that new strategies such as gene therapy, polytherapy by using scaffolds, healing promotive factors and stem cells, and finally three-dimensional printing are in their preliminary stages but may offer new exciting alternatives in the near future [61].

Tissue engineering is the "final" option in managing bone loss. Tissue engineering can involve the use of scaffolds, healing promotive factors (e.g., growth factors), and stem cells [61]. Since tissue engineering can be defined as a process that affects the structure and architecture of any viable and nonviable tissue with the aim of increasing the effectiveness of the construct in biologic environments [64], all the non-fresh grafts, which are processed for acellularization, belong to the tissue engineering category. In fact, acellularization is the basic tissue engineering technology described for allograft and xenografts [65]. This method of tissue engineering has been used to decrease the antigenicity of viable grafts for many years [65, 66].

Distler et al. [67] predicted that since they do not require bone donation, 3D-printed implants and biomaterial-based tissue engineering strategies promise to be the next generation therapies for bone regeneration. Polylactic acid (PLA)-bioactive glass (BAG) composite scaffolds were manufactured by fused deposition modeling (FDM) involving the fabrication of PLA-BG composite filaments, which are used to 3D-print controlled open-porous and osteoinductive scaffolds [67]. The printability of PLA-BG filaments as well as the bioactivity and cytocompatibility of PLA-BAG scaffolds using pre-osteoblast MC3T3E1 (which is an osteoblast precursor cell line derived from *Mus musculus* calvaria) cells was demonstrated. It was reported that gene expression analyses indicated the beneficial impact of BAG inclusions in FDM scaffolds regarding osteoinduction, as BAG inclusions lead to increased osteogenic differentiation of human adipose-derived stem cells in comparison to pristine PLA, confirming that FDM is a convenient additive manufacturing technology to develop PLA-BAG composite scaffolds suitable for bone tissue engineering.

9.2.2 Cell-based strategy

The extraordinary healing capacity of bone can be challenged by complex fractures (i.e., injuries above critical size) or health conditions (i.e., diabetes, genetic factors, poor lifestyle, and low HRQoL – health-related quality of living), resulting in non-union fractures that can lead to long-term disability and pain [68]. Bone reconstruction techniques are mainly based on the use of tissue grafts and artificial scaffolds. The former presents well-known limitations, such as restricted graft availability and donor site morbidity, while the latter commonly results in poor graft integration and fixation in the bone, which leads to the unbalanced distribution of loads, impaired bone formation, increased pain perception, and risk of fracture, ultimately leading to recurrent surgeries. In the past decade, research efforts have been focused on the development of innovative bone substitutes that not only provide immediate mechanical support, but also ensure appropriate graft anchoring by, for example, promoting de novo bone tissue formation. From the countless studies that were done in this direction, only few have made the big jump from the benchtop to the bedside, whilst most have perished along the challenging path of clinical translation. Sallent et al. [14] described some clinically successful cases of bone device development, including biological glues, stem cell-seeded scaffolds, and gene-functionalized bone substitutes and discussed the ventures that these technologies went through, the hindrances they faced and the common grounds among them, which might have been key for their success.

Cell-based strategies for bone tissue engineering have a long trajectory in the research stage but have minimally contributed to current clinical practices [69]. Indeed, the introduction of cells as a component in tissue engineering entails important economic and safety concerns; the former is related to the logistics, technology, and necessary human resources and the latter is related to possible immunogenicity, teratoma formation, and disease transmission risks [70]. As a result of the second, only those therapies that involve minimal ex vivo manipulation of autologous cells are FDA (US Food and Drug Administration)-approved, whilst those that follow the traditional tissue engineering paradigm (in vitro expansion of autologous/allogeneic cells and ex vivo development of a cell-based construct) present a more tortuous regulatory pathway that commonly results in the abandonment of the technology, in the best scenario, after clinical trials [71]. Taken together, tissue engineering approaches and, more precisely, the use of stem cells in combination with biomaterials has proven, in most of the clinical studies, to match or surpass the clinical outcomes of autografts. The extra step of in vitro cell expansion entails numerous risks and cost-related burdens but, if well designed, can increment the therapeutic outcomes. The major impediment in clinical translation of cell-based technologies is and will be the costs and risks associated, making the addressing of these issues, at an early stage of research, fundamental [14].

Regenerative medicine is the medical field that is involved in creating functional tissues to repair and replace damaged or malfunctioning tissues and organs [72]. Tissues or organs generated from a patient's own cells would allow transplants without tissue rejection. Furthermore, regenerative medicine treatments have the potential to replace organ transplants or artificial organs. As regenerative medicine generates tissues or organs using engineering technology, it is also called the tissue engineering [73]. To make regenerative medicine successful, three elements are required: stem cells, scaffolds, and growth factors [72]. Translational research, which takes results from the laboratory and translates them to the clinics, and industry-academic collaborations also play important roles in making regenerative medicine suitable for practical use [73].

In recent years, stem cell research has grown exponentially due to the recognition that stem cell-based therapies have the potential to improve the life of patients with several kinds of diseases, such as Alzheimer's disease and cardiac ischemia. These therapies have also a role in regenerative medicine, such as the repair of bone or tooth loss. Stem cells have the potential to differentiate into several cell types, including odontoblasts, neural progenitors, osteoblasts, chondrocytes, and adipocytes [73]. Mesenchymal stem cells (MSC) are multipotent progenitor cells that were originally isolated from various tissues, including adult bone marrow, adipose tissue, skin, umbilical cord, and placenta. Bone marrow-derived MSCs have been used in clinical trials for the effective treatment of osseous defects. However, bone marrow aspiration is an invasive and painful procedure for the donor and is a difficult procedure for a general practitioner. Furthermore, MSCs constitute heterogeneous cell types, and the potential for proliferation and differentiation of the MSCs depends on a patient's age, sex, or the presence of certain medical conditions, such as diabetes or hypertension [74]. Of late, mesenchymal stem cell-based therapies have the potential to provide an effective treatment for osseous defects. Jimi et al. [73] discussed both the current therapy for bone regeneration and the perspectives in the field of stem cell-based regenerative medicine, addressing the sources of stem cells and growth factors used to induce bone regeneration effectively and reproducibly. It was concluded that, although regenerative medicine has been tried in various fields, there is much demand for regenerative medicine in dentistry, particularly in bone regeneration. Depending on the state of periodontitis or jaw resection, it might take more than 6–2 months for occlusal reconstruction. Thus, the development of an efficient and high-quality bone derivation method is necessary. Cell-based therapy may pave the way to rejection-free regenerative treatment for bone defects. It is also likely that research concerning growth factor or cell-based therapies will continue to progress. However, there are also many problems, such as laws and costs of equipment that must be solved. Although it is unclear when the technology of regenerative medicine will be put into practical use, it is important to follow the current status of regenerative medicine to keep abreast with the progression of technology.

9.3 Concept and approach

Toward successful development of the third-generation bone-graft substitutes, which closely mimic the bone tissue microenvironment, Polo-Corrales et al. [75] listed several requirements, including: (1) a scaffold that matches the biomaterial degradation rate with the bone regeneration rate, (2) a multi-analyte growth factor delivery strategy with proper control over its release kinetics, (3) a strategy for stimulating scaffold vascularization (e.g., co-cultures, growth factor, etc.), (4) development of a bioreactor technology capable of preparing scaffolds for implantation, and (5) development of biophysical stimulation technologies that can be brought to the clinic. It was also mentioned that, at the same time, the following knowledge about the bone regeneration process is required; (i) temporal and multi-growth factor requirements, (ii) signaling pathway cross talk after growth factor stimulation, and (iii) signaling pathway cross talk after biophysical stimulation. All these challenges will surely be met through the close interdisciplinary collaboration of material scientists, engineers, medical doctors, chemists, and physicists [75].

Among various demanded properties for acceptable bone substitutes, there are needs for the careful structural design of new materials including considering vital biological parameters such as pore size, density, morphology, interconnectivity, and resorbability [17]. Additionally, there has also been increasing interest in the controlled time-release delivery of growth factors as a means of maintaining their bioactivity over the therapeutic window [76]. Another major challenge we face is the lack of research investigating the safety and efficacy of newer bone-grafting materials. Thus, the development of novel grafting materials should focus on incorporating as many ideal biological parameters as possible, whilst ensuring that such materials would be readily available, cost-effective, and clinically evidence-based [77]. Zhao et al. [77] furthermore described that (1) development of hybrid grafts, which utilize growth factors and living osteogenic cells capable of inducing bone regeneration, presents the future of dental bone-grafting and dental implants, (2) good examples include bone substitutes that can release bone morphogenic proteins or platelet-derived growth factors in a controlled manner, and (3) despite the progress highlighted in this review article more work is needed to develop dental biomaterials that have a porous structure, mechanical stability, controlled degradation, and remodeling ability, which is comparable with the rate of new bone formation. Jones [78] described that ideally, a nondestructive technique will be developed that can image cells within porous scaffolds in three dimensions, with a resolution high enough to image cellular processes. It would be very useful to be able to image cells and the formation of extracellular matrices in three-dimensional scaffolds, so that the migration of cells and formation of new tissue can be monitored throughout the scaffold without having to do serial sectioning and histology. Three-dimensional imaging of the migration of cells and matrix formation may soon be possible using high-resolution µCT.

Treatment of segmental bone defects following traumatic injury, Cierny-Mader Stage III and IV osteomyelitis, and en bloc tumor resections remain clinically challenging [79]. Without adequate vascularization, the inner regions of the graft either undergo cellular necrosis (can be the case with large segmental autografts) or failure of proper cellular infiltration and regeneration, leading to graft failure [80–82]. Based on this background, Valtanen et al. [79] mentioned that traditional bone tissue engineering has focused on designing a scaffold that is osteoconductive, osteoinductive, and osteogenic; however, current efforts have not been integrated with cutting-edge innovations in vascular tissue engineering. The translational future of vascularized bone tissue engineering scaffolds will be determined by our ability to integrate these two designs into one. Furthermore, composite rigid scaffolds will not make or break the success of synthetic bone-graft substitutes. Instead, this will depend on our ability to create an acute living tissue bed in the critical size bone defect to allow differentiation into vascularized bone. It is quite possible that the skillfully engineered rigid composite scaffolds may not even play a role in the final solution to this ongoing dilemma. Sergi et al. [83] proposed a unique hybrid composite consisting of BAG (bioactive glass) and natural polymer; characterizing that (i) collagen, gelatin, silk fibroin, hyaluronic acid, chitosan, alginate, and cellulose containing BAGs represent interesting materials for biomedical devices, for both hard and soft tissue, not least because of their biocompatibility and non-toxicity, (ii) the addition of nano-sized and micro-sized BAG particles has been shown to improve mechanical properties, to enhance bioactivity, and to promote the cells' viability, adhesion, proliferation, and differentiation, and (iii) the incorporation of BAGs offers a chance to adjust the elasticity of polymers and the cellular response. The elasticity of natural polymers represents a pivotal feature that affects the cellular response both in vitro and in vivo. It is furthermore claimed that (iv) overall, combining natural polymers with bioactive glass particles to produce composites for both hard and soft tissues can represent an efficient strategy to heal different damages of the body. Considering the in vitro behavior, further research involving the investigation of different cell lines needs to be considered, gaining more relevant data on the osteogenic, odontogenic, and angiogenic behavior of natural polymer composites. Furthermore, it is worth noting that preclinical studies in animals should be improved and become the focus of future studies. In addition, future studies should pay attention to investigate the most advantageous combination of natural polymers and BAGs to achieve good mechanical and biological performances of composites.

There is no single ideal graft material to choose in clinical practice; therefore researches are ongoing in all relevant fields to establish modern bone regeneration protocols that may lead to the innovation of ideal graft substitutes. Kheirallh et al. [3] reviewed articles on the current state of development of bone-graft substitutes that could be used for bone defect regeneration, as well as to analyze their efficacy for clinical use, and concluded that future trends may focus on the effective combinations of osteoinductive materials, osteoinductive growth factors, and cell-based tissue regeneration tactics using composite carriers. The development of biomaterials and techniques for alveolar bone augmentation applications is a challenge from an engineering, surgical, and biological perspective. The newer generation of biomaterials will have to be fine-tuned in terms of their physicochemical properties to have more predictable and improved graft resorption after implantation to be used as effective scaffolds for bone augmentation [84]. Various approaches discussed here separately or in combination have the potential for providing improved tissue regenerative results [85]. It is expected that the next generation of biomaterials will demonstrate vast improvements in graft and biological tissue interfacing based on the knowledge gained from recent research and allow clinicians to achieve more predictable clinical results with regards to vertical alveolar bone augmentation.

Huo et al. [86] conducted an extensive review on the state-of-art progress in the design, manufacturing, and assessment of the bone scaffold for fixing the large bone defects. It was concluded that (i) the microstructures of the bone scaffolds have evolved from the periodic regular unit to the nonperiodic irregular unit, which can be designed using some advanced optimization algorithms and (ii) the additive manufacturing technique has enabled the production of the scaffolds with complicated internal structures. Various techniques can be used to assess the performance of the scaffold, especially the emerging cell culture experiments; however, there are still many issues that remain to be solved, and to tackle these issues, further improvements and developments are still needed in the future, especially in the following perspectives: (1) to establish the optimization framework considering the dynamic interaction between the scaffold and surrounding tissues, especially the time-varying properties of both the degradable scaffolds and the bone tissues. In this challenge, the machine learning [87] algorithm can be utilized to efficiently and quickly predict the dynamic behaviors of both the scaffolds and the surrounding tissues. (2) To improve the quality of the scaffolds produced by additive manufacturing (AM), this challenge can be tackled first by improving the AM technique, second by improving the design of the scaffold, e.g., incorporating the AM defects, etc. into the design stage to minimize the defects in the products; (3) to comprehensively develop and apply the most advanced measurement techniques into the assessment of the performance of the scaffold. In recent years, the flexible measurement devices and many flexible and wireless sensors have been developed. These tools have enabled the fast and accurate measurements of some parameters (e.g., the surface strain, the muscle activation level) in the human body, which can be hardly measured previously. Therefore, the performance of the scaffolds implanted in the human body can be better assessed using these up-to-date measurement techniques in the future [86].

References

[1]   Nandi SK, Roy S, Mukherjee P, Kundu B, De DK, Basu D. Orthopedic applications of bone graft and graft substitutes: A review. Indian J Med Res. 2010, 132, 15–30.
[2]   Elsalanty ME, Genecov DG. Bone grafts in craniofacial surgery. Craniomaxillofac Trauma Reconstr. 2009, 2, 125–34.
[3]   Kheirallah M, Almeshaly M. Bone graft substitutes for bone defect regeneration. A collective review. Int J Dent Oral Sci. 2016, 3, 247–57.
[4]   Bigham AS, Dehghani SN, Shafiei Z, Torabi Nezhad S. Xenogenic demineralized bone matrix and fresh autogenous cortical bone effects on experimental bone healing: Radiological, histopathological and biomechanical evaluation. J Orthop Traumatol. 2008, 9, 73–80.
[5]   Scaglione M, Fabbri L, Dell'Omo D, Gambini F, Guido G. Long bone nonunions treated with autologous concentrated bone marrow-derived cells combined with dried bone allograft. Musculoskelet Surg. 2014, 98, 101–06.
[6]   Greenwald AS, Boden SD, Goldberg VM, Khan Y, Laurencin CT, Cato T, Randy RN. Bone-graft substitutes: Facts, fictions and applications. J Bone Joint Surg Am. 2001, 8398–103.
[7]   Kneser U, Schaefer DJ, Munder B, Klemt C, Andree C. Tissue engineering of bone. Min Invas Ther Allied Technol. 2002, 11, 107–16.
[8]   Kumar P, Saini M, Dehiya BS, Sindhu A, Kumar V, Kumar R, Lamberti L, Pruncu CI, Thakur R. Comprehensive survey on nanobiomaterials for bone tissue engineering applications. Adv Nanomater Biomed Nanomater. 2020, 10, 2019; https://doi.org/10.3390/nano10102019.
[9]   Oshida Y. Bioscience and Bioengineering of Titanium Materials. Elsevier, 2007.
[10] Oshida Y. Surface Engineering and Technology for Biomedical Implants. Momentum Press, 2014.
[11] McMahon RE, Wang L, Skoracki R, Mathur AB. Development of nanomaterials for bone repair and regeneration. J Biomed Mater Res B Appl Biomater. 2013, 101, 387–97.
[12] Zou D, Zhang Z, He J, Zhu S, Wang S, Zhang W, Zhou J, Xu Y, Huang Y, Wang Y, Han W, Zhou Y, You S, Jiang X. Repairing critical-sized calvarial defects with BMSCs modified by a constitutively active form of hypoxia-inducible factor-1α and a phosphate cement scaffold. Biomaterials. 2011, 32, 9707–18.
[13] Ishihara A, Bertone AL. Cell-mediated and direct gene therapy for bone regeneration. Expert Opin Biol Ther. 2012, 12, 411–23.
[14] Scheller EL, Villa-Diaz LG, Krebsbach PH. Gene therapy: Implications for craniofacial regeneration. J Craniofac Surg. 2012, 23, 333–37.
[15] Sallent I, Capella-Monsonís H, Procter P, Bozo IY, Deev RV, Zubov D, Vasyliev R, Perale G, Pertici G, Baker J, Gingras P, Bayon Y, Zeugolis DI. The few who made it: Commercially and clinically successful innovative bone grafts. Front Bioeng Biotechnol. 2020, 8, 952; https://doi.org/10.3389/fbioe.2020.00952.
[16] Buser Z, Brodke DS, Youssef JA, Meisel H-J, Myhre SL, Hashimoto R, Park J-B, Yoon ST, Wang JC. Synthetic bone graft versus autograft or allograft for spinal fusion: A systematic review. J Neurosurg Spine SPI. 2016, 25, 509–16.
[17] Haugen HJ, Lyngstadaas SP, Rossi F, Perale G. Bone grafts: Which is the ideal biomaterial?. J Clin Periodontol. 2019, 46, 92–102; doi: 10.1111/jcpe.13058.
[18] Stark JR, Hsieh J, Waller D. Bone graft substitutes in single- or double-level anterior cervical discectomy and fusion: A systematic review. Spine. 2019, 44, E618–8.
[19] Neufurth M, Wang X, Wang S, Steffen R, Ackermann M, Haep ND, Schröder HC, Müller WEG. 3D printing of hybrid biomaterials for bone tissue engineering: Calcium-polyphosphate microparticles encapsulated by polycaprolactone. Acta Biomater. 2017, 64, 377–88.
[20] Filardo G, Kon E, Tampieri A, Cabezas-Rodríguez R, Di Martino A, Fini M, Giavaresi G, Lelli M, Martínez-Fernández J, Martini L, Ramírez-Rico J, Salamanna F, Sandri M, Sprio S, Marcacci M. New

bio-ceramization processes applied to vegetable hierarchical structures for bone regeneration: An experimental model in sheep. Tissue Eng Part A. 2014, 20, 763–73.

[21] Gao F, Xu Z, Liang Q, Li H, Peng L, Wu M, Zhao X, Cui X, Ruan C, Liu W. Osteochondral regeneration with 3D-Printed biodegradable high-strength supramolecular polymer reinforced-gelatin hydrogel scaffolds. Adv Sci. 2019, 6, 1900867; doi: 10.1002/advs.201900867.

[22] Itoh S, Kikuchi M, Koyama Y, Takakuda K, Shinomiya K, Tanaka J. Development of an artificial vertebral body using a novel biomaterial, hydroxyapatite/collagen composite. Biomaterials. 2002, 23, 3919–26.

[23] Fratzl P, Gupta HS, Paschalis EP, Roschger P. Structure and mechanical quality of the collagen-mineral nano-composite in bone. J Mater Chem. 2004, 14, 2115–23.

[24] Dorozhkin SV. Biocomposites and hybrid biomaterials based on calcium orthophosphates. Biomatter. 2011, 1, 3–56.

[25] Burr DB. The contribution of the organic matrix to bone's material properties. Bone. 2011, 31, 8–11.

[26] Cui FZ, Li Y, Ge J. Self-assembly of mineralized collagen composites. Mater Sci Eng Rep. 2007, 57, 1–27.

[27] Chan CK, Kumar TSS, Liao S, Murugan R, Ngiam M, Ramakrishnan S. Biomimetic nanocomposites for bone graft applications. Nanomedicine (Lond). 2006, 1, 177–88.

[28] Weiner S, Wagner HD. The material bone: Structure-mechanical function relations. Annu Rev Mater Sci. 1998, 28, 271–98.

[29] Hench LL. Bioceramics. J Am Ceram Soc. 1998, 81, 1705–28.

[30] Vallet-Regi M, Arcos D. Nanostructured hybrid materials for bone tissue regeneration. Curr Nanosci. 2006, 2, 179–89.

[31] Bauer TW, Muschler GF. Bone graft materials. An overview of the basic science. Clin Orthop Relat Res. 2000, 371, 10–27.

[32] Athanasiou KA, Zhu CF, Lanctot DR, Agrawal CM, Wang X. Fundamentals of biomechanics in tissue engineering of bone. Tissue Eng. 2000, 6, 361–81.

[33] Doblaré M, Garcia JM, Gómez MJ. Modelling bone tissue fracture and healing: A review. Eng Fract Mech. 2004, 71, 1809–40.

[34] Huiskes R, Ruimerman R, van Lenthe GH, Janssen JD. Effects of mechanical forces on maintenance and adaptation of form in trabecular bone. Nature. 2000, 405, 704–06.

[35] Suchanek W, Yoshimura M. Processing and properties of hydroxyapatite-based biomaterials for use as hard tissue replacement implants. J Mater Res. 1998, 13, 94–117.

[36] Arcos D, Boccaccini AR, Bohner M, Diez-Perez A, Epple M, Gomez-Barrena E, Herrera A, Planell JA, Rodríguez-Mañas L, Vallet-Regí M. The relevance of biomaterials to the prevention and treatment of osteoporosis. Acta Biomater. 2014, 10, 1793–805.

[37] Schlickewei CW, Laaff G, Andresen A, Klatte TO, Rueger JM, Ruesing J, Epple M, Lehmann W. Bone augmentation using a new injectable bone graft substitute by combining calcium phosphate and bisphosphonate as composite – An animal model. J Orthop Surg and Res. 2015, 10, 116; https://doi.org/10.1186/s13018-015-0263-z.

[38] Einhorn TA. Can an anti-fracture agent heal fractures?. Clin Cases Miner Bone Metab. 2010, 7, 11–14.

[39] Boanini E, Gazzano M, Rubini K, Bigi A. Composite nanocrystals provide new insight on alendronate interaction with hydroxyapatite structure. Adv Mater. 2007, 19, 2499–502.

[40] Cattalini J, Boccaccini AR, Lucangioli S, Mourino V. Bisphosphonate-based strategies for bone tissue engineering and orthopedic implants. Tissue Eng Part B-Rev. 2012, 18, 323–40.

[41] Trabecular metal dental implant. ZimVie. https://www.zimvie.com/en/dental/dental-implant-systems/trabecular-metal-implant.html.

[42] Nouri A, Wen C. Introduction to surface coating and modification for metallic bio-materials. In: Wen C (Ed.)., Surface Coating and Modification of Metallic Biomaterials. Woodhead Publishing, Cambridge, UK, 2015, 3–60.

[43] Yue S, Pilliar RM, Weatherly GC. The fatigue strength of porous-coated Ti-6%Al-4%Vimplant alloy. J Biomed Mater Res. 1984, 18, 1043–58.

[44] Pilliar RM. Porous-surfaced metallic implants for orthopaedic applications. J Biomed Mater Res. 1987, 21, 1–33.

[45] Story BJ, Wagner WR, Gaisser DM, Cook SD, Rust-Dawicki AM. In vivo performance of a modified CSTi dental implant coating. Int J Oral Maxillofac Implants. 1998, 13, 749–57.

[46] Cook SD, Rust-Dawicki AM. In vivo evaluation of a CSTi dental implant: A healing time course study. J Oral Implantol. 1995, 21, 82–90.

[47] Shadanbaz S, Dias GJ. Calcium phosphate coatings on magnesium alloys for biomedical applications: A review. Acta Biomater. 2012, 8, 20–30.

[48] Oshida Y. Magnesium Materials. De Gruyter Pub, 2021.

[49] Zhang S, Li J, Song Y, Zhao C, Zhnag X, Xie C, Zhnag Y, Tao H, He Y, Jiang Y, Bian Y. In vitro degradation, hemolysis and MC3T3-E1 cell adhesion of biodegradable Mg–Zn alloy. Mater Sci Eng C. 2009, 29, 1907–12.

[50] Xu L, Zhang E, Yin D, Zeng S, Yang K. In vitro corrosion behaviour of Mg alloys in a phosphate buffered solution for bone implant application. J Mater Sci Mater in Med. 2008, 19, 1017–25.

[51] Wei J, Jia J, Wu F, Wei S, Zhou H, Zhang H, Shin J-W, Liu C. Hierarchically microporous/macroporous scaffold of magnesium-calcium phosphate for bone tissue regeneration. Biomaterials. 2010, 31, 1260–69.

[52] Witte F. The history of biodegradable magnesium implants: A review. Acta Biomater. 2010, 6, 1680–92.

[53] Staiger MP, Pietak AM, Huadmai J, Dias G. Magnesium and its alloys as orthopedic biomaterials: A review. Biomaterials. 2006, 27, 1728–34.

[54] Yamamoto A, Watanabe A, Sugahara K, Tsubakino H, Fukumoto S. Improvement of corrosion resistance of magnesium alloys by vapor deposition. Scripta Materialia. 2001, 44, 1039–42.

[55] Gray JE, Luan B. Protective coatings on magnesium and its alloys–a critical review. J Alloys Comp. 2002, 336, 88–113.

[56] Walker J, Shadanbaz S, Kirkland NT, Stace E, Woodfield T, Staiger MP, Dias GP. Magnesium alloys: Predicting in vivo corrosion with in vitro immersion testing. J Biomed Mater Res B. 2012, 100, 1134–41.

[57] Wang Y, Zhu Z, Xu X, He Y, Zhang B. Improved corrosion resistance and biocompatibility of a calcium phosphate coating on a magnesium alloy for orthopedic applications. Eur J Inflam. 2016, https://doi.org/10.1177/1721727X16677763.

[58] Saiz E, Zimmermann EA, Lee JS, Wegst UGK, Tomsia AP. Perspectives on the role of nanotechnology in bone tissue engineering. Dent Mater. 2013, 29, 103–15.

[59] Walmsley GG, McArdle A, Tevlin R, Momeni A, Atashroo D, Hu MS MD, Feroze AH, Wong VW MD, Lorenz PH, Longaker MT, Wan DC. Nanotechnology in bone tissue engineering. Nanomedicine. 2015, 11, 1253–63.

[60] Bruinink A, Bitar M, Pleskova M, Wick P, Krug HF, Maniura-Weber K. Addition of nanoscaled bioinspired surface features: A revolution for bone related implants and scaffolds?. J Biomed Mater Res A. 2014, 102, 275–94.

[61] Moshiri A, Oryan A. Role of tissue engineering in tendon reconstructive surgery and regenerative medicine: Current concepts, approaches and concerns. Hard Tissue. 2012, 1, 1–11.

[62] Oryan A, Alidadi S, Moshiri A, Maffulli N. Bone regenerative medicine: Classic options, novel strategies, and future directions. J Orthop Surg Res. 2014, 9; https://josr-online.biomedcentral.com/articles/10.1186/1749-799X-9-18.

[63]   Dimitriou R, Jones E, McGonagle D, Giannoudis PV. Bone regeneration: Current concepts and future directions. BMC Med. 2011, 9, 66; doi: 10.1186/1741-7015-9-66.

[64]   Rose FR, Oreffo RO. Bone tissue engineering: Hope vs hype. Biochem Biophys Res Commun. 2002, 292, 1–7.

[65]   Vavken P, Joshi S, Murray MM. Triton-X is most effective among three decellularization agents for ACL tissue engineering. J Orthop Res. 2009, 27, 1612–18.

[66]   Zhang AY, Bates SJ, Morrow E, Pham H, Pham B, Chang J. Tissue-engineered intrasynovial tendons: Optimization of acellularization and seeding. J Rehabil Res Dev. 2009, 46, 489–98.

[67]   Distler T, Fournier N, Grünewald A, Polley C, Seitz H, Detsch R, Boccaccini AR. Polymer-bioactive glass composite filaments for 3D scaffold manufacturing by fused deposition modeling: Fabrication and characterization. Front Bioeng Biotechnol. 2020, 8; https://doi.org/10.3389/fbioe.2020.00552.

[68]   Keating JF, Simpson AH, Robinson CM. The management of fractures with bone loss. J Bone Joint Surg Br. 2005, 87, 142–50.

[69]   Mishra R, Bishop T, Valerio IL, Fisher JP, Dean D. The potential impact of bone tissue engineering in the clinic. Regen Med. 2016, 11, 571–87.

[70]   Webber MJ, Khan OF, Sydlik SA, Tang BC, Langer R. A perspective on the clinical translation of scaffolds for tissue engineering. Ann Biomed Eng. 2015, 43, 641–56.

[71]   Jager M, Hernigou P, Zilkens C, Herten M, Fischer J, Krauspe R. Cell therapy in bone-healing disorders. Orthopade. 2010, 39, 449–62.

[72]   Gurtner GC, Callaghan MJ, Longaker MT. Progress and potential for regenerative medicine. Ann Rev Med. 2007, 58, 299–312.

[73]   Jimi E, Hirata S, Osawa K, Terashita M, Kitamura C, Fukushima H. The current and future therapies of bone regeneration to repair bone Defects. Intl J Dent. 2012, 148261. https://doi.org/10.1155/2012/148261.

[74]   Kern S, Eichler H, Stoeve J, Klüter H, Bieback K. Comparative analysis of mesenchymal stem cells from bone marrow, umbilical cord blood, or adipose tissue. Stem Cells. 2006, 24, 1294–301.

[75]   Polo-Corrales L, Latorre-Esteves M, Ramierz-Vick JE. Scaffold design for bone regeneration. J Nanosci Nanotechnol. 2014, 14, 15–56.

[76]   Oliveira É, Nie L, Podstawczyk D, Allahbakhsh A, Ratnayake J, Brasil D, Shavandi A. Advances in growth factor delivery for bone tissue engineering. Int J Mol Sci. 2021, 22, 903; doi: 10.3390/ijms22020903.

[77]   Zhao R, Yang R, Cooper PR, Khurshid Z, Shavandi A, Ratnayake J. Bone grafts and substitutes in dentistry: A review of current trends and developments. Molecules. 2021, 26, 3007; doi: 10.3390/molecules26103007.

[78]   Jones JR. Observing cell response to biomaterials. Mater Today. 2006, 9, 34–43.

[79]   Valtanen RS, Yang YP, Gurtner GC, Maloney WJ, Lowenberg DW. Synthetic and Bone tissue engineering graft substitutes: What is the future?. Injury. 2021, 52, S72–7.

[80]   Naito Y, Shinoka T, Duncan D, Hibino N, Solomon D, Cleary M, Rathore A, Fein C, Church S, Breuer C. Vascular tissue engineering: Towards the next generation vascular grafts. Adv Drug Deliv Rev. 2011, 63, 312–23.

[81]   Lovett M, Lee K, Edwards A, Kaplan DL. Vascularization strategies for tissue engineering. Tissue Eng – Part B Rev. 2009, 15, 353–70.

[82]   Elomaa L, Yang YP. Additive manufacturing of vascular grafts and vascularized tissue constructs. Tissue Eng – Part B Rev. 2017, 23, 436–50.

[83]   Sergi R, Bellucci D, Cannillo V. A review of bioactive glass/natural polymer composites: State of the art. Materials. 2020, 13, 5560; https://doi.org/10.3390/ma13235560.

[84]   Sheikh Z, Sima C, Glogauer M. Bone replacement materials and techniques used for achieving vertical alveolar bone augmentation. Materials (Basel). 2015, 8, 2953–93.

[85] Breitbart AS, Grande DA, Mason JM, Barcia M, James T, Grant RT. Gene-enhanced tissue engineering: Applications for bone healing using cultured periosteal cells transduced retrovirally with the BMP-7 gene. Ann Plast Surg. 1999, 42, 488–95.

[86] Huo Y, Lu Y, Meng L, Wu J, Gong T, Zou J, Bosiakov S, Cheng L. A critical review on the design, manufacturing and assessment of the bone scaffold for large bone defects. Front Bioeng Biotechnol. 2021, 9; https://doi.org/10.3389/fbioe.2021.753715.

[87] Oshida Y. Artificial Intelligence for Medicine – People, Society, Pharmaceuticals, and Medical Materials. De Gruyter Pub, 2022.

Postscript

Although the bone normally possesses the capability of self-regeneration and self-repair, in large and massive bone defect cases, bone healing and repair are hardly expected to be achieved, due to various factors including insufficient blood supply and infection of the injured bone or its surrounding tissues. Besides, systemic chronic disease could adversely affect negative bone healing activity, resulting in delayed unions or nonunions [1–3]. To overcome these issues, various bone-grafting materials have been developed to promote bone healing. The bone-grafting material can be used alone or in combination with other material(s) as hybrids or composites. Major biofunctions can be osteogenesis, osteoinduction, and osteoconduction, in combination or alone. In both dental and orthopedic implant treatments, mesh membrane and/or scaffold material are employed to support and ensure the structural integrity at healing areas. These supportive materials can be bioabsorbable or biodegradable under controlled resorption rate. From the last chapter, the common concept among many articles on the future perspective that is described is that tissue engineering (including bone tissue engineering as well as cell tissue engineering) is a new and developing option that had been introduced to reduce limitations of bone-grafts and improve the healing processes of the bone fractures and defects. The combined use of scaffolds, healing promoting factors together with gene therapy, and, more recently, three-dimensional printing of tissue-engineered constructs may open new insights in the near future. All these new concepts deal with the interfacial action between vital host hard tissue (i.e., bone) and foreign material (i.e., bone-grafts along with additional material), so that three important compatibilities (biological compatibility, biomechanical compatibility and morphological compatibility) should still be required to be fulfilled [4–6].

Magnesium is a very important macroelement for the human body and serves a lot of functions. Its deficiency can cause many disorders and health ailments. Magnesium is needed for more than 300 biochemical reactions in the body. At the biochemical level, magnesium is involved in energy metabolism and protein synthesis; it maintains normal muscle and nerve function, supports a healthy immune system, and keeps bones strong [7]. Magnesium also helps regulate blood sugar levels and promotes normal blood pressure, playing important roles in preventing and managing disorders such as hypertension, cardiovascular disease, and diabetes [8]. Dietary magnesium is absorbed in the small intestine and is excreted through the kidneys [8]. The kidneys are efficient at excreting excess magnesium and it is unlikely that the mineral will accumulate to toxic levels, although there is a risk of renal dysfunction with an overdose of magnesium, due to precipitation of magnesium salts. A high intake of magnesium can compete with calcium and lead to impairment of its absorption [9]. Symptoms of magnesium overload include diarrhea, difficulty in breathing, and depression of the central nervous system, causing muscle weakness, lethargy, sleepiness

https://doi.org/10.1515/9783111136691-010

or even hyperexcitability [10]. Accordingly, magnesium is widely used in therapy, primarily for the treatment of heart disease, cardiovascular system, or the respiratory system. This became a premise for the use of magnesium and its alloys in medicine as a potential biomaterial for medical implants. The concept of using magnesium alloys as resorbable medical implants assumes that it will be nontoxic to the human body. The alloy components will also be elements present in the human body. The resorbable biomaterial of a magnesium alloy would be an alternative to the previously used implants, mainly orthopedic ones [11]. Amorphous and crystalline magnesium alloys developed for medical applications – especially implantology – present the characteristics of biocompatible magnesium alloys (Mg-Zn, Mg-Zn-Ca, Mg-Ca etc.) [12]. Biocorrosion resistance (in other words, bioresorbability) can be affected by various material factors including crystalline vs. amorphous (aka, polycrystalline vs. monocrystalline), unavoidable impurities and effective alloying elements, heat treatment, and surface modifications including oxidation/anodization, depositing/coating, or shot/laser peening [4, 13]. As regards magnesium alloys for resorbable implants with a crystalline structure, the following groups of alloys have been examined: Mg-Ca, Mg-Zn, Mg-Zn-Ca, Mg-Mn, Mg-Si, Mg-Zr, Mg-Zn-Zr, Mg-Zn-Y, Mg-Zn-Zr-Y, and Mg-Zn-Mn [6, 13]. The mechanical and corrosion properties of the alloy can be regulated by the structural and chemical composition of the alloy. Compared to their crystalline counterparts, magnesium-based metallic glasses may be more resistant to corrosion due to their single-phase structure, which may result in more uniform alloy corrosion. In the process of designing new degradable biomaterials, elements with potential toxicological problems should be omitted whenever possible and, if they are absolutely necessary, they should be reduced to the minimum. Calcium and zinc are essential elements in the human body; therefore, these elements should be the first choice for alloying additives in biomedical magnesium alloys. The concentration of calcium should not exceed 2 wt%, and zinc 6 wt%, due to the corrosive properties of these magnesium alloys [11].

In chapter 8, we have reviewed various techniques for bone-grafting, including onlay bone block technique, osteoperiosteal flap technique, bone expansion technique, sandwich osteotomy, alveolar ridge splitting technique, minimally invasive tunnel technique, or sausage technique. We also described that each of these techniques possesses merits and demerits. The following describes effectiveness of bioabsorbable magnesium wire network in promoting new bone generation. In Figure 1, (1) depicts placed implant fixture at left maxillary No.3 area, (2) shows placement of 0.2-mm diameter pure Mg wire network, followed by (3) suturing without any additional application such as resorbable membranes.

Figure 2 shows a scan image of Mg wire placement area after the two-month placement, indicating that (i) placed fine Mg wire cannot be recognized due to bioabsorption on the X-ray image and (ii) at the same time, newly generated bone tissue was indicated. It can be considered that although Mg fine wire network was not co-applied with bony substances, such wire network as a scaffold-like spacer played

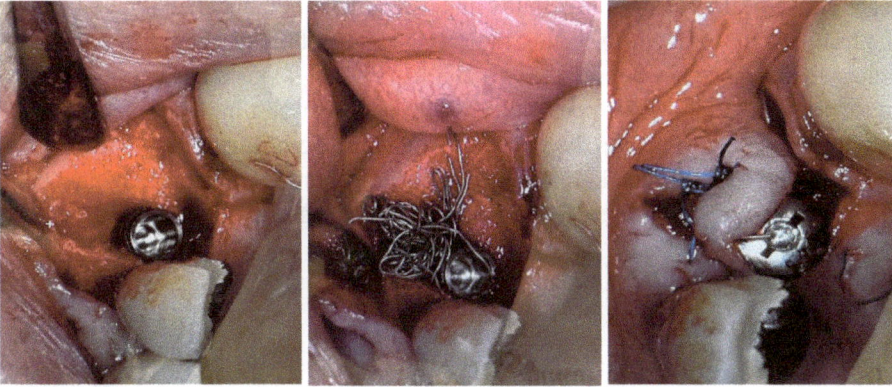

Figure 1: Placement implant and bioresorbable pure magnesium wire network.

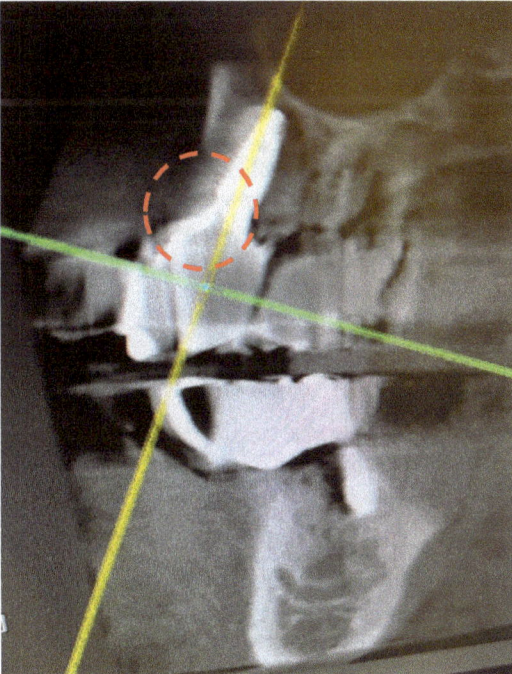

Figure 2: X-ray image, taken two months after Mg wire placement, where broken red circle indicates the area where Mg wire network was placed.

an important and effective role in maintaining sufficient space, and as a result, new bony structure can be generated and grown in a relatively short period of healing time.

In implant dentistry, the implant stability quotient (ISQ) value is considered as an important parameter to judge the success of placed implant. The ISQ is the value on a scale that indicates the level of stability and osseointegration in dental implants, ranging from 1 to 100, with higher values indicating greater stability. Normally, the acceptable stability range lies between 55 ISQ and 85 ISQ [14]. It was indicated that the ISQ of this case is as follows; at operation time, it was 66 at post-implantation and 73 at initial loading (which was 45 days post-implantation). Accordingly, herein-introduced Mg wire network scaffold-like spacer technique appears to be a promising new bone growth promoting technique, which is simpler and easier than other conventional techniques.

References

[1]   Oryan A, Alidadi S, Moshiri, Maffulli N. Bone regenerative medicine: Classic options, novel strategies, and future directions. J Orthop Surg Res. 2014, 9; https://josr-online.biomedcentral.com/articles/10.1186/1749-799X-9-18.

[2]   Hegde C, Shetty V, Wasnik S, Ahammed I, Shetty V. Use of bone graft substitute in the treatment for distal radius fractures in elderly. Eur J Orthop Surg Traumatol. 2013, 23, 651–56.

[3]   Scaglione M, Fabbri L, Dell'omo D, Gambini F, Guido G. Long bone nonunions treated with autologous concentrated bone marrow-derived cells combined with dried bone allograft. Musculoskelet Surg. 2014, 98, 101–06.

[4]   Oshida Y. Bioscience and Bioengineering of Titanium Materials. Elsevier, 2007.

[5]   Oshida Y. Surface Engineering and Technology for Biomedical Implants. Momentum Press, 2014.

[6]   Oshida Y, Miyazaki T. Biomaterials and Engineering for Implantology. De Gruyter Pub, 2022.

[7]   Chen Q, Thouas GA. Metallic implant biomaterials. Mater Sci Eng R. 2015, 87, 1–57.

[8]   Torshin IY, Cromova OA. The biological roles of magnesium. In: Magnesium and Pyridoxine: Fundamental Studies and Clinical Practice. Nova Science Pub, New York, 2009, 1–17.

[9]   Stahlmann R. J Antimicrob Chemother. 2001, 4, 11–12.

[10]   Kim WS. Magnesium toxicity and its use for prediction of calcium deficiency in apple fruit before harvest. Acta Hortic. 1997, 448, 359–60.

[11]   Park JB, Bronzino JD. Biomaterials: Principles and Applications. CRC Press, Florida, 2003.

[12]   Cesarz-Andraczke K, Kania A, Młynarek K, Babilas R. Amorphous and crystalline magnesium alloys for biomedical applications. In: Tański T, et al. (Ed.)., Magnesium Alloys Structure and Properties. 2022, https://www.intechopen.com/chapters/74199.

[13]   Oshida Y. Magnesium Materials. De Gruyter Pub, 2021.

[14]   Miyazaki T, Yutani T, Murai N, Kawata A, Shimizu H, Uejima N, Miyazaki Y, Oshida Y. Early osseointegration attained by UV-Photo treated implant into piezosurgery-prepared site. Report III. influence of surface treatment by hydrogen peroxide solution and determination of early loading timing. Int J Oral Health. 2021, 7(7); dx.doi.org/10.16966/2378-7090.381.

Acknowledgments

In closing, we would like to express our sincere appreciations to faculty members of Indiana University School of Dentistry (particularly, Department of Prosthodontics and Periodontics, Institute of Dental Implantology) and Indiana University School of Medicine (Department of Anatomy, Department of Orthopedics, and Bone Study Club) for their direct and indirect involvement in implantology research and lecturing. We should not forget to thank individual authors and institutes of articles cited in this book. As importantly, our sincere gratitude goes to excellent editorial team of De Gruyter Brill. Thank you all!

https://doi.org/10.1515/9783111136691-011

Index

https://doi.org/10.1515/9783111136691-012

www.ingramcontent.com/pod-product-compliance
Lightning Source LLC
Chambersburg PA
CBHW061405210326
41598CB00035B/6103